ISBN 978-0-265-98610-3
PIBN 10919492

GREENLEAF'S ARITHMETIC,
Improved Stereotype Edition.

THE

NATIONAL ARITHMETIC,

ON THE

INDUCTIVE SYSTEM;

COMBINING THE

ANALYTIC AND SYNTHETIC METHODS,

IN WHICH

THE PRINCIPLES OF ARITHMETIC ARE EXPLAINED IN A
PERSPICUOUS AND FAMILIAR MANNER;

CONTAINING ALSO

PRACTICAL SYSTEMS OF MENSURATION, GAUGING, GEOMETRY, AND

BOOK-KEEPING;

FORMING A

COMPLETE MERCANTILE ARITHMETIC.

DESIGNED FOR COMMON SCHOOLS AND ACADEMIES.

By BENJAMIN GREENLEAF, A. M.,
PRINCIPAL OF BRADFORD TEACHERS' SEMINARY.

BOSTON:

ROBERT S. DAVIS, AND GOULD, KENDALL, & LINCOLN.
N. YORK: ROBINSON, PRATT, & Co., AND COLLINS, BROTHER, & Co.
PHILADELPHIA: THOMAS, COWPERTHWAIT, & Co.
BALTIMORE: CUSHING & BROTHER.
And sold by the trade generally.

1843.

GREENLEAF'S INTRODUCTORY ARITHMETIC,

Designed for Younger Scholars.

PUBLISHED BY **ROBERT S. DAVIS**, BOSTON,

And sold by the principal Booksellers throughout the United States.

☞ This small and cheap work, having been prepared at the urgent solicitation of many distinguished teachers, in various parts of the United States, who have adopted the Author's *larger* volume, the Publisher trusts it will be found worthy of the extensive patronage which has been extended to the National Arithmetic.

From the many commendatory letters, already received from those best qualified to judge of its merits, there is space here for only the following.

From W. R. Ellis, Esq., Principal of Sandwich Academy.

Mr. R. S. DAVIS. — Dear Sir : Before introducing Greenleaf's "Introduction to the National Arithmetic" into my school, I gave it a careful and thorough examination, and was at that time much pleased with it. And since its introduction, I have become convinced, from comparing the improvement of my pupils in this branch, with that of previous terms, that as an introduction to the study of arithmetic, teachers cannot place in the hands of their pupils a better book. Very respectfully, yours,
        (Signed)   W. R. ELLIS.

Sandwich, Mass., March 8, 1843.

Extract from a Letter from A. Mackie, Esq., of Grove School, New Bedford.

"Teachers will find in this book precisely that which they would desire placed in the hands of a scholar, who attends school but for three or four months in the year ; they will find that it goes into the subject as far as is necessary to prepare the young for the common avocations of life. They will find it no less practical than the Author's larger work, — while the several questions and rules succeed each other in such a manner, as will not perplex and embarrass the student, but, on the contrary, in that inductive and progressive manner, which cannot fail to encourage and stimulate him onward ; and, above all, they will find it is calculated to unfold to the scholar the powers of his own mind, which he only needs to know and feel in order to bring into action. The work is printed with a beautiful type on good paper, and is indeed got up in a style, which reflects great credit on the publisher.   (Signed) "ADAM MACKIE.

"New Bedford, Dec. 6, 1842."

From James K. Bollough, Esq., Teacher, Savannah, Ga.

"I have with care and much satisfaction, examined Greenleaf's Arithmetic, and deem it a work of peculiar merit. — It is in every respect well calculated to facilitate the improvement of pupils in that highly useful and important science, the science of numbers. It is a book much needed in our elementary schools. I shall most cordially recommend it to others of my profession, and shall also use my utmost endeavors to introduce it into my own school.   (Signed) JAMES K. BOLLOUGH.

"Savannah, Feb. 1, 1843."

From O. M. Randall, Esq., Teacher, Lynn, Mass.

"I have examined the "Introduction to the National Arithmetic," and think it admirably adapted to the use of common schools. I have long felt the need of something different from what we have had in our town schools, to give the scholar a practical knowledge of Arithmetic, without retarding his proficiency by a multiplicity of questions so arranged as to confuse the mind, rather than unfold to it the principles of the science. I think the 'Introduction' meets the exigency of the case, and cannot fail to secure that patronage which it so richly merits.   (Signed) O. M. RANDALL.

"Lynn, Mass., Jan. 9, 1843,"

☞ *The reader is referred to the Advertising Sheet at the end of this volume.* ☜

CAMBRIDGE:
METCALF, KEITH, AND NICHOLS,

PREFACE.

THE author of the following work is far from flattering himself, that he is about to present to the public any considerable number of new principles in the science of arithmetic. But from thirty years' experience in the business of teaching, he has been led to suppose, that some improvement might be made in the arrangement and simplification of the rules of the science. How far he has succeeded in his attempt at making this improvement, the public must judge.

An opinion has prevailed among some teachers, that the pupil should have no *rule* to perform his questions by, but should form all his rules himself, by mere induction. This plan might do very well, could it be carried into *effect*. But, if the experience of the author has been of any service to him, one thing, it has taught him, is, that, in a given time, a student will acquire more knowledge of arithmetic by having some plain rules given him, with examples, than he will without them ; especially, if he be required to give an analysis of a suitable number of questions under each rule.

A few of the rules, which some arithmeticians of the present day have laid aside as *useless*, the author has thought best to retain; as Practice, Progression, Position, Permutation, etc. For though some of these rules are not of much practical utility, yet, as they are well adapted to improve the reasoning powers, they ought not, in the author's judgment, to be laid aside by any, who wish to become thorough arithmeticians.

The author does not think it expedient, that the pupil should spend a long time on *mental* arithmetic without the use of the slate. He has introduced, as much mental arithmetic, as he thinks necessary for students *generally;* unless they are very young, and for such he would recommend the "*Introduction to the National Arithmetic,*" recently published.

The circumstances of some students may be such, that
they have not time to study all the rules.   Such can make
a selection of those likely to be most useful to them, many
of the rules having no necessary connexion with the rest,
or dependence on them.

Those teachers, who have a number of pupils to instruct
in the same rule or rules, will do well to class them in
arithmetic, as in reading ; and require the whole class to
go on together.   This will enable the teacher to be more
thorough in giving his instructions, and will excite the pupils
to greater diligence.

Every student should be able and be required to give a
thorough analysis of every question he performs.   At least
he should be required to do this, till he has proved himself
perfectly familiar with all the principles involved in the rule,
and with their application.

Every class should review often, and should be exercised
frequently, in the *elements* of the science.

This will teach them not only to be more accurate, but
enable them to solve questions with greater ease.

In preparing this work, the author has consulted most of
the standard writers on the subject in the English language;
from some he has quoted, as he has found occasion, and
from many of which, he has received profitable *hints* and
suggestions.   On the article of Exchange, he is under particular obligations to that very able work, Kelley's British
Cambist, to which he has had access through the politeness
and favor of the gentlemen of the Boston Atheneum.   And
to such as wish to go more extensively into the subject
than he has, he would recommend Grund's Merchants' As-
sistant, as the only thorough work on the subject, published
in this country.

As some parts of the work were prepared, while the author was much engaged in teaching, several errors escaped
detection in the present edition, which, it is hoped, will not
be the case in another, should it be called for.

With these prefatory remarks, the author commends his
work to the blessing of God, and to the candor of an en
lightened *public*.

THE AUTHOR.

Bradford Academy, Nov. 12, 1835.

ADVERTISEMENT

TO THE

SECOND (STEREOTYPE) EDITION.

THE rapid and extensive sale of the first edition of the National Arithmetic, together with its flattering reception in various sections of our country, has induced the author thoroughly to revise, and improve the work, which he trusts will give it additional merit.

The author believes that not an error or inaccuracy of essential importance will be found in the present (stereotype) edition, which could not be wholly avoided in the first.

It has been deemed expedient, that the new edition should embrace more of the inductive plan than the former, with the addition of much important and valuable matter.

The author has availed himself of the assistance of several experienced teachers, among whom he would acknowledge his obligations particularly to Mr. CHARLES H. ALLEN, one of the associate Principals of the Franklin Academy, Andover; and to Mr. DAVID P. PAGE, Principal of the English High School, Newburyport; also to several mercantile gentlemen, who have afforded valuable suggestions of a practical nature.

B GREENLEAF.

Bradford, Nov. 5th, 1836.

☞ A Key, containing the operation of the more difficult questions is now published for the use of teachers *only*.

1*

INDEX.

CHARACTERS USED IN THIS WORK.

$ Contraction, for U. S. United States' currency, and is prefixed to dollars and cents.

= Sign of equality; as 12 inches =1 foot, signifies, that 12 inches are equal to one foot.

+ Sign of addition; as 8+6=14, signifies, that 8 added to 6 is equal to 14.

— Sign of subtraction ; 8—6=2, that is, 8 less 6 is equal to 2.

× Sign of multiplication; as 7×6=42, that is, 7 multiplied by 6, is equal to 42.

÷ Sign of division ; as 42÷6=7, that is, 42 divided by 6 is equal to 7.

$\frac{7}{3}$ Numbers placed in this manner imply, that the *upper* line is to be divided by the *lower* line.

: :: : Signs of proportion ; thus, 2 : 4 :: 6 : 12, that is, 2 has the same ratio to 4, that 6 has to 12 ; and such numbers are called *proportionals*.

$\overline{12-3}+4=13$. Numbers placed in this manner show, that 3 is to be taken from 12, and 4 added to the remainder. The line at the top is called a *vinculum*, and connects all the numbers over which it is drawn.

9^2 Implies, that 9 is to be raised to the second power ; that is, multiplied by itself.

8^3 Implies, that 8 is to be multiplied into its square.

√ This sign prefixed to any number shows, that the square root is to be extracted.

$\sqrt[3]{}$ This sign prefixed to a number, shows, that the cube root is to be extracted.

ARITHMETIC.

SECTION I.

ARITHMETIC is the art of computing by numbers. Its five principal rules are Numeration, Addition, Subtraction, Multiplication, and Division.

NUMERATION.

Numeration teaches to express the value of numbers either by words or characters.

The numbers in Arithmetic are expressed by the following ten characters, or Arabic numeral figures, which the Moors introduced into Europe about nine hundred years ago; viz. 1 one, 2 two, 3 three, 4 four, 5 five, 6 six, 7 seven, 8 eight, 9 nine, 0 cipher, or nothing.

The first nine are called *significant figures*, as distinguished from the cipher, which is of itself insignificant.

Besides this value of those figures, they have also another, which depends on the place in which they stand, when connected together; as in the following table.

Hundreds of Millions.	Tens of Millions.	Millions.	Hundreds of Thousands.	Tens of Thousands.	Thousands.	Hundreds.	Tens.	Units.
9	8	7	6	5	4	3	2	1
	9	8	7	6	5	4	3	2
		9	8	7	6	5	4	3
			9	8	7	6	5	4
				9	8	7	6	5
					9	8	7	6
						9	8	7
							9	8
								9

Here any figure in the first place, reckoning from right to left, denotes only its simple value; but that in the second place, denotes ten times its simple value; and that in the third place a hundred times its simple value; and so on; the value of any successive place being always ten times its former value.

Thus in the number 1834, the 4 in the first place denotes only four units, or simply 4; 3 in the second place signifies three tens, or thirty; 8 in the third place signifies eighty tens or eight hundred; and the 1, in the fourth place, one thousand; so that the whole number is read thus,—one thousand eight hundred and thirty-four.

As to the cipher, 0, though it signify nothing of itself, yet being joined to the right hand of other figures, it increases their value in a tenfold proportion; thus 5 signifies only five, but 50 denotes 5 tens or fifty; 500 is five hundred; and so on.

NOTE. — The idea of number is the latest and most difficult to form. Before the mind can arrive at such an abstract conception, it must be familiar with that process of classification, by which we successively remount from individuals to species, from species to genera, from genera to orders. The savage is lost in his attempts at numeration, and significantly expresses his inability to proceed, by holding up his expanded fingers or pointing to the hair of his head. See Lacroix.

NUMERATION TABLE.

817,897;431,032;639,864;361,816;461,315;123,675;816,131;123,456;128,614;315,131;398,832;563,871;851,615;123,561.

Thousands.
Tridecillions.
Thousands.
Duodecillions.
Thousands.
Undecillions.
Thousands.
Decillions.
Thousands.
Nonillions.
Thousands.
Octillions.
Thousands.
Septillions.
Thousands.
Sextillions.
Thousands.
Quintillions.
Thousands.
Quatrillions.
Thousands.
Trillions.
Thousands.
Billions.
Thousands.
Millions.
Thousands.
Units.

In order to enumerate any number of figures, they must be separated by semicolons into *divisions* of six figures each, and each division by a comma, as in the annexed table. Each division will be known by a different *name*. The first three figures in each division will be so many thousands of that name, and the next three will be so many of *that* name, that is over its unit place. The value of the numbers in the annexed table, expressed in words, is Three hundred and seventeen thousand, eight hundred and ninety-seven tridecillions; four hundred and thirty-one thousand, thirty-two duodecillions; six hundred thirty-nine thousand, eight hundred sixty-four undecillions; three hundred sixty-one thousand, three hundred sixteen decillions; four hundred sixty-one thousand, three hundred fifteen nonillions; one hundred twenty-three thousand, six hundred seventy-five octillions; eight hundred sixteen thousand, one hundred thirty-one septillions; one hundred twenty-three thousand, four hundred fifty-six sextillions; one hundred twenty-three thousand, six hundred fourteen quintillions; three hundred fifteen thousand, one hundred thirty-one quatrillions; three hundred ninety-eight thousand, eight hundred thirty-two trillions; five hundred sixty-three thousand, eight hundred seventy-one billions; three hundred fifty-one thousand, six hundred fifteen millions; one hundred twenty-three thousand five hundred sixty-one.

NOTE.—The student must be familiar with the names, from Units to Tridecillions, and from Tridecillions to Units, so that he may repeat them with facility either way.

The, following Table is the French method of Enumeration.

809,317,123,586,315,961,831,561,786,811,123,456,617,831,515.

Tridecillions.
Duodecillions.
Undecillions.
Decillions.
Nonillions.
Octillions.
Septillions.
Sextillions.
Quintillions.
Quatrillions.
Trillions.
Billions.
Millions.
Thousands
Units.

By the annexed table it will be seen, that every three figures have a different name. Their value would be expressed thus:— Eight hundred nine tridecillions, three hundred and seventeen duodecillions, one hundred twenty-three undecillions, five hundred eighty-six decillions, three hundred fifteen nonillions, eight hundred sixty-one octillions, eight hundred thirty-one septillions, five hundred sixty-one sextillions, seven hundred eighty-six quintillions, eight hundred eleven quatrillions, one hundred twenty-three trillions, four hundred fifty-six billions, six hundred seventeen millions, eight hundred thirty-one thousand, five hundred and fifteen.

NOTE.—It is very doubtful, whether the French method is so good as that of the English.

Let the following numbers be written in words.

76
319
74,679
864,093
27;623,781
367,832;732,567
216;746,230;095,623
56,000;000,602;007,020
7;065,182;347,627;907,640
802;678,327;006,279;206,040
6,070;360,217;642,176;400,567
4;332,705;607,204;062,376;407,592
3;469,200;000,000;111,111;100,000;001,101
7;006,437;576,589;081,828;384,859;091,929;394,332
71,326;436,035;769,846;012,131;415,161;718,192;021,223
712,642;976,403;796,411;604,220;404,263;575,068;076,001
170,907;642,376;756,809;000,604;020,760;307,000;100,000

Let the following numbers be written in figures.

1. Twenty-nine. `
2. Four hundred and seven.
3. Twenty-three thousand and twenty-seven.
4. Five million two thousand and five.
5. Seventeen trillion two hundred million and six.
6. Fifteen billions twenty-seven thousand million nine thousand two hundred and five.
7. Seven billions, five millions, six thousand, five hundred and twenty-five.
8. Ninety-nine trillions, seventy-nine thousand six hundred billions, one hundred and twenty-four millions, three hundred and twenty-nine.
9. Fifteen quintillions, thirty-three thousand millions, seventy-six thousand and five.
10. Eight thousand five hundred and forty-three septillions, five quintillions, seven hundred twenty-nine thousand three hundred and forty-six quatrillions, three hundred fifty-seven thousand two hundred sixty-one trillions, four hundred and two thousand, twenty-three billions, seven millions and forty-six.
11. Nine nonillions, forty-seven trillions, ten thousand seven billions, two million and seventy-two.
12. Three hundred and twenty-thousand and fifteen duodecillions, two thousand and ten trillions, one hundred and twenty-seven billions, twenty-six millions, three hundred and twenty thousand four hundred and twenty-six.

SECTION II.

MENTAL OPERATIONS IN ADDITION.

1. How many are 3 and 4? How many are 3 and 6? How many are 3 and 7? How many are 9, 7 and 3? 4 and 10? 7 and 6? 5 and 11? 3 and 15? 6 and 5? 4 and 8? 10 and 5? 3 and 12? 6 and 6? 6 and 8? 6 and 7? 6 and 3? 6 and 10? 6 and 9? 6 and 12? 6 and 11? 6 and 13? 5 and 3? 5 and 5? 5 and 4? 5 and 7? 5 and 9? 5 and 8? 5 and 6? 5 and 12? 5 and 15? 5 and 13? 7 and 3? 7 and 5? 7 and 7? 7 and 9? 7 and 8? 8 and 2? 8 and 5? 8 and 4? 8 and 8? 8 and 7? 8 and 9? 8 and 12? 8 and 10? 9 and 5? 9 and 2? 9 and 4? 9 and 9? 9 and 7? 9 and 8? 10 and 5? 10 and 7? 10 and 4? 10 and 3? 10 and 10?

2. Bought an orange for 3 cents and some nuts for 6 cents; what did they both cost?

3. Bought a pound of figs for 8 cents and a pint of cherries for 7 cents; what was the price of both?

4. Bought a book for 12 cents and some paper for 9 cents; what did they both cost?

2

5. Gave 6 dollars for a cow and 9 dollars for a load of hay ; what did I give for both?

6. A boy gave 12 cents for a penknife and 10 cents for a bunch of quills ; what did he give for both?

7. A boy gave 8 cents for a top and 9 cents for some apples ; what was the price of the whole?

8. A lady gave 11 dollars for a silk cloak and 7 dollars for a bonnet ; what was the price of both?

9. How many are 3 and 2 and 5? How many are 7 and 8 and 10? How many are 11 and 5 and 2? 6 and 5 and 4? 7 and 5 and 8? 12 and 3 and 9? 15 and 2 and 3? 16 and 3 and 4? 10 and 9 and 3? 12 and 5 and 2? 13 and 10 and 2? 15 and 5 and 10? 25 and 2 and 3? 30 and 2 and 8? 1 and 50 and 2? 50 and 50 and 1? 12 and 12 and 10? 13 and 7 and 5? 30 and 20 and 10? 10 and 60 and 30? 15 and 5 and 50? 100 and 50 and 60? 17 and 9 and 7? 19 and 9 and 8? 99 and 6 and 10? 29 and 8 and 8? 15 and 15 and 15?

10. James gave 30 cents for an arithmetic, 10 cents for a slate, and 15 cents for a writing book ; what did he give for the whole?

11. A merchant has due from one creditor 20 dollars, from another 10, from another 25. What is the sum due from the whole?

12. A gentleman gave 17 dollars for a coat, 12 dollars for pantaloons, 5 dollars for a vest, 6 dollars for a hat. What did they all cost him?

13. A drover has three flocks of sheep ; the first has 25, the second 35, and the third 50. How many in the whole?

14. William is 12 years old, John is 9, and Thomas is 8. What is the sum of their ages?

15. Samuel has 5 apples, Enoch has 18, and Levi has 20. How many in the whole?

16. A beggar received of one man 14 cents, of another 12, of another 8, and of another 7. How many has he received?

17. Samuel has 12 marbles in one pocket, 12 in another, and 2 in each hand ; how many has he in all?

18. John has 7 birds in one cage, and 12 in another ; how many has he in both cages?

19. A farmer has 11 calves in one pasture, 6 in another, and 5 in another ; how many has he in the whole?

20. Thomas caught one afternoon 7 pickerel, 9 trout, and 10 perch ; how many had he in all?

21. A boy at service earned the first week 9 shillings, the second week 11 shillings, and the third week 12 shillings how much money had he?

22. A lady gave 15 dollars for a bonnet, 20 dollars for a silk dress, and 12 dollars for other articles ; what was the amount of her bill?

23. In one school room are 30 scholars, in another 20, and in another 50 ; how many in all ?

24. My shoes cost me 2 dollars, my books 5, my hat 5, my cloak 20, and my other garments 10 ; what are they all worth?

25. The distance from Bradford to Salem is 22 miles, from Salem to Marblehead 4 miles ; what is the whole distance?

26. From Lowell to Haverhill is 20 miles, from Haverhill to Newburyport is 14 miles ; what is the whole distance?

27. A boy has 15 cents in each of his two pockets, and 5 in each hand ; how many has he?

28. A girl gave 17 cents for cherries, 5 for apples, and 7 for peaches ; what did they all cost?

29. A man gave 11 dollars for a plough, 15 dollars for a cart, 5 dollars for a wheelbarrow, and 10 dollars for a wagon ; what did they all cost?

30. A butcher gave 12 dollars for an ox, 9 dollars for a cow, and 7 dollars for three pigs ; what did they all cost?

31. A farmer has 25 apple trees, 10 pear trees, and 15 peach trees ; how many in all?

32. A merchant receives 18 barrels of flour by one vessel, 12 by another, and 12 by another ; how many has he?

33. A boy at one time lost 7 marbles, at another 5, at another 8, at another 12 ; how many has he lost?

34. Thomas gave 17 cents for a whip, 13 cents for a top, 12 for a knife, and 10 for marbles ; how much did he lay out?

35. Gave 60 dollars for a horse, 30 for a wagon ; what did they both cost?

36. A boy at a Sabbath school committed at one lesson 25 verses, at another 15 verses, at another 20 verses ; how many did he commit in all?

37. I owe a merchant 7 dollars for sugar, 5 dollars for coffee, 17 for flour, 3 dollars for tea, and 2 dollars for salt ; what is the whole sum due?

38. A laborer worked 9 days for one man, 11 days for another, 7 for another, 6 for another, 10 for another, and 3 for another ; how many days has he labored?

Let the pupil perform the last 10 questions on his slate, and hence notice, that

ADDITION is the collecting of numbers together to find their sum.

RULE.

Write units under units, tens under tens, &c. Then add upwards the units, and find how many tens are contained in their sum. Write down directly below what remains more than those tens ; or, if nothing remains, write down a cipher and carry as many units to the next column, as there are tens. Then add up the second column together with the number carried, in the same manner as before. Proceed in this way, till the whole is finished, writing down the total amount in the last column.

PROOF.

Begin at the top, and add together all the columns of numbers *downwards*, in the same manner, as they were before added *upwards*; then if the two sums agree, the work is right.

EXAMPLES FOR THE SLATE.

1.	2.	3.	4.	5.	6.	7.
1	12	123	98765	367856	1234567	678123
2	23	456	32145	789123	8912345	456789
3	34	789	67811	478956	7856627	567128
4	61	234	21346	367123	4478536	889998
5	32	567	12389	678910	7813791	123456
6	58	890	99678	123456	8951372	789138
7	96	345	12345	895567	3467854	767799
8	89	678	78912	347890	9514673	135798
9	32	357	34567	643634	1234578	765432
45	437	4439				

8.	9.	10.	11.	12.
123456	876543	789012	987654	678953
789012	789112	345678	456112	467631
345678	345678	901234	222533	117777
901234	965887	789037	456789	888888
567890	445566	891133	987654	444444
987654	788743	477666	321178	667679
321032	399378	557788	123456	998889
765437	456789	888878	789561	671236

13. Add the following numbers, 763, 4663, 37, 49763, 6173, and 671.   Ans. 62075.

14. A butcher sold to A 369 lbs. of beef, to B 169 lbs., to C 861 lbs., to D 901 lbs., to E 71 lbs., and to F 8716 lbs.; what did they all receive?   Ans. 11087 lbs.

15. A owes to one creditor 596 dollars, to another 3961, to another 581, to another 6116, to another 469, to another 506, to another 69381, and to another 1261. What does he owe them all?   Ans. $ 82871.

16. If a boy earn 17 cents a day, how much will he earn in a week?   Ans. 119 cts.

17. If a man's wages be 19 dollars per month, what are they per year?   Ans. $ 228.

18. If a boy receive a present every new year's day of 1783 dollars, how much money will he possess, when he is 21 years old?   Ans. $ 37443.

19. How many inhabitants are there in Essex county? Haverhill having 3912, Bradford 1856, Methuen 2011, Andover 4540, Boxford 937, Newburyport 6338, Newbury 3603, West Newbury 1586, Rowley 2044, Ipswich 2951, Danvers 4228,

Essex 1345, Topsfield 1011, Lynn 6138, Saugus 960, Middleton 607, Marblehead 5150, Salem 13886, Lynnfield 617, Gloucester 7513, Beverly 4079, Hamilton 743, Wenham 612, Salisbury 2519, Manchester 1238, and Amesbury 2445.   Ans. 82869.

20. Of how many members does the British House of Commons consist? England sending county members, 143; Isle of Wight, 1; universities, 4; cities, boroughs, and cinque ports, 323; Wales, county members, 15; boroughs, 14; Scotland, county members, 30; cities and boroughs, 23; Ireland, county members, 64; university, 2; cities and boroughs, 30?
Ans. 649.

21. How many are the members of Congress? There being 2 Senators from each State; and New Hampshire sending 5 Representatives; Massachusetts 12, Maine 8, Connecticut 6, Rhode Island 2, Vermont 5, New York 40, New Jersey 6, Pennsylvania 28, Delaware 1, Maryland 8, North Carolina 13, South Carolina 9, Georgia 9, Kentucky 13, Tennessee 13, Ohio 19, Alabama 5, Mississippi 2, Missouri 2, Illinois 3, Virginia 21, Louisiana 3, Indiana 7.   Ans. 288.

22. According to the census of 1830, Maine had 399437 inhabitants, New Hampshire 269367; Massachusetts 610014; Connecticut 297513; Rhode Island 97210; Vermont 280679; New York 1913508; New Jersey 320779; Pennsylvania 1347672; Delaware 76739; Maryland 446913; Virginia 1211272; North Carolina 738470; South Carolina 581458; Georgia 516567; Ohio 937679; Kentucky 688844; Indiana 341582; Illinois 157575; Missouri 140192; Tennessee 684833; Louisiana 215762; Alabama 308997; Mississippi 136806; Florida Territory 34723; Michigan Territory 31260; Arkansas Territory 30383; District of Columbia 39858. What is the whole number of inhabitants?
Ans. 12,856,092.

SECTION III.

MENTAL OPERATIONS IN SUBTRACTION.

1. If 3 be taken from 4, how many will be left? If 2 be taken from 5, how many will be left? If 5 be taken from 9, how many will be left? What will remain, if 7 be taken from 10? 8 from 13? 7 from 11? 13 from 16? 3 from 20? 5 from 15? 10 from 19? 7 from 15? 7 from 14? 3 from 10? 4 from 12? 5 from 19? 19 from 21? 7 from 16? 7 from 19? 4 from 20? 13 from 17? 18 from 23? 3 from 27? 7 from 18? 6 from 16? 13 from 21? 16 from 19? 16 from 23? 16 from 26? 11 from 19? 13 from 19? 13 from 14? 27 from 29? 21 from 23? 12 from 17? 12 from 19? 12 from 21? 13 from 17? 14 from 19? 14 from 15? 14 from 21?
2*

2. Sold an apple for 6 cents, and a knife for 13 cents. How much more did I receive for my knife, than for the apple?

3. Gave 7 dollars for an umbrella, and 10 dollars for a coat. How much more did the coat cost than the umbrella?

4. Bought a horse for 21 dollars, and sold it for 30 dollars. How much did I gain?

5. Bought a book for 35 cents and sold it for 25 cents. How much did I lose?

6. Bought a bunch of quills for 26 cents and sold it for 30 cents. How much did I gain?

7. Bought a chaise for 90 dollars, and a horse for 80 dollars; how much more did the chaise cost than the horse?

8. Received 25 dollars from one man and 15 dollars from another; how much more did I receive from the first than the last?

9. James has 17 nuts, he gave 6 to Benjamin; how many has he left?

10. A boy went to market with 50 cents, he gave 20 cents for butter, and 15 cents for cheese; how much has he left?

11. Fifteen black birds are on a tree, a man fired and killed 12 of them; how many were left on the tree?

12. Samuel had 20 cents, he spent 3 for raisins, and 4 for apples; how many has he left?

13. John has 15 apples, he gave 4 to Thomas, and 7 to Dick; how many has he left? and how many more has Dick than Thomas?

14. William is 17 years old, and Enoch is 9; how much older is William than Enoch? and what is the sum of their ages?

15. Mary lost 12 cents, and Jane 15 cents; how many did Jane lose more than Mary? and what did they both lose?

16. Seth has 25 nuts, he gave 4 to Abel, 5 to John, and 6 to Samuel; how many has he left?

17. If a barrel of pork is worth 15 dollars, and a barrel of beef is worth 12 dollars; how much is the pork worth more than the beef?

18. The distance from Haverhill to Boston is 30 miles, to Salem is 22; how much farther is it to Boston than Salem?

19. I have a book of 75 pages, I have read 60 of them; how many remain to be read?

20. I have a field of 48 rows of corn, I have hoed 30 of them; how many remain?

21. Benjamin has a basket of quinces, 41 in number; he gave 7 to James, 8 to Moses, and 5 to his sister; how many has he left?

22. A farmer has 18 turkeys; 3 the foxes killed, and 7 he carried to market; how many has he left?

23. Bought a whip for 50 cents, and sold it for 75 cents; what did I gain?

24. Bought a load of hay for 25 dollars, and sold it for 17 dollars; what did I lose?

25. Found 63 dollars, and lost 5 at one time, and 10 at another; how many have I left?

26. James has 43 cents, and Thomas 20; James gave Thomas 3 of his; how many more has James than Thomas?

27. I wish to pay a debt of 50 dollars; John gave me 12 dollars, and Samuel gave 15; how much do I lack of having enough?

28. I have 80 cents; I wish to give 20 to Peter, 8 to John, and 5 to Luke; how many shall I have left?

29. John had 30 marbles; he lost half of them, and then gave away 5; how many had he left?

30. A farmer had 46 sheep, the wolves killed 15, and the dogs 20; how many had he left?

Let the pupil perform the last 12 questions on the slate, and hence notice, that

Subtraction teaches to take one number from another to find the difference.

RULE.

Place the less number under the greater; units under units, and tens under tens. Begin with the units, and take the lower figure from the upper figure, and place the difference between them immediately under the units; but, if the upper figure be less than the lower figure, add ten to the upper one, and place the difference between them under the units as before, and carry one to the next number at the bottom, and proceed thus till all the numbers are subtracted.

Note. — The upper line is called the Minuend, and the lower one the Subtrahend. The *result* of the question is called the Remainder.

PROOF.

Add the Remainder to the Subtrahend, and if their sum be like the Minuend, the work is right.

EXAMPLES.

1. £.	2. Cwt.	3. Tons.	4. Miles.	5. Inches.	6. Rods.
79	86	469	876315123	11630078	56785111
24	25	188	177897638	1919179	38819566
55	61	286	698417485		

7. Gallons.	8. Bushels.	9. Minutes.
10203400	100200300400500	600700800900
5617851	90807060504039	191318917185

10. Hours.	11. Dollars.
61567101	100000
91678	1

13. From   6767851 tak
14. From   761619161 tak
15. From   31671675 tak
16. From   16781321 tak
17. From   1002007000 tak
18. From   91611237 tak
19. From   4637561 tak
20. From   88895651 tak
21. From   1111111 tak
22. From   7163878 tak
23. From   8887771 tak
24. From   1379156 tak
25. From   3671652 tak

Sum of the rer

26. Sir Isaac Newton was born in the in 1727 ; how old was he at the time of

27. Gunpowder was invented in the was this before the invention of printin

28. The mariners' compass was inv year 1302 ; how long was this before th by Columbus, which happened in 1492

29. What number is that, to which sum will be one million?

30. A man bought an estate for seven dred and sixty-five dollars, and sold it f three hundred and seventy-five dollars and how much ?

31. Bought a pair of oxen for 85 doll lars, three cows at 25 dollars apiece ; three hundred dollars ; how much did I

32. Bonaparte was declared emper years is it since ?

33. The union of the government of was in the year 1603 ; how long was it time of the declaration of the indep States ?

34. Jerusalem was taken and destro 70 ; how long was it from this period Crusade, which was in the year 1096.

SECTION IV.

MENTAL OPERATIONS IN MULTIPLICATION.

1. What will 2 oranges cost at 3 cents a piece ?
2. What will 3 apples cost at 2 cents a piece ?
3. What will 7 bushels of wheat cost at 2 dollars a bushel ?
4. At 6 cents a quart, what will 5 quarts of cherries cost?
5. At 4 cents a pound, what will 3 pounds of raisins cost ?
6. At 9 dollars per acre, what will 8 acres of land cost ?
7. At 3 shillings a bushel, what will 9 bushels of oats cost ?
8. At 5 shillings a bushel, what will 9 bushels of corn cost ?
9. At 8 cents per yard, what will 5 yards of cloth cost ?
10. What will 10 loads of hay cost at 9 dollars a load ?
11. At 8 dollars per ton, what will 12 tons of lead cost ?
12. At 7 dollars per box, what will 9 boxes of lemons cost ?
13. What will 12 pounds of rice cost at 7 cents per pound ?
14. What will 11 pounds of sugar cost at 9 cents a pound ?
15. What will 6 pounds of pepper cost at 12 cents a pound ?
16. What will 9 pounds of coffee cost at 11 cents a pound ?
17. What will 10 pounds of mace cost at 11 cents a pound ?
18. What will 8 yards of ribbon cost at 8 cents per yard ?
19. What will 8 ounces of peppermint cost at 6 cents per oz.?
20. What will 7 pair of hose cost at 3 shillings a pair ?
21. What will 6 papers of pins cost at 12 cents a paper ?
22. What will 11 cwt. of sugar cost at 9 dollars per hundred weight ?
23. What will 7 yards of buckram cost at 12 cents per yard ?
24. What will 10 oranges cost at 9 cents a piece ?
25. What will 9 pair of boots cost at 4 dollars a pair ?
26. What will 6 dozen of hoes cost at 11 dollars a dozen ?
27. What will 15 doz. of penknives cost at 8 dollars a dozen ?
28. What will 7 sheep cost at 5 dollars a piece ?
29. What will 6 cwt. of copperas cost at 6 dollars a cwt. ?
30. What will 25 oxen cost at 50 dollars each ?
31. What will 15 calves cost at 3 dollars each ?
32. What cost 13 shad at 13 cents each ?
33. At 20 dollars each, what cost 20 cows ?
34. At 15 dollars each, what cost 14 bales of cloth ?
35. At 25 dollars each, what cost 10 bureaus ?
36. At 15 cents per pound, what cost 9 lbs. of starch ?
37. At 11 cents per oz., what cost 11 oz. of cloves ?
38. What cost 6 lbs. of logwood at 9 cents per lb. ?
39. What cost 7 cwt. of iron at 8 dollars per cwt. ?
40. If one peck of meal weigh 15 lbs., what will 12 pecks weigh ?
41. If a man travel 7 miles in one hour, how far will he travel in 5 hours ?　In 6 ?　In 7 ?　In 8 ?　In 9 ?　In 10 ?　In 11 ?

42. If one pound of sugar cost 9 cents, what will be the cost of 4 lbs.? 5 lbs.? 6 lbs.? 7 lbs.? 8 lbs.? 9 lbs.? 10 lbs.? 11 lbs.?

43. If in 1 shilling there be 12 pence, how many pence in 5 shillings? In 6? In 7? In 8? In 9? In 10? In 11? In 12?

44. If in 1 mile there are 8 furlongs, how many furlongs in 3 miles? In 4? In 5? In 6? In 7? In 8? In 9? In 10?

45. If 12 oz. make a pound troy, how many ounces in 10 lbs.?

46. If 5 quarters make one ell English, how many quarters in 7 ells English? In 8? In 9? In 10? In 12? In 14?

47. If there be 9 square feet in one square yard, how many square feet in 7 yards? In 8? In 9? In 10? In 11? In 12?

48. If there be 16 ounces in one pound avoirdupois, how many ounces in 5 lbs.? In 7? In 8? In 9? In 10? In 12?

Let the pupil perform the last 12 questions on the slate, and notice, that

MULTIPLICATION is a compendious way of performing Addition.

It consists of three parts, the number to be multiplied or the multiplicand, the number to multiply by or multiplier, and the result or product, which is the answer to the question.

RULE.

Place the larger number uppermost, and then set the multiplier under it, so that units may be under units, &c., and multiply each figure in the multiplicand by the multiplier, beginning at the unit's place and carry for ten, as in Addition. When the multiplier consists of more places than one, multiply each figure in the multiplicand by every figure in the multiplier, beginning with the units, and placing the first figure of each product directly under its multiplier; then add all these several products together in the same order as they stand, and their sum will be the true product required.

If there be ciphers at the right hand of the multiplier or multiplicand, they may be neglected in the operation, but their number must be affixed to the product.

PROOF.

The regular way to prove questions in Multiplication is by Division; but a more expeditious way is by casting out the 9's, thus; cast the 9's from the multiplicand and place the remainder at the *right* of a cross, then cast the 9's from the multiplier and place the remainder at the *left* of the cross; then cast the 9's from the product and set the remainder at the *top* of the cross. Multiply together the numbers on each side of the cross, and cast the '9s from their product, and if the remainder be like the number at the *top* of the cross, it may be presumed the work is right. See Ex. 11.

TABLE OF PYTHAGORAS.

1	2	3	4	5	6	7	8	9	10	11	12	13	14	15	16	17	18	19	20	21	22	23	24
1	2	3	4	5	6	7	8	9	10	11	12	13	14	15	16	17	18	19	20	21	22	23	24
2	4	6	8	10	12	14	16	18	20	22	24	26	28	30	32	34	36	38	40	42	44	46	48
3	6	9	12	15	18	21	24	27	30	33	36	39	42	45	48	51	54	57	60	63	66	69	72
4	8	12	16	20	24	28	32	36	40	44	48	52	56	60	64	68	72	76	80	84	88	92	96
5	10	15	20	25	30	35	40	45	50	55	60	65	70	75	80	85	90	95	100	105	110	115	120
6	12	18	24	30	36	42	48	54	60	66	72	78	84	90	96	102	108	114	120	126	132	138	144
7	14	21	28	35	42	49	56	63	70	77	84	91	98	105	112	119	126	133	140	147	154	161	168
8	16	24	32	40	48	56	64	72	80	88	96	104	112	120	128	136	144	152	160	168	176	184	192
9	18	27	36	45	54	63	72	81	90	99	108	117	126	135	144	153	162	171	180	189	198	207	216
10	20	30	40	50	60	70	80	90	100	110	120	130	140	150	160	170	180	190	200	210	220	230	240
11	22	33	44	55	66	77	88	99	110	121	132	143	154	165	176	187	198	209	220	231	242	253	264
12	24	36	48	60	72	84	96	108	120	132	144	156	168	180	192	204	216	228	240	252	264	276	288
13	26	39	52	65	78	91	104	117	130	143	156	169	182	195	208	221	234	247	260	273	286	299	312
14	28	42	56	70	84	98	112	126	140	154	168	182	196	210	224	238	252	266	280	294	308	322	336
15	30	45	60	75	90	105	120	135	150	165	180	195	210	225	240	255	270	285	300	315	330	345	360
16	32	48	64	80	96	112	128	144	160	176	192	208	224	240	256	272	288	304	320	336	352	368	384
17	34	51	68	85	102	119	136	153	170	187	204	221	238	255	272	289	306	323	340	357	374	391	408
18	36	54	72	90	108	126	144	162	180	198	216	234	252	270	288	306	324	342	360	378	396	414	432
19	38	57	76	95	114	133	152	171	190	209	228	247	266	285	304	323	342	361	380	399	418	437	456
20	40	60	80	100	120	140	160	180	200	220	240	260	280	300	320	340	360	380	400	420	440	460	480
21	42	63	84	105	126	147	168	189	210	231	252	273	294	315	336	357	378	399	420	441	462	483	504
22	44	66	88	110	132	154	176	198	220	242	264	286	308	330	352	374	396	418	440	462	484	506	528
23	46	69	92	115	138	161	184	207	230	253	276	299	322	345	368	391	414	437	460	483	506	529	552
24	48	72	96	120	144	168	192	216	240	264	288	312	336	360	384	408	432	456	480	504	528	552	576

EXAMPLES.

1.	2.	3.	4.
678956324	36785678	123456789	987654321
3	7	6	9
2036868972	257499746	740740734	8888888889

5.	6.	7.	8.
678954	616783	789563	789567
24	36	57	98
2715816	3700698	5526941	6316536
1357908	1850349	3947815	7106103
16294896	22204188	45005091	77377566

9.	10.	11.	PROOF.
3678543	67854	612346	0
4567	10234	230049	0 ✕ 4
25749801	271416	5511114	0
22071258	203562	2449384	
18392715	135708	1837038	
14714172	67854	1224692	
16799905881	694417836	140869584954	

12.	13.	14.	15.
678567	4567895	6785000	478763000
8007	60004	32000	12000

16. Multiply 75432 by 47.   Ans. 3545304.
17. Multiply 76785316 by 7615.   Ans. 584720181340.
18. Multiply 67853000 by 8765.   Ans. 594731545000.
19. Multiply 38123450 by 31243.   Ans. 1191090948350.
20. Multiply 40670007 by 10002.   Ans. 406781410014.
21. Multiply 31235678 by 10203.   Ans. 318697622634.
22. Multiply 76786321 by 30070.   Ans. 2308964672470.
23. Multiply 317160070 by 700500. Ans. 222170629035000.
24. Multiply 467325812 by 167000.   Ans. 78043410604000.
25. Multiply 6176777 by 22222.   Ans. 137260338494.
26. Multiply 123456789 by 987654321.
                                      Ans. 121932631112635269.
27. What will 27 oxen cost at 35 dollars each ?
                                                       Ans. $ 945.
28. What will 365 acres of land cost at 73 dollars per acre ?
                                                       Ans. $ 26645.
29. What will 97 tons of iron cost at 57 dollars a ton ?
                                                       Ans. $ 5529.
30. What will 397 yards of cloth cost at 7 dollars per yard ?
                                                       Ans. $ 2779.
31. What will 569 hogsheads of molasses cost at 37 dollars
per hogshead ?   Ans. $ 21053.
32. If a man travel 37 miles in one day, how far will he travel
in 365 days ?   Ans. 13505 miles.
33. If one quire of paper have 24 sheets, how many sheets
are in a ream, which consists of 20 quires ?   Ans. 480 sheets.

34. If a vessel sails 169 miles in one day, how far will she sail in 144 days ?   Ans. 24336 miles.

35. What will 698 barrels of flour cost at 7 dollars a barrel ?
                                                    Ans. $4886.

36. What will 376 lbs. of sugar cost at 13 cents a pound ?
                                                    Ans. 4888 cts.

37. What will 97 lbs. of tea cost at 93 cents a pound ?
                                                    Ans. 9021 cts.

38. If a regiment of soldiers consists of 1128 men, how many men are there in an army of 53 regiments ?   Ans. 59784.

39. What will an ox weighing 569 pounds amount to at 8 cents a pound ?   Ans. 4552 cts.

40. If a barrel of cider can be bought for 93 cents, what will 75 barrels cost ?   Ans. 6975 cts.

41. If in a certain factory 786 yards of cloth are made in one day, how many will be made in 313 days ?   Ans. 246018 yds.

42. A certain house contains 87 windows, and each window has 32 squares of glass, how many squares are there in the whole house ?   Ans. 2784 squares.

43. There are 407 wagons each loaded with 30009 pounds of coal, how many pounds are there in the whole ?
                                            Ans. 12213663 pounds.

44. Multiply three hundred and seventy-five millions two hundred and ninety-six thousand three hundred and twenty-one, by seventy-nine thousand and twenty-four.
                                            Ans. 29657416470704.

45. What would be the cost of 687 fother of lead at 73 dollars a fother ?   Ans. $50151.

SECTION V.

MENTAL OPERATIONS IN DIVISION.

1. How many times 2 are there in 4?   In 6?   In 8?   In 10?
   How many times 3 are there in 6?   In 9?   In 12?   In 15?
   How many times 2 are there in 12?   In 14?   In 16?   In 18?
   How many times 3 are there in 18?   In 21 ?   In 24?   In 27 ?
   How many times 2 are there in 10?   In 13?   In 14?   In 15?
   How many times 4 are there in 12?   In 16?   In 20?   In 24?
   How many times is 2 contained in 3? In 7?   In 9?   In 11?
   How many times is 5 contained in 10 ?   2 is contained in 12, how many times?   4 in 12 how many times?   3 in 12 how many times?   6 in 12 how many times?

2. James had 8 apples, and John had half as many, how many had he ?

3. How many pears at 2 cents a piece can you buy for 4 cents?

3

4. If you have 8 apples to give to 4 boys, how many can you give to each?

5. Joseph has 12 pears, which he wishes to distribute equally between 6 of his companions, how many can he give them apiece?

6. If a man travel 3 miles in an hour, how many hours will it take him to travel 12 miles?

7. Fourteen are how many times 2?

8. At 6 cents apiece, how many oranges can you buy for 18 cents?

9. If 1 pencil cost 2 cents, how many can you buy for 20 cents?

10. If you can buy 1 slate for 6 cents, how many can you buy for 30 cents?

11. If you had 60 cents, how many quires of paper could you buy at 20 cents per quire?

12. If one pound of raisins cost 12 cents, how many pounds may be bought for 60 cents?

13. If 10 dollars will buy 1 ton of hay, how many tons may be bought for 30 dollars?

14. 24 are how many times 4? are how many times 6? are how many times 8? are how many times 12?

15. 27 are how many times 3? how many times 9?

16. 28 are how many times 4? how many times 7?

17. 30 are how many times 3? how many times 5? how many times 15? how many times 6? how many times 10?

18. 32 are how many times 4? 8? 16? 5? 9? 11? 13? 17?

19. 36 are how many times 4? 6? 9? 12? 5? 7? 11? 30? 35?

20. 40 are how many times 4? 5? 8? 10? 3? 6? 7? 9? 11?

21. 42 are how many times 6? 7? 14? 21? 8? 9? 10? 11?

22. 48 are how many times 4? 6? 8? 12? 5? 6? 7? 9? 10?

23. 56 are how many times 4? 7? 8? 9? 10? 11? 12? 13?

24. 60 are how many times 3? 4? 5? 6? 10? 12? 7? 8? 9?

25. 64 are how many times 4? 8? 16? 5? 6? 7? 9? 10? 11?

26. 66 are how many times 2? 3? 6? 22? 33? 4? 5? 7? 8?

27. 72 are how many times 8? 9? 12? 6? 7? 8? 17? 18?

28. 84 are how many times 7? 12? 8? 9? 10? 11? 13? 14?

29. A man buys 6 sheep for 48 dollars; what did he give apiece?

30. James gave 72 cents for 8 quarts of cherries; what were they a quart?

31. Thomas gave 96 cents for 8 penknives; what was the price of each?

32. Benjamin had 37 nuts; he gave 7 to James, and divided the remainder equally between himself and two sisters; how many will be his share?

33. Thomas has 98 cents; he gave his sister Nancy 18 and divided the remainder between himself and 9 others; how many will each have?

34. Daniel has 47 apples; he gave William 7 and Samuel 10; how many will he have, if he share the remainder with himself and 2 brothers?

Let the pupil perform the last 16 questions on the slate, and hence notice, that

DIVISION is a short or compendious way of performing Subtraction.

Its object is to find how many times one number is contained in another. It consists of four parts; the dividend or number to be divided; the divisor or the number to be divided by; the quotient, which shows how many times the divisor is contained in the dividend; and the remainder, which is always less than the divisor, and of the same name of the dividend.

When the divisor is less than 13, the question should be performed by

SHORT DIVISION.

EXAMPLE.

1. Divide 948 dollars equally among 4 men.

Dividend.   In performing this question, inquire how
Divisor 4)948   many times 4, the divisor, is contained in 9,
———   which is 2 times, and 1 remaining; write the
Quotient 237   2 under the 9 and suppose 1, the remainder,
to be placed before the next figure of the dividend, 4, and the number would be 14. Then inquire how many times 4, the divisor, is contained in 14. It is found to be 3 times and 2 remaining. Write the 3 under the 4, and suppose the remainder, 2, to be placed before the next figure of the dividend, 8, and the number would be 28. Inquire again how many times 28 will contain the divisor. It is found to be 7 times, which we place under the 8. Thus we find each man receives 237 dollars.

From the above illustration, we deduce the following

RULE.

See how many times the divisor may be had in the first figure or figures of the dividend, and place the result immediately under that figure, and what remains suppose to be placed directly before the next figure of the dividend, and then inquire how many times these two figures will contain the divisor, and place the result as before; and so proceed until the question is finished.

2.	3.	4.
3)67856336	5)123456789	6)98786356

5.	6.	7.
7)126711001	8)33445567	9)1234567

8.
10)178985

9.
11)1667789

10.
12)9167856

			Quotients.	Rem.
11.	Divide 67893536	by 2	33946768	
12.	Divide 316789311	by 3	105596437	
13.	Divide 567895326	by 4	141973831	2
14.	Divide 123456789	by 5	24691357	4
15.	Divide 671678953	by 6	111946492	1
16.	Divide 166336711	by 7	23762387	2
17.	Divide 161331793	by 8	20166474	1
18.	Divide 161677678	by 9	17964186	4
19.	Divide 363895678	by 11	33081425	3
20.	Divide 164378956	by 12	13698246	4

When the divisor exceeds 12, the operation is performed by

LONG DIVISION,

as in the following   EXAMPLE.

1. A prize, valued at $ 3978, is to be equally divided among 17 men. What is the share of each?

OPERATION.
Dividend.
Divisor. 17) 3978 (234 Quotient.
    34   17
    ——   ——
    57   1638
    51   234
    ——   ——
    68 3978 Proof.
    68
    ——
    00 Remainder.

The object of this question is to find, how many times 3978 will contain 17, or how many times must 17 be subtracted from 3978, until nothing shall remain. We first inquire, how many times the first two figures of the dividend will contain the divisor; that is, how many times 39 (thousand) will contain 17. Having found it to be 2 times, we write 2 in the quotient and multiply it by the divisor, 17, and place their product 34 under 39, from which we subtract it, and find the remainder to be 5, to which we annex the next figure of the dividend, 7. And having found that 57 will contain the divisor 3 times, we write 3 in the quotient, multiply it by 17, and place the product 51 under 57, from which we subtract it, and to the remainder, 6, we annex the next figure of the dividend, 8, and inquire how many times 68 will contain the divisor, and find it to be 4 times. And having placed the product of 4 times 17 under 68, we find there is no remainder, and that 3978 will contain 17, the divisor, 234 times; that is, each man will receive 234 dollars. To prove our work is right, we reason thus. If one man receives 234 dollars, 17 men will receive 17 times as much, and 17 times 234 are 3978, the same as the dividend; and this operation is effected by multiplying the divisor by the quotient. The student will now see the propriety of the following

RULE.

Place the divisor before the dividend, and inquire how many times it is contained in a competent number of figures in the dividend, and place the result in the quotient ; multiply the figure in the quotient by the divisor, and place the product under those figures in the dividend, in which it was inquired how many times the divisor was contained ; subtract this product from the dividend, and, to the remainder, bring down the next figure of the dividend, and then inquire how many times this number will contain the divisor, and place the result in the quotient, and proceed as before, until all the figures in the dividend are brought down.

NOTE 1. — It will sometimes happen, that, after a figure is brought down, the number will not contain the divisor ; a cipher is then to be placed in the quotient, and another figure is to be brought down, and so continue until it will contain the divisor, placing a cipher each time in the quotient.

NOTE 2. — The remainder in all cases is less than the divisor, and of the same denomination of the dividend ; and, if at any time, we subtract the product of the figure in the quotient and the divisor from the dividend, and the remainder is more than the divisor, the figure in the quotient is not large enough.

PROOF.

Division may be proved by Multiplication, Addition, casting out the 9's, or by Division itself.

To prove it by Multiplication, the divisor must be multiplied by the quotient, and to the product, the remainder must be added, and if the result be like the dividend, the work is right See Example 1.

To prove it by Addition — Add up the several products of the divisor and quotient with the remainder, and if the result be like the dividend the work is right.　See Example 2.

To prove it by casting out the 9's — Cast the 9's out of the divisor, and place the remainder at the left hand of a *cross ;* then cast them out of the quotient, and place the remainder at the right hand of the cross, and lastly subtract the remainder from the dividend, and cast the 9's out of what may remain, and place the result at the top of the cross ; and if it be like the product of the figures at the sides of the cross (after the 9's are cast out of their products) the work is right.　See Example 3.

To prove it by Division itself, subtract the remainder from the dividend, and divide this number by the quotient, and the quotient found by this division, will be equal to the former divisor, when the work is right.　See Example 4.

3*

EXAMPLES.

```
            2.                              3.
97)147856(1524              328)678767(2069
   97*                          656
 ─────                        ─────
   508                         2276
   485*                        1968                 Proof.
 ─────                        ─────                   5
   235                         3087                 4×8
   194*                        2952                   5
 ─────                        ─────
   416                          135
   388*                       ─────
 ─────                        678632
    28*
 ─────
 147856
```

* NOTE. — The asterisms show the numbers to be added.

```
            4.                          Proof.
72)37895(526       -                    37895
   360                                     23
 ─────                                  ─────
   189                             526)37872(72
   144                                  3682
 ─────                                ─────
   455                                  1052
   432                                  1052
 ─────                                ─────
    23
```

				Quotients.	Remainders.
5.	Divide	6756785	by	35	193051	
6.	Divide	789636	by	46	17166	
7.	Divide	7967843	by	52	153227	44
8.	Divide	16785675	by	61	275175	
9.	Divide	675753	by	39	17327	
10.	Divide	5678910	by	82	69255	
11.	Divide	6716394	by	94	71451	
12	Divide	1167861	by	135	8650	111
13	Divide	7861783	by	87	90365	28
14.	Divide	1678567	by	365	4598	297
15.	Divide	87635163	by	387	226447	174
16.	Divide	34567890	by	6789	5091	5091
17.	Divide	78911007	by	36712	2149	16919
18.	Divide	78963167	by	45671	1728	43679
19.	Divide	671616589	by	61476	10924	52765
20.	Divide	471361876	by	36789	12812	21208
21.	Divide	300700801	by	10037	29959	2318
22.	Divide	10000000	by	9999	1000	1000
23.	Divide	199999999	by	123456		
24.	Divide	6716789513	by	7816789		
25.	Divide	1613716131	by	3151638		
26.	Divide 1219326311112635269	by	123456789			

CASE II.

To divide by any number with ciphers annexed.

Cut off the ciphers from the divisor, and the same number of digits from the right hand of the dividend; then divide the remaining figures as in the last case, and the quotient is the answer; and what remains written before the figures cut off is the true remainder.

EXAMPLE.

1. Divide 36378967 by 31000.

$$31,000)36378,967(1173$$

```
             31
             ─
             53
             31
             ─
            227
            217
            ───
            108
             93
            ───
```

15967 Remainder.

				Quotients.	Remainders.
2. Divide	32100	by	6000	5	2100
3. Divide	3167810	by	160000	19	127810
4. Divide	12345678	by	1400000	8	1145678
5. Divide	1637851	by	500000	3	137851
6. Divide	3678953	by	326100	11	91853
7. Divide	41111111	by	1100000	37	411111

CASE III.

To divide by an unit with ciphers annexed.

Cut off as many figures from the right hand of the dividend, as there are ciphers in the divisor, and the figures on the left hand of the separation will be the quotient, and those on the right hand the remainder.

1. Divide	123456789	by	10	12345678	9
2. Divide	987654321	by	100	9876543	21
3. Divide	122112347800	by	1000	122112347	800
4. Divide	89765432156	by	1000000	89765	432156

CASE IV.

To divide by a composite number, that is, a number, which is produced by the multiplying of two or more numbers.

Divide the dividend by one of those numbers, and the quotient thence arising by another, and so continue; and the last quotient will be the answer.

NOTE. — To find the true remainder, we multiply the last remainder by the last divisor but one, and, to the product, add the next preceding remainder ; we multiply this sum by the next preceding divisor, and to the product add the next preceding remainder; and so on, till we have gone through all the divisors, and remainders to the first.

This rule will be better understood by the pupil, after he has become acquainted with fractions.

EXAMPLES.

1. Divide 47932 by 72.

9)47932

8)5325 — 7

665 — 5 = 52

As 72 is equal to 9 times 8, we first divide the dividend by 9, and the quotient, thence arising by 8 ; and to find the true remainder, we multiply the last remainder, 5, by the first divisor, 9 ; and to the product add the first remainder, 7 ; and find the amount to be 52.

2. Divide 5371 by 192.

4)5371

6)1342 — 3

8)223 — 4

27 — 7 = 187

We find 192 equal to the product of 4 times 6 times 8. = 4 × 6 × 8 = 192. We therefore divide by these *factors*, as in the last example. To find the true remainder, we multiply the last remainder, 7, by the last divisor but one, 6 ; and to the product add the last remainder but one, 4 ; this sum we multiply by the first divisor, 4 ; and to the product add the first remainder, 3 ; and find the amount to be 187.

					Quotients.	Remainders.
3. Divide	7691 by	24 =	4 × 6		320	11
4. Divide	8317 by	27 =	3 × 9		308	1
5. Divide	3116 by	81 =	9 × 9		38	38
6. Divide	61387 by	121 =	11 × 11		507	40
7. Divide	19917 by	144 =	12 × 12		138	45
8. Divide	91746 by	336 =	6 × 7 × 8		273	18
9. Divide	376785 by	315 =	5 × 7 × 9		1196	45

MISCELLANEOUS EXAMPLES.

1. What number multiplied by 1728 will produce 1705536 ?
Ans. 987.

2. If a garrison of 987 men are supplied with 175686 pounds of beef, how much will there be for each man ? Ans. 178 lbs.

3. In one dollar there are 100 cents, how many dollars in 697800 cents ?   Ans. $ 6978

4. In one pound there are 16 ounces, how many pounds are in 111680 ounces ?   Ans. 6980 lbs.

5. A dollar contains 6 shillings, how many dollars are in 5868 shillings ?   Ans. $ 978.

6. The President of the United States receives a salary of $ 25,000 ; what does he receive per month ?   Ans. $ 2083⅓.

7. A man receiving $ 96 for 8 months' labor ; what does he receive for 1 month ?   Ans. $ 12.

8. The distance from Haverhill to Boston is 30 miles ; and if a man travel 6 miles an hour, how long will he be in going this distance ?   Ans. 5 hours.

9. The annual revenue of a gentleman being $ 8395, how much per day is that equivalent to, there being 365 days in a year ?   Ans. $ 23.

10. The car on the Liverpool railroad goes at the rate of 65 miles an hour ; how long would it take to pass round the globe, the distance being about 25,000 miles ?   Ans 384$\frac{8}{13}$ hours.

11. How much sugar at $ 15 per cwt. may be bought for $ 405 ?   Ans. 27 cwt.

12. In 6789560 shillings how many pounds, there being 20 shillings in a pound ?   Ans. 339478 pounds.

13. The Bible contains 31,173 verses ; how many must be read each day, that the book may be read through in a year ?
     Ans. 85$\frac{148}{365}$ verses.

14. In 123456720 minutes how many hours ?
     Ans. 2057612 hours.

15. A gentleman possessing an estate of $ 66,144, bequeathed one-fourth to his wife, and the remainder was to be divided between his 4 children ; what was the share of each ?
     Ans. $ 12,402.

16. A man disposed of a farm, containing 175 acres at $ 87 per acre ; of the avails he distributed $ 1234 for charitable purposes ; $ 197 was expended for the purchase of a horse and chaise ; the remainder was divided between 6 gentlemen and 8 ladies, and each lady was to receive twice as much as a gentleman ; what was the share of each ?
     Ans. $ 627 for a gentleman, and $ 1254 for a lady.

17. If there are 160 square rods in an acre, how many acres are in 1086240 square rods?   Ans. 6789 acres.

18. If 144 square inches make one square foot, how many square feet in 14222160 square inches?   Ans. 98765 feet.

19. What number is that, which being multiplied by 24, the product divided by 10, the quotient multiplied by 2, 32 subtracted from the product, the remainder divided by 4, and 8 subtracted from the quotient, the remainder shall be 2 ?   Ans. 15.

20. What is the difference between half a dozen dozen, and six dozen dozen ?   Ans. 792.

SECTION VI.

TABLES OF MONEY, WEIGHTS, AND MEASURES.

UNITED STATES MONEY.

10 Mills	make	1 Cent,	marked	c.
10 Cents	"	1 Dime,	"	d.
10 Dimes	"	1 Dollar,	"	$.
10 Dollars	"	1 Eagle,	"	E.

Mills.		Cents.		Dimes.		Dollars.		Eagle.
10	=	1						
100	=	10	=	1				
1000	=	100	=	10	=	1		
10000	=	1000	=	100	=	10	=	1

ENGLISH MONEY.

4 Farthings	make	1 Penny,	marked	d.
12 Pence	"	1 Shilling,	"	s.
20 Shillings	"	1 Pound,	"	£.

qrs.		d.		s.		£.
4	=	1				
48	=	12	=	1		
960	=	240	=	20	=	1

FRENCH MONEY.

100 Centimes   make   1 Franc   =   .1875 dollar.

TROY WEIGHT.

24 Grains	make	1 Pennyweight,	marked	dwt.
20 Pennyweights	"	1 Ounce,	"	oz.
12 Ounces	"	1 Pound,	"	lb.

grs.		dwt.		oz.		lb.
24	=	1				
480	=	20	=	1		
5760	=	240	=	12	=	1

By this weight are weighed gold, silver, and jewels.

NOTE.—"The original of all weights, used in England, was a grain or corn of wheat, gathered out of the middle of the ear ; and, being well dried, 32 of them were to make one pennyweight, 20 pennyweights one ounce, and 12 ounces one pound. But in later times, it was thought sufficient to divide the same pennyweight into 24 equal parts, still called grains, being the least weight now in common use ; and from hence the rest are computed."

APOTHECARIES' WEIGHT.

20 Grains	make	1 Scruple,	marked	sc. or Ɗ
3 Scruples	"	1 Dram,	"	dr. or ℨ
8 Drams	"	1 Ounce,	"	oz. or ℥
12 Ounces	"	1 Pound,	"	lb. or ℔

gr.		sc.		dr.		oz.		lb.
20	=	1						
60	=	3	=	1				
480	=	24	=	8	=	1		
5760	=	288	=	96	=	12	=	1

Apothecaries mix their medicines by this weight; but buy and sell by Avoirdupois. The pound and ounce of this weight are the same as in Troy Weight.

AVOIRDUPOIS WEIGHT.

16 Drams	make	1 Ounce,	marked	oz.
16 Ounces	"	1 Pound,	"	lb.
28 Pounds	"	1 Quarter,	"	qr.
4 Quarters	"	1 Hundred Weight,	"	cwt.
20 Hundred Weight	"	1 Ton,	"	ton.

dr.		oz.		lb.		qr.		cwt.		ton.
16	=	1								
256	=	16	=	1						
7168	=	448	=	28	=	1				
28672	=	1792	=	112	=	4	=	1		
573440	=	35840	=	2240	=	80	=	20	=	1

By this weight are weighed almost every kind of goods, and all metals except gold and silver. By a late law of Massachusetts, the cwt. contains 100 lbs. instead of 112 lbs.

LONG MEASURE.

3 Barleycorns	make	1 Inch,	marked	in.
12 Inches	"	1 Foot,	"	ft.
3 Feet	"	1 Yard,	"	yd.
6 Feet	"	1 Fathom,	"	fth.
5½ Yards, or 16½ feet	"	1 Rod, or pole,	"	rd.
40 Rods	"	1 Furlong,	"	fur.
8 Furlongs	"	1 Mile,	"	m.
3 Miles	"	1 League,	"	lea.
69½ Miles nearly	"	1 Degree,	" Deg. or °
360 Degrees	"	1 Circle of the Earth.		

in.		ft.		yd.		rd.		fur.		m.
12	=	1								
36	=	3	=	1						
198	=	16½	=	5½	=	1				
7920	=	660	=	220	=	40	=	1		
63360	=	5280	=	1760	=	320	=	8	=	1

CLOTH MEASURE.

2¼ Inches	make	1 Nail,	marked	na.
4 Nails	"	1 Quarter of a Yard,	"	qr.
4 Quarters	"	1 Yard,	"	yd.
3 Quarters	"	1 Ell Flemish,	"	E. F.
5 Quarters	"	1 Ell English,	"	E. E.
4 Quarters 1½ inch	"	1 Ell Scotch,	"	E. S.

SQUARE MEASURE.

144 Square inches	make	1 Square foot,	marked	ft.
9 Square feet	"	1 Square yard,	"	yd.
30¼ Square yards	"	1 Square rod or pole,	"	p.
272¼ Square feet	"	1 Square rod or pole,	"	p.
40 Square rods or poles	"	1 Rood,	"	R.
4 Roods	"	1 Acre,	"	A.
640 Acres	"	1 Square mile,	"	S. M.

in.	ft.				
144 =	1	yd.			
1596 =	9 =	1	p.		
39204 =	272¼ =	30¼ =	1	R.	
1568160 =	10890 =	1210 =	40 =	1	A.
6272640 =	43560 =	4840 =	160 =	4 =	1 S.M.
4014489600 =	27878400 =	3097600 =	102400 =	2560 =	640 = 1

DRY MEASURE.

2 Pints	make	1 Quart,	marked	qt.
4 Quarts	"	1 Gallon,	"	gal.
2 Gallons	"	1 Peck,	"	pk.
4 Pecks	"	1 Bushel,	"	bu.
36 Bushels	"	1 Chaldron,	"	ch.

pts.		gal.					
8	=	1		pk.			
16	=	2	=	1		bu.	
64	=	8	=	4	=	1	ch.
2304	=	288	=	144	=	36	= 1

NOTE.—This measure is applied to all Dry Goods, as Corn, Fruit, Salt, Coals, &c. A Winchester Bushel is 18½ inches in diameter, and 8 inches deep. The standard Gallon Dry Measure contains 268⅜ cubic inches.

ALE AND BEER MEASURE.

2 Pints	make	1 Quart,	marked	qt.
4 Quarts	"	1 Gallon,	"	gal.
36 Gallons	"	1 Barrel,	"	bar.
54 Gallons	"	1 Hogshead,	"	hhd.
2 Hogsheads	"	1 Butt,	"	butt.
2 Butts	"	1 Tun,	"	tun.

pts.		qt.									
2	=	1		gal.							
8	=	4	=	1		bar.					
288	=	144	=	36	=	1		hhd.			
432	=	216	=	54	=	1½	=	1		butt.	
864	=	432	=	108	=	3	=	2	=	1	

NOTE. — By a law of Massachusetts, the Barrel for Cider and Beer shall contain 32 Gallons. The Ale Gallon contains 282 cubic or solid inches.

WINE MEASURE.

4 Gills	make	1 Pint,	marked	pt.
2 Pints	"	1 Quart,	"	qt.
4 Quarts	"	1 Gallon,	"	gal.
42 Gallons	"	1 Tierce,	"	tier.
63 Gallons, or 1½ Tierces	"	1 Hogshead,	"	hhd.
2 Tierces	"	1 Puncheon,	"	pun.
2 Hogsheads	"	1 Pipe or Butt,	"	pi.
2 Pipes, or 4 Hhds.	"	1 Tun,	"	tun.

pts.		qt.											
2	=	1		gal.									
8	=	4	=	1		tier.							
336	=	168	=	42	=	1		hhd.					
504	=	252	=	63	=	1½	=	1		pun.			
672	=	336	=	84	=	2	=	1⅓	=	1		pi.	
1008	=	504	=	126	=	3	=	2	=	1½	=	1	tun.
2016	=	1008	=	252	=	6	=	4	=	3	=	2	= 1

NOTE. — The Wine Gallon contains 231 cubic inches.

OF TIME.

60 Seconds, or 60″	make	1 Minute,	marked	m.
60 Minutes	"	1 Hour,	"	h.
24 Hours	"	1 Day,	"	d.
7 Days	"	1 Week,	"	w.
4 Weeks	"	1 Month,	"	mo.
13 Months, 1 day, 6 hours, or 365 days, 6 hours,	}	1 Julian Year,		y.
12 Calendar months	"	1 Year,		y

sec.		m.									
60	=	1		h.							
3600	=	60	=	1		d.					
86400	=	1440	=	24	=	1		w.			
604800	=	10080	=	168	=	7	=	1		mo.	
2419200	=	40320	=	672	=	28	=	4	=	1	y.
31557600	=	525960	=	8766	=	365¼				=		1

NOTE. — The true solar year is the time measured from the sun's leaving either equinox or solstice, to its return to the same again. A periodical year

4

is the time the earth revolves round the sun, and is 365 d. 6 h. 9 m. 14¼ sec. and is often called the Siderial year. The civil year is that which is in common use among the different nations of the world, and contains 365 days for three years in succession, but every fourth year contains 366 days. When any year can be divided by four, without any remainder, it is *leap year*, and has 366 days. The days in each month are stated in the following distich : —

> Thirty days hath September,
> April, June, and November ;
> All the rest have thirty-one,
> Except February alone,
> Which hath but twenty-eight,
> Except leap year, when it hath twenty-nine.

	w.	d.	h.		mo.	d.	h.		
Or,	52	1	6	=	13	1	6	=	1 Julian Year.

	day.		h.		m.		sec.		
But,	365		5		48		57	=	1 Solar Year.

	day.		h.		m.		sec.		
And,	365		6		9		14½	=	1 Siderial Year.

CIRCULAR MOTION.

60	Seconds	make	1 Prime minute,	marked	$''$
60	Minutes	"	1 Degree,	"	$\circ$
30	Degrees	"	1 Sign,	"	s.

12 Signs, or 360 Degrees, the whole great circle of the zodiac.

$''$.		$'$.				s.			
60	=	1							
3600	=	60	=	1					
108000	=	1800	=	30	=	1			
1296000	=	21600	=	360	=	12	=	zodiac.

MEASURING DISTANCES.

7 $\frac{92}{100}$	Inches	make	1 Link.
25	Links	"	1 Pole.
100	Links	"	1 Chain.
10	Chains	"	1 Furlong
8	Furlongs	"	1 Mile.

Inches.		Link.		Pole.		Chain.		Furlong.		Mile
7 $\frac{92}{100}$	=	1								
192	=	25	=	1						
792	=	100	=	4	=	1				
7920	=	1000	=	40	=	10	=	1		
63360	=	8000	=	320	=	80	=	8	=	1

SOLID MEASURE.

1728 Inches	make	1 Foot.
27 Feet	"	1 Yard.
40 Feet of round timber, or 50 feet of hewn timber,	"	1 Ton.
128 Feet, i. e. 8 in length, 4 in breadth, and 4 in height,	"	1 Cord of wood.

MISCELLANEOUS TABLE.

A gallon of train oil	weighs	7½	pounds.
A stone of butcher's meat	"	8	"
A gallon of molasses	"	11	"
A stone of iron	"	14	"
A tod	"	28	"
A firkin of butter	"	56	"
A firkin of soap	"	94	"
A quintal of fish	"	100	"
A weigh	"	182	"
A sack	"	364	"
A puncheon of foreign prunes	"	1120	"
A last	"	4368	"
A fother of lead	"	19½	cwt.
A barrel of anchovies	"	30	pounds.
" raisins	"	112	"
" flour	"	196	"
" pork or beef	"	200	"
" soap	"	256	"
" shad or salmon in Conn. or New York,	"	200	"
" fish in Massachusetts,	is	30	gallons.
" cider and beer	"	32	"
" herrings in England	"	32	"
" salmon, or eels do.	"	42	"
8 bushels of salt, measured on board the vessel	"	1	hogshead
7½ do. measured on shore,	"	1	"
3 hoops	make	1	cast.
40 casts	"	1	hundred.
10 hundred	"	1	thousand.
12 units, or things,	"	1	dozen.
12 dozen	"	1	gross.
144 dozen	"	1	great gross.

COMPOUND ADDITION.

COMPOUND ADDITION is the adding together of two or more numbers of different denominations.

RULE.

Write all the given numbers of the same denomination under each other; then add the numbers of the first denomination together, and divide the sum by so many as make one of the next denomination; set the remainder under its column, and add the quotient to the next column; which add together and divide as before; thus proceed to the last denomination, under which, place its whole sum.

EXAMPLES.

UNITED STATES' MONEY.

$	cts.	m.		$	cts.	m.		E.	$	cts.	m.
325	67	3		28	15	6		71	3	41	5
186	35	8		16	16	3		61	6	82	6
161	89	9		63	81	5		16	1	96	2
987	15	8		14	61	6		41	7	82	1
891	61	6		38	74	5		54	3	36	3
176	81	3		16	16	8		41	9	48	5
2729	51	7									

ENGLISH MONEY.

£.	s.	d.		£.	s.	d.		£.	s.	d.	qr.
471	16	9		28	6	9¼		31	17	9	2
147	17	8		15	16	11¼		16	16	6	1
613	13	11		31	13	10¼		16	11	11	1
115	11	7		14	16	9		19	19	9	3
41	19	6		17	17	7¼		61	17	1	3
48	12	2		32	18	8¼		14	14	4	2
1439	11	7									

TROY WEIGHT.

lb.	oz.	dwt.	grs.		lb.	oz.	dwt.	grs.
16	11	19	23		123	9	7	13
31	10	18	16		92	11	17	14
63	9	12	15		49	7	13	21
17	8	13	12		13	10	10	20
61	7	12	16		47	9	19	23
17	6	17	22		51	5	15	15
209	7	15	8					

APOTHECARIES' WEIGHT.

9.

lb.	℥.	ℨ.	Ə.	gr
27	11	7	2	19
16	10	6	1	13
41	9	3	2	16
38	10	5	2	14
41	4	4	1	11
16	6	6	2	6
183	6	3	1	19

10.

lb.	℥.	ℨ.	Ə.	gr
37	9	6	1	18
14	4	4	2	11
61	6	3	2	6
41	4	7	2	16
39	8	4	1	12
51	11	7	2	19

AVOIRDUPOIS WEIGHT.

11.

Ton.	cwt.	qr.	lb.	oz.	dr.
61	19	3	27	15	15
63	13	3	16	11	11
51	12	3	17	7	6
61	16	1	11	12	12
13	13	3	12	13	15
71	18	2	13	14	14
324	15	2	16	12	9

12.

cwt.	qr.	lb.	oz.	dr.
61	2	11	11	14
16	3	15	15	11
41	3	13	9	9
38	2	11	10	10
42	1	9	8	13
31	3	27	11	12

LONG MEASURE.

13.

Deg	m.	fur.	rd.	ft.	in.	br.
17	69	7	39	16	11	2
61	62	3	17	12	9	1
16	16	6	16	13	10	2
48	19	3	15	15	6	1
17	58	6	33	14	7	1
33	35	5	19	9	9	2
195	54	5	24	1	1	0

14.

M.	fur.	rd.	yd.	ft.	in.	br.
69	7	31	5	2	11	1
16	6	16	4	1	6	2
61	7	32	3	2	10	1
73	3	16	4	2	9	2
19	4	14	1	1	8	2
75	5	25	5	2	7	1

CLOTH MEASURE.

15.

yd.	qr.	na.	in.
37	3	3	2
61	3	1	1
13	2	2	2
32	1	1	1
61	2	2	2
22	1	3	0
229	3	3	1¼

16.

EE.	qr.	na.	in.
671	1	1	1
161	3	3	2
617	3	1	2
178	3	2	1
717	2	1	2
166	3	2	1

4*

LAND OR SQUARE MEASURE.

17.

A.	R.	p.	ft.	in.
761	3	37	260	125
131	2	16	135	112
613	1	14	116	131
161	3	13	116	123
321	2	31	97	96
47	3	19	91	48
2038	1	13	2	95

18.

A.	R.	p.	ft.
38	1	39	272
61	3	38	167
35	3	19	198
47	3	16	271
86	2	13	198
46	1	14	269

SOLID MEASURE.

19.

Ton.	ft.	in.
29	36	1279
69	19	1345
67	18	1099
71	14	1727
43	35	916
53	17	1719
335	23	1173

20.

Cord.	ft.	in.
61	127	1161
37	89	1711
61	98	1336
43	56	1678
91	119	1357
81	115	1129

WINE MEASURE.

21.

Tun.	hhd.	gal.	qt.	pt.
61	3	62	3	1
39	2	16	1	1
68	3	57	2	1
87	3	45	3	1
47	2	59	3	1
47	3	39	2	1
354	0	30	1	0

22.

hhd.	gal.	qt.	pt.
67	15	3	1
16	16	3	0
39	16	3	0
47	62	1	1
43	57	3	0
71	61	3	1

ALE AND BEER MEASURE.

23.

Tun.	hhd.	gal.	qt.
46	3	50	3
91	2	48	3
17	3	18	0
81	3	38	2
41	1	47	1
37	2	29	3
317	2	17	0

24.

hhd.	gal.	qt.	pt.
161	53	3	1
371	52	3	1
98	19	1	0
47	43	1	0
61	43	1	1
42	27	3	1

DRY MEASURE.

25.

Bu.	pk.	qt.	pt.
37	3	5	1
61	3	7	1
32	2	2	0
71	1	6	1-
61	1	3	1
32	3	3	1
298	0	4	1

26.

ch.	bu.	pk.	qt.
16	31	3	3
39	31	3	1
14	16	3	1
55	15	3	0
71	17	3	1
42	14	3	1

TIME.

27.

y.	mo.	w.	d.	h.	m.	s.
57	11	3	6	23	29	55
31	11	1	3	19	19	39
46	9	2	2	17	28	56
43	10	1	1	18	17	48
32	9	1	3	16	23	28
14	1	1	5	22	28	16
227	2	0	3	21	28	2

28.

y.	m.	d.	h.
13	5	29	17
61	11	17	21
15	9	19	16
61	10	25	23
41	4	16	17
18	5	9	6

CIRCULAR MOTION.

29.

s.	°.	'.	".
4	29	59	59
6	17	17	29
11	16	56	58
9	13	46	51
5	27	16	42
2	25	17	17
5	10	35	-16

30.

s.	°.	'.	".
11	11	16	51
6	6	6	16
9	14	56	56
3	29	29	49
9	17	18	58
6	13	13	52

MEASURING DISTANCES.

31.

m.	fur.	ch.	p.	l.
17	7	9	3	24
16	3	4	1	15
27	4	6	2	17
18	6	3	3	21
61	7	7	2	16
17	1	8	2	19
160	0	1	1	12

32.

m.	fur.	ch.	p.	l.
27	4	3	1	21
29	3	1	3	23
67	3	3	1	19
21	7	1	3	16
16	7	9	3	13
31	4	8	1	20

COMPOUND SUBTRACTION.

COMPOUND SUBTRACTION teaches to find the difference between numbers of different denominations.

RULE.*

Write those numbers under each other, which are of the same denomination, the less compound number under the greater; begin with the lowest denomination, and, if it exceeds the number over it, add as many units to the upper number, as it takes of that denomination to make the next higher; then subtract and carry one to the next number of the subtrahend.

EXAMPLES.

FEDERAL MONEY.

	1.				2.	
$.	cts.	m.		$.	cts.	m.
169	81	3		681	16	7
85	93	8		189	43	8
83	87	5				

ENGLISH MONEY.

	3.				4.	
£.	s.	d.		£.	s.	d.
87	16	3¼		617	11	5½
19	17	9½		181	15	8¼
67	18	5¾				

TROY WEIGHT.

	5.				6.	
lb.	oz.	dwt.	gr.	lb.	oz.	dwt.	gr.
71	3	12	15	58	5	12	10
16	10	17	20	19	9	17	21
54	4	14	19				

* The reason of this rule will readily appear, from what was said in Simple subtraction; for the adding to the upper line depends on the same principle, and is only different, as the numbers to be subtracted are of different denomitions.

APOTHECARIES' WEIGHT.

		7.		
lb.	℥.	ℨ.	Ϩ.	gr.
71	1	3	1	13
18	6	7	2	19
52	6	3	1	14

		8.		
lb.	℥.	ℨ.	Ϩ.	gr.
15	2	2	0	15
9	9	1	1	18

AVOIRDUPOIS WEIGHT.

		9.			
T.	cwt.	qr.	lb.	oz.	dr.
71	18	1	13	1	13
19	19	2	16	8	5
51	18	2	24	9	8

	10.		
cwt.	qr.	lb.	oz.
73	1	15	13
19	1	19	15

CLOTH MEASURE.

	11.		
yd.	qr.	na.	in.
67	1	1	1
18	2	2	2
48	2	2	1½

	12.	
EE.	qr.	na.
51	2	3
19	3	1

LONG MEASURE.

		13.			
m.	fur.	rd.	ft.	in.	bar.
16	7	18	3	2	1
9	7	19	16	8	2
6	7	38	2	11	2

			14.				
deg.	m.	fur.	rd.	yd.	ft.	in.	bar.
38	41	3	29	2	1	7	2
29	36	5	31	3	1	9	1

LAND OR SQUARE MEASURE.

		15.		
A.	R.	p.	ft.	in.
56	1	19	119	110
17	3	13	127	113
38	2	5	264	33

		16.			
A.	R.	p.	yd.	ft.	in.
13	1	15	19	1	17
9	3	16	30	5	19

SOLID MEASURE.

	17.	
Tons.	ft.	in.
49	13	1611
18	15	1719
30	37	1620

	18.	
Cords.	ft.	in.
361	47	1178
197	121	1617

WINE MEASURE

| | | 19 | | | | | | 20. | | |
Tun.	hhd.	gal.	qt.	pt.			hhd.	gal.	qt.	pt.
79	3	19	1	1			16	1	1	0
11	1	28	2	1			9	2	2	1
68	1	53	3	0						

ALE AND BEER MEASURE.

| | | 21. | | | | | | 22. | |
Tun.	hhd.	gal.	qt.	pt.			hhd.	gal.	qt.
63	1	15	1	0			769	18	1
19	3	16	3	1			191	19	3
43	1	52	1	1					

DRY MEASURE.

| | 23. | | | | | | 24. | | |
ch.	bu.	pk.	qt.			ch.	bu.	pk.	qt.
56	2	1	1			39	12	2	1
38	3	1	2			12	25	3	5
17	34	3	7						

TIME.

| | | 25. | | | | | | | 26. | | | |
mo	d.	h.	m.	s.		y.	m.	w.	d.	h.	m.	s.
6	16	13	27	19		48	9	2	5	19	27	31
1	22	16	41	37		19	10	3	7	21	38	56
4	23	20	45	42								

CIRCULAR MOTION.

| | | 27. | | | | | 28. | |
S.	°.	'.	''.		S.	°.	'.	''.
6	11	12	48		4	19	41	22
9	8	15	56		1	22	19	28
9	2	56	52					

MEASURING DISTANCES.

| | | 29. | | | | | | 30. | | |
m.	fur.	ch.	p.	l.		m.	fur.	ch.	p.	l.
21	1	3	2	19		28	6	1	2	18
19	2	1	3	21		15	7	3	1	19
1	7	1	2	23						

EXERCISES IN COMPOUND ADDITION AND SUBTRACTION.

1. What is the sum of 16£. 5s. 8d. 2qr. — 31£. 16s. 11d. 3qr. — 21£. 11s. 1qr. — 19£. 0s. 10d. 3qr. — 13£. 13s. 7d. 3qr. and 28£. 17s. 5d. 1qr. ?   Ans. 131£. 5s. 8¼d.

2. Bought of a London tailor a vest for 1£. 13s. 4d., a coat for 7£. 12s. 9d., pantaloons for 2£. 3s. 9d., and surtout for 9£. 8s. 0d. ; what was the whole amount ?
Ans. 20£. 17s. 10d.

3. Bought a silver tankard, weighing 1lb. 8oz. 17dwt. 14gr., a silver can, weighing 1lb. 2oz. 12dwt., a porringer, weighing 11oz. 19dwt. 20gr. and three dozen of spoons, weighing 1lb. 9oz. 15dwt. 10gr. ; what was the whole weight ?
Ans. 5lb. 9oz. 4dwt. 20gr.

4. What is the weight of a mixture of 3℔. 4℥. 2ʒ. 2Ә. 14gr. of aloe, 2℔. 7℥. 6ʒ. 1Ә. 13gr. of picra, and 1℔. 10℥. 1ʒ. 2Ә. 17gr. of saffron ?   Ans. 7℔. 10℥. 3ʒ. 1Ә. 4gr.

5. Add 32℔. 9℥. 1ʒ. 2Ә. 14gr. ; 13℔. 7℥. 6ʒ. 1Ә. 13gr. ; and 16℔. 11℥. 7ʒ. 1Ә. 12 gr. together.
Ans. 63℔. 4℥. 7ʒ. 2Ә. 19gr.

6. Sold 4 loads of hay ; the first weighed 27cwt. 3qr. 18lb. ; second 31cwt. 1qr. 15lb. ; third 19cwt. 1qr. 15lb. ; and fourth 38cwt. 2qr. 27lb. ; what is the weight of the whole ?
Ans. 117cwt. 1qr. 19lb.

7. Bought 5 pieces of broadcloth ; the first contained 17yd. 3qr. 2na. ; second, 13yd. 2qr. 1na. ; the third 37yd. 1qr. 3na. ; the fourth 27yd. 1qr. 2na. ; and the fifth 29yd. 1qr. 2na. ; what was the whole quantity purchased ?   Ans. 175yd. 2qr. 2na.

8. A pedestrian travelled, the first week, 371m. 3fur. 37rds. 5yd. 2ft. 10in. ; the second week, 289m. 2fur. 18rds. 3yds. 1ft. 9in. ; and the third week he travelled 399m. 7 fur. 3ft. 11in. ; how many miles will he have travelled ?
Ans. 1060m. 5fur. 16rds. 5yds. 1ft.

9. A man has 3 farms ; the first contains 186A. 3R. 14p. ; the second 286A. 17p. ; and the third, 115A. 2R. ; how much do they all contain ?   Ans. 588A. 1R. 31p.

10. The Moon is 5s. 18°. 14'. 17". east of the Sun ; Jupiter is 7s. 10°. 29'. 28". east of the Moon ; Mars is 11s. 12°. 11'. 56". east of Jupiter ; and Herschell is 7s. 18°. 38'. 15". east of Mars ; how far is Herschell from the Sun ?   Ans. 7s. 29°. 33'. 56".

11. I have 4 piles of wood ; the first contains 7 cords, 76ft. 1671in. ; the second 16c. 28ft. 56in. ; the third 29c. 127ft. 1000in. ; and the fourth 29c. 10ft. 1216in. ; how much is there in all ?   Ans. 82c. 115ft. 487in.

12. A vintner sold at one time 73hhd. 43gal. 3qt. 1pt. of wine; at another, 27hhd. 3gal.; at another, 15hhd. 3qts. 1pt.; and, at another, 161hhd. and 2qts.; how much did he sell in all?
                                   Ans. 276hhd. 48gal. 1qt.

13. A man has 3 sons; the first is 14y. 3m. 2w. 5d. old; the second is 9y. 10mo. 3w. 4d. 23h. 12m. 15sec.; and the third is 2y. 1mo. 3w. 2d. 7m.; what is the sum of their ages? And how much older is the first than the second?
                     Ans. 26y. 3mo. 1w. 4d. 23h. 19m. 15sec.
                      "   4y. 5mo. 3w. 0d. 0h. 47m. 45sec.

4. I have 73A. of land; if I should sell 5A. 3R. 1p. 7ft., how much should I have left?   Ans. 67A. 0R. 38p. 265¼ft.

15. A. owes B. £100; what will remain due after he has paid him 3s. 6¼d.?   Ans. £99, 16s. 5¼d.

16. It is about 25000 miles round the globe; if a man shall have travelled 43m. 17rds. 9in. how much will remain to be travelled?   Ans. 24956m. 7fur. 22rds. 15ft. 9in.

17. Bought 7 cords of wood; and 2 cords 78ft. having been stolen, how much remained?   Ans. 4c. 50ft.

18. I have 15 yards of cloth; having sold 3yds. 2qr. 1na., what remains?   Ans. 11yds. 1qr. 3na.

19. If a wagon loaded with hay weighs 43cwt. 2qr. 18lbs., and the wagon is afterwards found to weigh 9cwt. 3qr. 23lbs., what is the weight of the hay?   Ans. 33cwt. 2qr. 23lbs.

20. Bought a hogshead of wine, and by an accident 8gal. 3qts. 1pt. leaked out; what remains?   Ans. 54gal. 1pt.

21. I have 10A. 3R. 10p. of land; I have sold two house lots, one containing 1A. 2R. 13p., the other, 2A. 2R. 5p.; how much have I remaining?   Ans. 6A. 2R. 32p.

22. The Moon moves 13°. 10′. 35″. in a solar day, and the Sun 59′. 8″. 20‴.; now supposing them both to start from the same point in the heavens, how far will the Moon have gained on the Sun in 24 hours?   Ans. 12°. 11′. 26″. 40‴.

23. A farmer raised 136bush. of wheat; if he sells 49bu. 2pk 7qt. 1pt., how much has he remaining? Ans. 86bu. 1pk. 0qt. 1pt.

24. If from a stick of round timber, containing 2T. 18ft. 1410in., there be taken 38ft. 1720in., how much will be left?
                                   Ans. 1T. 19ft. 1418in.

25. If from 1℔. of ipecacuanna there be taken at one time 4ʒ. 2ℨ. 13gr. and at another, 3℥. 1ℨ. 2 Э. 14gr., how much will be left?   Ans. 4ʒ. 3ℨ. 2Э. 13gr.

26. A brewer has in one cellar 18bbls. 3gal. 2qt. of beer, and in another, 13bbls. 1p.; what is the whole quantity, and how much more is in one cellar than the other?
                              Ans. 31bbls. 3gal. 2qt. 1pt.
                               "   5bbls. 3gal. 1qt. 1pt.

27. If from $100.00 there be paid at one time $17.28.5, at another time, $10.00.5, and at another $37.15, how much will remain?
                                   Ans. $35.56.

SECTION VII.

MENTAL OPERATIONS IN REDUCTION.

1. In 8 farthings how many pence ? In 12 ? In 16 ? In 20 ? In 24 ? In 36 ? In 40? In 48 ? In 60 ? In 90 ? In 80 ?

2. How many farthings in 2 pence ? In 3 ? In 4 ? In 6 ? In 8 ? In 9 ? In 12 ? In 14 ? In 15 ? In 16 ? In 18 ?

3. In 3 shillings how many pence ? In 4 ? In 5 ? In 6 ? In 8 ? In 9 ? In 10 ? In 11 ? In 12 ? In 15 ? In 17 ? In 20 ?

4. In 24 pence how many shillings ? In 36 ? In 48 ? In 60 ? In 72 ? In 84 ? In 30 ? In 40 ? In 50 ? In 60 ? In 70 ?

5. In 2 pounds how many shillings ? In 4 ? In 5 ? In 6 ? In 8 ? In 10 ? In 20 ? In 12 ? In 14 ? In 15 ? In 20 ?

6. In 40 shillings how many pounds ? In 60 ? In 80 ? In 120 ? In 160 ? In 200 ? In 70 ? In 90 ? In 95 ? In 130 ? 135 ? In 150 ? In 155 ? In 170 ? In 199 ? In 500 ? In 1000 ?

7. In 3s. 6d. how many pence ? In 3s. 2d. ? In 4s. 2d. ? In 4s. 6d. ? In 4s. 9d. ? In 5s. 7d. ? In 5s. 9d.? In 6s. 3d. ? In 6s. 7d. ? In 6s. 11d. ? In 7s. 6d. ? In 8s. 3d. ? In 9s. 10d. ?

8. In $2\frac{1}{2}$ pence how many farthings ? In $3\frac{1}{4}$? In $4\frac{1}{4}$? In $5\frac{1}{4}$? In $6\frac{1}{4}$? In $7\frac{1}{4}$? In $9\frac{1}{4}$? In $10\frac{1}{4}$? In $11\frac{1}{4}$? In $11\frac{1}{2}$?

9. In 2£. 3s. how many shillings ? In 2£. 6s. ? In 5£. 5s. ? In 5£. 8s. ? In 6£. 8s. ? In 10£. 12s. ? In 7£. 8s. ?

10. In 2lbs. how many ounces ? In 3lbs. ? In 4lbs. ? In 6 ? In 8 ? In 10 ? In 12 ? In 15 ? In 20 ? In 24 ? In 25 ?

11. In 32 ounces how many pounds ? In 64 ? In 80 ? In 90 ?

12. In 3cwt. how many quarters ? In 5 ? In 7 ? In 8 ? In 10 ? In 12 ? In 15 ? In 16 ? In 20 ? In 24 ? In 25 ? In 30 ?

13. In 13 quarters how many cwt. ? In 16 ? In 20 ? In 32 ? In 36 ? In 60 ? In 66 ? In 68 ? In 70 ? In 72 ? In 80 ?

14. In 5 tons how many cwt. ? In 6 ? In 7 ? In 9 ? In 12 ? In 15 ? In 24 ? In 26 ? In 28 ? In 30 ? In 32 ? In 36 ?

15. In 40cwt. how many tons ? In 60 ? In 120 ? In 160 ? In 200 ? In 220 ? In 230 ? In 240 ? In 250 ? In 260 ?

16. In 57cwt. how many tons ? In 69 ? In 75 ? In 82 ? In 105 ? In 115 ? In 125 ? In 130 ? In 135 ? In 140 ?

17. In 5lbs troy how many ounces ? In 6 ? In 8 ? In 10 ? In 12 ? In 15 ? In 20 ? In 24 ? In 25 ? In 28 ? In 30 ?

18. In 36 ounces how many pounds troy ? In 60 ? In 72 ? In 84 ? In 90 ? In 95 ? In 98 ? In 100 ? In 110 ? In 120 ?

19. In 7 ounces how many pennyweights ? In 8 ? In 10 ? In 12 ? In 14 ? In 16 ? In 28 ? In 30 ? In 35 ? In 40 ?

20. In 40 grains how many pennyweights ? In 50 ? In 60 ? In 70 ? In 90 ? In 120 ? In 130 ? In 140 ? In 150 ? In 160 ?

21. In 7 scruples how many grains ? In 8 ? In 9 ? In 10 ?

22. In 40 grains how many scruples ? In 50 ? In 60 ? In 70 ? In 80 ? In 100 ? In 110 ? In 120 ? In 130 ? In 140 ?

5

23. In 2 yards how many quarters? In 3? In 5? In 10? In 12? In 15? In 16? In 20? In 24? In 26? In 30?

24. In 3 ells English how many quarters? In 7? In 12?

25. In 12 quarters how many yards? In 16? In 19? In 24? In 30? In 36? In 38? In 40? In 41? In 47?

26. In 6 quarters how many nails? In 6? In 7? In 8? In 12? In 14? In 16? In 20? In 22? In 30?

27. In 24 nails how many quarters? In 30? In 36? In 40? In 46? In 50? In 70? In 100? In 150? In 160?

28. In 100 pence how many shillings? In 144? In 160?

29. In 140 shillings how many pounds? In 150? In 160?

30. In 96 ounces how many lbs.? In 100? In 200? In 300?

31. In 56lbs. how many quarters? In 84? In 112? In 200?

32. In 8 quarters how many cwt.? In 12? In 16? In 28? In 44? In 48? In 50? In 60? In 66? In 70? In 80?

33. In 40cwt. how many tons? In 80? In 100? In 120? In 180? In 200? In 800? In 850? In 866? In 1000?

34. In 48qrs. how many dwt.? In 72? In 96? In 144?

35. In 9 scruples how many drams? In 12? In 18?

36. In 48 ounces, troy, how many lbs.? In 96? In 144?

37. In 12 quarters how many lbs.? In 16? In 28? In 40? In 48? In 56? In 60? In 100? In 120? In 130? In 150?

38. In 25 quarters how many ells English? In 30? In 60? In 70? In 100? In 110? In 120? In 125? In 150?

39. In 36 inches how many feet? In 48? In 60? In 144? In 180? In 190? In 100? In 120? In 130? In 140? In 150?

40. In 12 feet how many yards? In 15? In 24? In 30? In 48? In 60? In 66? In 90? In 100? In 120? In 140?

41. In 80 rods how many furlongs? In 120? In 720?

42. In 24 furlongs how many miles? In 32? In 48? In 72? In 80? In 90? In 96? In 100? In 200? In 300? In 400?

43. In 3 weeks how many days? In 4? In 5? In 6? In 7? In 8? In 9? In 12? In 14? In 20? In 25? In 30? In 50?

44. In 5 years how many months? In 3? In 4? In 8? In 9? In 10? In 12? In 24? In 30? In 32? In 36? In 40?

45. In 2 quarts how many pints? In 4? In 6? In 8? In 10? In 20? In 50? In 60? In 190? In 120? In 150?

46. In 5 years how many months? In 3? In 4? In 8? In 9? In 10? In 12? In 24? In 26? In 28? In 30? In 32?

47. In 2 quarts how many pints? In 4? In 6? In 8? In 10? In 20? In 50? In 60? In 70? In 75? In 80? In 90?

48. In 3 gallons how many quarts? In 4? In 5? In 6? In 8? In 9? In 10? In 20? In 40? In 50? In 100? In 200?

49. In 2 hogsheads how many gallons? In 3? In 4? In 5? In 6? In 7? In 8? In 10? In 12? In 14? In 20? In 24?

50. In 3 pipes how many hogsheads? In 5? In 7? In 8? In 10? In 12? In 20? In 24? In 26? In 30? In 40?

51. In 2 tuns how many hogsheads? In 5? In 7? In 8? In 9? In 10? In 12? In 14? In 16? In 20? In 24?

52. In 2 bushels how many pecks ? In 3 ? In 5 ? In 6 ? In 3 ? In 10 ? In 15 ? In 20 ? In 60 ? In 70 ? In 80 ?

53. In 3 pecks how many quarts ? In 4 ? In 5 ? In 10 ? In 12 ? In 15 ? In 20 ? In 25 ? In 30 ? In 40 ? In 50 ? In 60 ?

54. In 2 feet how many square inches ? In 4 ? In 5 ? In 10 ? In 20 ? In 50 ? In 60 ? In 70 ? In 80 ? In 90 ? In 100 ?

55. In 3 square yards how many feet ? In 4 ? In 6 ? In 7 ? In 8 ? In 10 ? In 20 ? In 30 ? In 40 ? In 50 ? In 60 ?

56. In 5 roods how many rods ? In 10 ? In 20 ? In 30 ?

57. In 2 acres how many roods ? In 5 ? In 10 ? In 20 ? In 30 ? In 40 ? In 50 ? In 100 ? In 120 ? In 150 ? In 160 ?

58. In 3 acres how many rods ? In 5 ? In 10 ? In 50 ?

59. In 5 minutes how many seconds ? In 10 ? In 20 ?

60. In 2 hours how many minutes ? In 3 ? In 4 ? In 5 ? In 8 ? In 12 ? In 15 ? In 20 ? In 30 ? In 35 ? In 40 ?

61. In 3 days how many hours ? In 5 ? In 6 ? In 8 ? In 10 ? In 15 ? In 20 ? In 25 ? In 50 ? In 100 ? In 120 ?

62. In 2 cents how many mills ? In 3 ? In 4 ? In 6 ? In 8 ? In 9 ? In 12 ? In 16 ? In 18 ? In 25 ? In 35 ? In 45 ?

63. In 20 mills how many cents ? In 30 ? In 40 ? In 60 ? In 70 ? In 80 ? In 90 ? In 100 ? In 120 ? In 150 ? In 160 ?

64. In 3 dimes how many cents ? In 5 ? In 7 ? In 9 ?

65. In 30 cents how many dimes ? In 80 ? In 100 ? In 200 ?

66. In 2 dollars how many cents ? In 5 ? In 6 ? In 7 ? In 10 ? In 20 ? In 30 ? In 50 ? In 100 ? In 120 ? In 130 ?

67. In 300 cents how many dollars ? In 400 ? In 500 ? In 600 ? In 800 ? In 900 ? In 1000 ? In 10000 ? In 100000 ?

68. In 5 eagles how many dollars ? In 6 ? In 10 ? In 20 ? In 25 ? In 35 ? In 75 ? In 100 ? In 120 ? In 200 ? In 300 ?

69. In 20 dollars how many eagles ? In 30 ? In 50 ? In 70 ? In 100 ? In 200 ? In 1000 ? In 1500 ? In 100000 ?

70. In 4 yards how many nails ? In 8 ? In 10 ? In 12 ?

71. In 2 Ells English how many yards ? In 3 ? In 4 ? In 5 ?

72. What is the difference between 3 square feet, and 3 feet square ?

73. How many inches round one foot square ? 2 feet ? 3 feet ?

The student will now perceive that the object of

REDUCTION is to change numbers of one denomination to another without losing their value.

It consists of two parts, Descending and Ascending. The former is performed by Multiplication, the latter by Division.

Reduction Descending teaches to bring numbers of a higher denomination to a lower ; as, to bring pounds into shillings, or tons into hundred weights.

Reduction Ascending teaches to bring numbers of a lower denomination into a higher ; as, to bring farthings into pence, or shillings into pounds.

REDUCTION DESCENDING.

EXAMPLE.

1. In 48£. 12s. 7d. 2qr, how many farthings?

£.   s.   d. qr.
48 12 7 2
20
—————
972 + 12
12
—————
11671 + 7
4
—————
46686 + 2

In this example, we multiply the 48£. by 20, because it takes 20 shillings to make a pound; and, to this product, we add the 12s. in the question. Then multiply by 12, because it takes 12 pence to make one shilling; and, to the product, we add the 7 pence in the question. We then multiply by 4, the number of farthings in a penny, and, to the product, we add the 2 farthings, and the work is done.

From the above example and illustration, we deduce the following

RULE.

Multiply the highest denomination given by so many of the next less, as will make one of that greater; and so proceed, until it is brought to the denomination required, observing to bring in the lower denominations to their respective places.

NOTE 1.— To multiply by a ½, we divide the multiplicand by 2, and to multiply by a ¼, we divide by 4.

NOTE 2.— The answers to Reduction Descending will be found in the questions of Reduction Ascending.

2. In 127£. 15s. 8d. how many farthings?
3. In 28£. 19s. 11d. 3qr. how many farthings?
4. In 378£. how many pence?
5. How many grains in 28lb. 11oz. 12dwt. 15gr. troy?
6. In 17lb. 12dwt. troy, how many pennyweights?
7. If a silver tankard weigh 3lb. 11oz. how many grains will it be?
8. How many scruples in 23lbs., apothecaries' weight?
9. If a load of hay weigh 3 T. 16cwt. 2qr. 18lbs. how many ounces will it be?
10. Required the number of drachms in a hogshead of sugar, weighing 2T. 17cwt. 3qr. 16lb. 15oz. 13dr.?
11. In 57yds. how many nails?
12. In 83947 E.E. 4qr. how many nails?
13. In 2263 E.F. 2qr. how many quarters?
14. How many feet in 79 miles?
15. How many inches in 396 furlongs?
16. How many inches from Haverhill to Boston, the distance being 30 miles?

17. How many barleycorns will it take to reach round the world ?

18. In 403m. 7fur. 35rd. 2yd. 0ft. 0in. 1bar. how many barleycorns ?

19. In 413Le. 2m. 2fur. 33rd. 1yd. 0ft. 7in. how many inches ?

20. In 144m. 1fur. 3rd. 1yd. 1ft. how many feet ?

21. How many inches in 1051yd. 2ft. 5in. ?

22. In 3576fur. 12rd. 3yd. how many yards ?

23. How many square feet in 25 acres ?

24. How many square rods in 365 square miles ?

25. The surface of the earth contains 196563942 square miles. What would it be in square inches ?

26. Required the number of feet in 10A. 3R. 38p. 6yd. 5ft. 72in.

27. In 2R. 0p. 24yd. 3ft. how many inches ?

28. In 1A. 3R. 34p. 27yd. 4ft. 54in. how many inches ?

29. In 17 cords of wood how many inches ?

30. In 19 tons of round timber how many inches ?

31. How many cubic feet of wood in 123 cords ?

32. In 4899hhd. 4galls. 3qt. how many quarts ?

33. In 1224 tuns, 1p. 1hhd. 19galls. 1qt. 0pt. 1gi. how many gills ?

34. How many pints in 790p. 0hhd. 58galls. 0qt. 1pt. ?

35. In 460 butts, 1hhd. 31galls. of beer, how many galls. ?

36. In 36hhd. 26galls. 3qt. 1pt. how many pints ?

37. In 16 tons of round timber, how many inches ?

38. How many seconds since the deluge, it being 2348 years B. C. ?

39. How many days did the last war continue, it having commenced June 18, 1812, and ended Feb. 17, 1815 ?

40. How many pecks in 676 chaldrons ?

41. In 657 cents how many mills ?

42. In 3165 dimes how many mills ?

43. In 63 dollars how many cents ?

44. In 27 eagles, how many mills ?

REDUCTION ASCENDING.

EXAMPLE.

1. In 76789 farthings, how many pounds ?

Ans. 79£. 19s. 9¼d.

4)76789

12)19197 — 1

20)1599 — 9

79£. 19s. 9¼d.

We first divide by 4, because it takes 4 farthings to make a penny. We then divide by 12, because pence is the next higher denomination. Lastly we divide by 20, it being the number of shillings in a pound.

5*

From the preceding illustration and example we deduce the following

RULE.

Divide the lowest denomination given by that number, which it takes of that denomination to make one of the next higher ; so proceed, until it is brought to the denomination required.

NOTE 1. — To divide by 5½, we multiply the multiplicand by 2, and divide the product by 11 ; to divide 16½, we multiply by 2 and divide by 33 ; and to divide by 272¼, we multiply by 4 and divide by 1089.

NOTE 2. — The answers to Reduction Ascending are the questions in Reduction Descending.

2. In 122672 farthings, how many pounds ?
3. In 27839 farthings, how many pounds ?
4. In 90720 pence, how many pounds ?
5. In 166863 grains, how many pounds troy ?
6. How many pounds in 4692 pennyweights ?
7. How many pounds troy in 22560 grains ?
8. In 6624 scruples, how many pounds ?
9. In 137376 ounces, how many tons ?
10. In 1660157 drachms, how many tons ?
11. How many yards in 912 nails ?
12. Required the ells English in 1678956 nails.
13. Required the ells Flemish in 6791 quarters.
14. Required the miles in 417120 feet.
15. Required the furlongs in 3136320 inches.
16. Required the miles in 1900800 inches.
17. How many degrees in 4755801600 barleycorns ?
18. How many miles in 76789567 barleycorns ?
19. How many leagues in 78653167 inches ?
20. Required the miles in 761116 feet.
21. Required the yards in 37865 inches.
22. Required the furlongs in 786789 yards.
23. How many acres in 1089000 square feet ?
24. How many square miles in 37376000 square rods ?
25. How many square miles in 789,103,900,894,003,200 square inches ?
26. In 478675 square feet, how many acres ?
27. In 3167856 square inches, how many roods ?
28. How many acres in 12345678 square inches ?
29. How many cords in 3760128 cubic inches ?
30. How many tons of round timber in 1313280 cubic inches ?
31. How many cords in 16384 cubic feet ?
32. How many hogsheads of wine in 1234567 quarts ?
33. How many tuns of wine in 9877001 gills ?
34. In 796785 pints of wine how many pipes ?
35. How many butts of beer in 49765 gallons ?
36. In 15767 pints, how many hogsheads of beer ?

37. In 1105920 inches, how many tons of round timber?
38. In 132005440800 seconds, how many years?
39. In 974 days how many years and calendar months?
40. How many chaldrons in 97344 pecks?
41. How many cents in 6570 mills?
42. How many dimes in 316500 mills?
43. How many dollars in 6300 cents?
44. How many eagles in 270000 mills?

COMPOUND REDUCTION.

1. In 57£. 15s. how many dollars?   Ans. $192.50 cts.
2. In 67£. 14s. 9d. how many crowns at 6s. 7d. each?
           Ans. 205cr. 5s. 2d.
3. How many pounds, shillings, and pence in 678 dollars?
           Ans. 203£. 8s.
4. How many ells English in 761 yards?
           Ans. 608 E.E. 4qrs.
5. How many yards in 61 ells Flemish?   Ans. 45yds. 3qrs.
6. How many bottles, that contain 3 pints each, will it take to hold a hogshead of wine?   Ans. 168 bottles.
7. How many steps, 2ft. 8in. each, will a man take in walking from Bradford to Newburyport, the distance being 15 miles?
           Ans. 29700.
8. How many spoons, each weighing 2oz. 12dwt. can be made from 5lbs. 2oz. 8dwt. of silver.   Ans. 24 spoons.
9. How many times will the wheel of a coach revolve, whose circumference is 14ft. 9in. in passing from Boston to Washington, the distance being 436 miles?   Ans. 156073 $\frac{2}{177}$.
10. I have a field of corn, consisting of 123 rows, and each row contains 78 hills, and each hill has 4 ears of corn; now if it take 8 ears of corn to make a quart, how many bushels does the field contain?   Ans. 149bu. 3pks. 5qts. 0pts.
11. If it take 5yds. 2qrs. 3na. to' make a suit of clothes, how many suits can be made from 182 yards?   Ans. 32.
12. A goldsmith wishes to make a number of rings, each weighing 5dwt. 10gr. from 3lb. 1oz. 2dwt. 2grs. of gold; how many will there be?   Ans. 137 rings.
13. How many shingles will it take to cover the roof of a building, which is 60 feet long and 56 feet wide, allowing each shingle to be 4 inches wide and 18 inches long, and to lay one third to the weather?   Ans. 20160 shingles.
14. There is a house 56 feet long; and each of the two sides of the roof is 25 feet wide. How many shingles will it take to cover it, if it require 6 shingles to cover a square foot?
           Ans. 16800 shingles.

15. If a man can travel 22m. 3fur. 17rd. a day, how long would it take him to walk round the globe, the distance being about 25000 miles? Ans. 1114$\frac{6873}{7177}$ days.

16. If a family consume 7lbs. 10oz. of sugar in a week, how long would 10cwt. 3qrs. 16lbs. last them? Ans. 160 weeks.

17. Sold 3 tons, 17cwt. 3qrs. 18lbs. of lead at 7d. a pound, what did the lead amount to? Ans. 254£. 10s. 2d.

18. What will 5cwt. 1qr. 10lbs. of tobacco cost, at 4½d. a pound? Ans. 11£. 4s. 3d.

19. What will 7 hogsheads of wine cost at 9 cents a quart? Ans. $158.76.

20. What will 15 hogsheads of beer cost, at 3 cents a pint? Ans. $194.40.

21. What will 73 bushels of meal cost, at 2 cents a quart? Ans. $46.72.

22. A merchant has 29 bales of cotton cloth; each bale contains 57 yards. What is the value of the whole at 15 cents a yard? Ans. $247.95.

23. A merchant bought 4 bales of cotton; the first contained 6cwt. 2qrs. 11lbs.; the second, 5cwt. 3qrs. 16lbs.; the third, 7cwt. 0qr. 7lbs.; the fourth, 3cwt. 1qr. 17lbs. He sold the whole at 15 cents a pound. What did it amount to? Ans. $385.65.

24. A merchant having purchased 12cwt. of sugar, sold at one time 3cwt. 2qrs. 11lbs. and at another time he sold 4cwt. 1qr. 15lbs. What is the remainder worth at 15 cents per pound? Ans. $67.50.

25. Bought 4 chests of hyson tea. The weight of the first was 2cwt. 1qr. 7lbs.; the second, 3cwt. 2qrs. 15lbs.; the third, 2cwt. 0qr. 20lbs.; the fourth, 5cwt. 3qrs. 17lbs. What is the value of the whole at 37½ a pound? Ans. $589.12½.

26. Purchased a cargo of molasses, consisting of 87 hogsheads. What is the value of it at 33 cents a gallon? Ans. $1808.73.

27. From a hogshead of wine, 10galls. 1qt. 1pt. 3 gills leaked out. The remainder was sold at 6 cents a gill. To what did it amount? Ans. $100.86.

28. A man has 3 farms; the first containing 100A. 3R. 15rds.; the second, 161A. 2R. 28rds.; the third, 360A. 3R. 5rds. He gave his oldest son a farm of 112A. 3R. 30rds.; his second son a farm of 316A. 1R. 18rds.; his youngest son a farm of 168A. 3R. 13rds.; and sold the remainder of his land at 1 dollar and 35 cents a rod. To what did it amount? Ans. $5436.45.

29. A grocer bought a hogshead of molasses, containing 87galls. 1qt. from which 13 gallons leaked out. What is the remainder worth at 1 cent a gill? Ans. $23.76.

30. A man bought 4 loads of hay; the first weighing 25cwt. 0qr. 17lbs.; the second, 37cwt. 2qrs. 17lbs.; the third, 18cwt. 3qrs. 14lbs.; and the fourth, 37cwt. 1qr. 17lbs. What is the value of the whole, at 2 cents a pound? Ans. $266.74.

REDUCTION OF THE OLD NEW ENGLAND CURRENCY TO FEDERAL MONEY.

RULE.

To the pounds annex three ciphers, and divide the number by 3, and the quotient will be the answer in cents ; but if there be pounds and skillings, annex to the pounds half the number of shillings and two ciphers ; and if the shillings be an odd number, instead of two ciphers, annex a 5 and a cipher, and divide the number by 3, and the quotient will be the answer in cents, as before ; if there be pounds, shillings, pence and farthings, let the pounds and shillings be arranged as before, then reduce the pence and farthings into farthings, and add 1 if their number exceed 12, and 2 if it exceed 36, and let them be so annexed to the pounds, that the units of the farthings shall possess the third place from the pounds ; then divide by 3, as before, and the quotient is cents.

NOTE.—As a dollar is six-twentieths of a pound, or, which is the same thing, three-tenths, therefore a pound is ten-thirds of a dollar. It is also evident, that if pounds be multiplied by 10, their product will be thirds of a dollar; and if thirds of a dollar be divided by 3, the quotient will be dollars. Again, as shillings are twentieths of a pound, it is evident that if they be halved, they become tenths or decimals of a pound; and if 1 be added to 24 farthings, or 2 be added to 48 farthings, and they be made to occupy the third place from pounds, they become decimals also; for twenty-four farthings are twenty-four nine hundred sixtieths of a pound; but if 1 be added to the numerator of this fraction, then must 40 be added to the denominator, to make a fraction of equal value, for twenty-four nine hundred sixtieths is equal to twenty-five one thousandths; now, if a pound be considered the unit, twenty-five thousandths will be expressed thus, ,025£. Therefore, as the number of farthings may be less or greater than 24, so a greater or less number than a unit must be added to their number. Q. E. D.

EXAMPLES.

1. Reduce 162£. to Federal Money.

$$3)162000$$

$$54000 \qquad \text{Ans. \$ 540.}$$

2. Change 319£. 17s. to Federal Money.

$$3)319850$$

$$1066_{,}16\tfrac{2}{3} \qquad \text{Ans. \$-1066.16}\tfrac{2}{3}$$

3. Change 176£. 17s. 8¾d. to Federal Money.

$$176850$$
$$36$$

$$3)176886$$

$$58962 \qquad \text{Ans. \$ 589.62}$$

4.	Reduce	315£.	to	Federal	Money.	Ans.	$1050.00
5.	"	619£.	to	"	"	"	2063.33⅓
6.	"	166£.	to	"	"	"	553.33⅓
7.	"	318£.	to	"	"	"	1060.00
8.	"	101£.	to	"	"	"	336.66⅔
9.	"	144£.	to	"	"	"	480.00
10.	"	161£. 18s.	to	"	"	"	539.66⅔
11.	"	361£. 17s.	to	"	"	"	1206.16⅔
12.	"	99£. 11s.	to	"	"	"	331.83⅓
13.	"	100£. 9s.	to	"	"	"	334.83⅓
14.	"	661£. 7s.	to	"	"	"	2204.50
15.	"	47£. 11s.	to	"	"	"	158.50
16.	"	109£. 1s.	to	"	"	"	363.50
17.	"	16£. 17s. 6½d.	to	"	"	"	56.25⅔
18.	"	69£. 1s. 3½d.	to	"	"	"	230.21½
19.	"	87£. 16s. 11d.	to	"	"	"	292.82
20.	"	14£. 7s. 7½d.	to	"	"	"	47.93¾
21.	"	73£. 3s. 4½d.	to	"	"	"	243.89⅚
22.	"	47£. 12s. 10d.	to	"	"	"	158.80⅝
23.	"	187£. 5s. 0¾d.	to	"	"	"	624.17¾
24.	"	10£. 0s. 3¼d.	to	"	"	"	33.38

SECTION VIII.

MULTIPLICATION OF FEDERAL MONEY.

RULE.

When the price is in cents, or dollars and cents, multiply the quantity by the price, and the answer is in cents, which may be reduced to dollars and cents by cutting off TWO *figures from the right hand, which are cents; and those at the left are dollars. If the price be in mills, or cents and mills, or dollars, cents and mills, multiply as before, and cut off* THREE *figures from the right hand, the unit of which is mills, the other two are cents; and those figures at the left hand are dollars.*

EXAMPLES.

1. What will 365 barrels of Genessee flour cost, at $5.75 a barrel?

```
     365
     575
    ----
    1825
   2555
  1825
  ------
Ans. $2098.75
```

2. What will 128 pounds of sugar cost, at 13 cents, 7 mills a pound ?

$$128$$
$$137$$
$$\overline{896}$$
$$384$$
$$128$$
$$\overline{\$ 17.53.6}$$

3. What will 126 pounds of butter cost, at 13 cents a pound ?
Ans. $ 16.38.

4. What will 63 pounds of tea cost, at 93 cents a pound ?
Ans. $ 58.59.

5. What will 43 tons of hay cost, at 13 dollars 75 cents a ton ?
Ans. $ 591.25.

6. If 1 pound of pork is worth 7 cents 3 mills, what is 46 pounds worth ? Ans. $ 3.35.8.

7. If 1cwt. of beef cost 3 dollars 28 cents, what is 76cwt. worth ? Ans. $ 249.28.

8. What will 96,000 feet of boards cost, at 11 dollars 67 cents a thousand ? Ans. $ 1120.32.

9. If a barrel of cider be sold for 2 dollars 12 cents, what will be the value of 169 barrels ? Ans. $ 358.28.

10. What will be the value of a hogshead of wine, containing 63gals., at 1 dollar 63 cents a gallon ? Ans. $ 102.69.

11. Sold a sack of hops weighing 396 pounds, at 11 cents 3 mills a pound ; to what did it amount ? Ans. $ 44.74.8.

12. Sold 19 cords of wood, at 5 dollars 75 cents a cord ; to what did it amount ? Ans. $ 109.25.

13. Sold 169 tons of timber, at 4 dollars 68 cents a ton ; what did I receive ? Ans. $ 790.92.

14. Sold a hogshead of sugar weighing 465 pounds ; to what did it amount, at 14 cents a pound ? Ans: $ 65.10.

15. What will be the amount of 789 pounds of leather, at 18 cents a pound ? Ans. $ 142.02.

16. What will be the expense of 846 pounds of sheet lead, at 5 cents 7 mills a pound ? Ans. $ 48.22.2.

17. When potash is sold for 132 dollars 55 cents a ton, what will be the price of 369 tons ? Ans. $ 48910.95.

18. What will 365 pounds of bees' wax cost, at 18 cents 4 mills a pound ? Ans. $ 67.16.

19. If 1 pound of tallow cost 7 cents 3 mills, what is 968 pounds worth ? Ans. $ 70.66.4.

20. What will a chest of souchong tea be worth, containing 69 pounds, at 29 cents 9 mills a pound ? Ans. $ 20.63.1.

21. If 1 drum of figs cost 2 dollars 75 cents, what will be the price of 79 drums ? Ans. $ 217.25.

22. If 1 box of oranges cost 6 dollars 71 cents, what will be the price of 169 boxes? Ans. $ 1133.99.

23. Purchased 796 pounds of cocoa, at 11 cents 4 mills a pound; what did I have to pay? Ans. $ 90.74.4.

24. A farmer sold 691 bushels of wheat, at 1 dollar 25 cents a bushel; what did he receive for it? Ans. $ 863.75.

25. What is 97 pounds of madder worth, at 17 cents 6 mills a pound? Ans. $ 17.07.2.

26. A merchant sold 73 hogsheads of molasses, each containing 63 gallons, for 44 cents a gallon; how much money did he receive? Ans. $ 2023.56.

27. A drover has 169 sheep, which he values at 2 dollars 69 cents a head; what is the value of the whole drove? Ans. 454.61.

28. A farm containing 144 acres, is valued at 69 dollars, 74 cents 8 mills an acre; what is the amount of the whole? Ans. $ 10043.71.2.

29. An auctioneer sold 48 bags of cotton, each containing 397 pounds, at 13 cents 7 mills a pound; what is the value of the whole? Ans. $ 2610.67.2.

30. If 1 yard of broadcloth cost 5 dollars 67 cents, what will be the value of 48 yards? Ans. $ 272.16.

31. A wool grower has 179 sheep, each producing 4 pounds of wool; what will be its value, at 59 cents 3 mills a pound? Ans. $ 424.58.8.

32. A house having 17 rooms, requires 6 rolls of paper for each room; now if each roll costs 1 dollar 17 cents, what will be the expense for all the rooms? Ans. $ 119.34.

33. What will 89 yards of brown sheeting cost, at 17 cents a yard? Ans. $ 15.13.

34. What will 47,000 of shingles cost, at 3 dollars 75 cents a thousand? Ans. $ 176.25.

35. Bought 47 hogsheads of salt, each containing 7 bushels, for 1 dollar 12 cents a bushel; what did it cost? Ans. $ 368.48.

36. What will a ton of hay cost, at 1 dollar 17 cents a hundred weight? Ans. $ 23.40.

37. If 1 foot of wood cost 63 cents, what will 39 cords cost? Ans. $ 196.56.

38. What will 163 buckets cost, at 1 dollar 21 cents a bucket? Ans. $ 197.23.

39. What will 78 barrels of shad cost, at 3 dollars 89 cents a barrel? Ans. $ 303.42.

40. If 1 pound of salmon cost 17 cents 5 mills, what will 789 pounds cost? Ans. $ 138.07.5.

41. Bought 163 grindstones, at 6 dollars 79 cents each; what did the whole cost? Ans. $ 1106.77.

42. Sold 49 green hides, at 1 dollar 95 cents each; what did they all amount to? Ans. $ 95.55.

43. If a man's wages be 1 dollar 19 cents a day, what are they for a year?   Ans. $ 434.35.

44. Hired a horse and chaise to go a journey of 146 miles, at 16 cents a mile; what did it cost?   Ans. $ 23.36.

45. If it be worth 3 dollars 68 cents to plough one acre of land, what would it be worth to plough 79 acres?

Ans. $ 290.72.

46. What will 148 tons of plaster of Paris cost, at 2 dollars 28 cents a ton?   Ans. $ 337.44.

47. Bought 79 tons of logwood, at 49 dollars 75 cents a ton; what did it cost?   Ans. $ 3930.25.

48. Bought 5 gross bottles of castor oil, at 37 cents a bottle; what did it cost?   Ans. $ 266.40.

49. Sold 19 dozen pair of men's gloves, at 47 cents a pair; to what did they amount?   Ans. $ 107.16.

50. Bought a hogshead of wine for 97 cents a gallon, and sold it for 1 dollar 75 cents a gallon; what did I gain?

Ans. $ 49.14.

51. Bought 75 barrels of flour at 5 dollars 75 cents a barrel, and sold it at 6 dollars 37 cents; what did I gain?

Ans. $ 46.50.

52. Bought 17 score of penknives, at 17 cents each; what did they all cost?   Ans. $ 57.80.

53. What will 17½ tons of coal cost at $ 9.62 per ton?

Ans. 168.35.

54. What will 19 barrels of cider cost at $ 1.37½ per barrel?

Ans. $ 26.12.5.

BILLS.

Boston, July 4, 1835.

Mr. James Dow,

Bought of Dennis Sharp,

17 yds.	Flannel,	at	.45	cts.
19 "	Shalloon,	"	.37	"
16 "	Blue Camlet,	"	.46	"
13 "	Silk Vesting,	"	.87	"
9 "	Cambric Muslin,	"	.63	"
25 "	Bombazin,	"	.56	"
17 "	Ticking,	"	.31	"
19 "	Striped Jean,		.16	"

$ 61.33

Received payment,

Dennis Sharp.

Haverhill, May 5, 1835.

Mr. Samuel Smith,   Bought of David Johnson,

		at	.98 cts.
13 lbs.	Tea,	"	.15 "
16 "	Coffee,	"	.13 "
36 "	Sugar,	"	.09 "
47 "	Cheese,	"	.19 "
12 "	Pepper,	"	.17 "
7 "	Ginger,	"	.61 "
13 "	Chocolate,		

$ 35.45.

Received payment,   David Johnson.

Salem, February 29, 1835.

Mr. John Dow,   Bought of Richard Fuller,

		at	$ 5.25
17 yds.	Broadcloth,	"	1.62
29 "	Cassimere,	"	.17
60 "	Bleached Shirting,	"	.27
49 "	Ticking,	"	3.19
18 "	Blue Cloth,	"	2.75
27 "	Habit do.	"	.61
75 "	Flannel,	"	.75
36 "	Plaid Prints,	"	.18
49 "	Brown Sheeting,		

$ 372.90.

Received payment,   Richard Fuller.

Baltimore, January 20, 1835.

Mr. John Rilley,   Bought of James Somes,

		at	$ 2.75
10 pair	Boots,	"	1.25
19 "	Shoes,	"	1.29
83 "	Hose,	"	.17
47 lbs	Ginger,	"	.39
91 "	Chocolate,	"	.23
47 "	Pepper,	"	.13
68 "	Flour,	"	1.39
27 pair	Gloves,		

$ 258.98.

Received payment,   James Somes.

Philadelphia, June 11, 1835.

Mr. Moses Thomas,

Bought of Luke Dow,

27 National Spelling Books,	at	$ 0.19
25 Young Readers,	"	.27
17 Liber Primus,	"	.75
9 Greek Lexicons,	"	3.75
8 Ainsworth's Dictionaries,	"	4.50
27 Greek Readers,	"	2.25
18 Folio Bibles,	"	9.87
75 Testaments,	"	.31
67 National Readers,	"	.75
15 Pocket Bibles,	"	1.12

$ 423.09.

Received payment,

Luke Dow, by
Timothy True.

Boston, June 26, 1835.

Dr. Enoch Cross,

Bought of Maynard & Noyes,

14 oz.	Ipecacuanha,	at	$ 0.67
23 "	Laudanum,	"	.89
17 "	Emetic Tartar,	"	1.25
25 "	Cantharides,	"	2.17
27 "	Gum Mastic,	"	.61
56 "	Gum Camphor	"	.27

$ 136.94.

Received payment,

Maynard & Noyes,
by Timothy Jones.

Newburyport, June 5, 1835.

Mr. John Somes,

Bought of Samuel Gridley,

7½ yds.	Broadcloth,	at	$ 4.50
16¾ lbs.	Coffee,	"	.16
18¼ "	Candles,	"	.25
30 "	Soap,	"	.17
3 "	Pepper,	"	.19
7½ "	Ginger,	"	.18

$ 48.01¼.

Received payment,

Samuel Gridley.

Boston, May 1, 1835.

Mr. Benjamin Treat,

Bought of John True,

37 Chests Green Tea,	at	$25.50
41 " Black do.	"	16.17
40 Casks Wine,	"	97.75
13 Crates Liverpool Ware,	"	169.37

$7718.23.

Received payment,

John True.

New York, July 11, 1835.

Mr. John Cummings,

Bought of Lord & Secomb,

97 bls.	Genessee Flour,	at	$6.25
167 "	Philadelphia do.	"	5.95
87 "	Baltimore do.	"	6.07
196 "	Richmond do.	"	5.75
275 "	Howard st. do.	"	7.25
69 bu.	Rye,	"	1.16
136 "	Virginia Corn,	"	.67
68 "	North river do.	"	.76
169 "	Wheat,	"	1.37
76 Ton	Lehigh Coal,	"	9.67
89 "	Iron,	"	69.70
49	Grindstones,	"	3.47
39	Pitchforks,	"	1.61
197	Rakes,	"	.17
86	Hoes,	"	.69
78	Shovels,	"	1.17
187	Spades,	"	.85
91	Ploughs,	"	11.61
83	Harrows,	"	17.15
47	Handsaws,	"	3.16
35	Millsaws,	"	18.15
47 cwt.	Steel,	"	9.47
57 "	Lead,	"	6.83

$17315.32.

Received payment,

Lord & Secomb.

SECTION IX.

COMPOUND MULTIPLICATION.

CASE I.

1. If 1 quire of paper cost 8d., what will 2 quires cost? 3? 4?
2. If 1 bushel of corn cost 4s. 6d., what will 2 bushels cost?
3. What is the width of 2 boards, each of which is 1ft. 6in. wide?
4. What is the width of 3 boards? 4 boards? 5? 6? 7? 8?
5. If a man drink 2qt. 1pt. of water a day, how much will he drink in 2 days? In 3? In 4? In 5? In 6? In 7? In 8?
6. If 1 spoon weigh 2oz. 10dwt. what will 2 spoons weigh?
7. If a man walk 3m. 4fur. in an hour, how far will he walk in 2 hours? In 3? In 4? In 5? In 6? In 7? In 8? In 9?
8. What will 6 bales of cloth cost, at 7£. 12s. 7d. per bale?

```
£    s.   d.
7    12   7
          6
----------------
45   15   6
```

In this question, we multiply 7d. by 6 and find the product to be 42d. This we divide by 12, the number of pence in a shilling, and find it contains 3s., and 6d. remaining, which we write under the pence, and carry 3 to the product of 6 times 12, and find the amount to be 75s. which we reduce to pounds by dividing them by 20, and find them to be 3£. 15s. We write down the shillings under the shillings and carry 3 to the product of 6 times 7£.; and we thus find the answer to be 45£. 15s. 6d.

From the above illustration, we deduce the following

RULE.

When the quantity is less than 12, multiply the price by the quantity, and carry as in Compound Addition.

NOTE. — For the answers in Multiplication, see Section 10, in Division.

9. Multiply 18£. 16s. 7¼d. by 4.	Ans. 75£. 6s. 6d.
10. Multiply 15£. 11s. 8¼d. by 8.	Ans. 124£. 13s. 10d.
11. Multiply 27£. 19s. 11½d. by 9.	Ans. 251£. 19s. 7¼d.
12. Multiply 19£. 5s. 7¼d. by 11.	Ans. 212£. 1s. 7¾d.
13. Multiply 81£. 14s. 9d. by 8.	Ans. 653£. 18s. 0d.
14. Multiply 15£. 18s. 5d. by 7.	Ans. 111£. 8s. 11d.
15. Multiply 13£. 5s. 4¼d. by 12.	Ans. 159£. 4s. 9d.
16. Multiply 17lb. 7oz. 13dwt. 13gr. by 9.	
17. Multiply 15lb. 11oz. 19dwt. 15gr. by 7.	
18. Multiply 16T. 12cwt. 3qr. 13lb. 12oz. by 11.	
19. Multiply 13T. 3cwt. 1qr. 14lb. 13oz. by 8.	
20. Multiply 2℔. 5℥. 5ʒ. 1Ɖ. 16½gr. by 8.	

6*

21. Multiply 47yds. 3qr. 2na. 2in. by 7.
22. Multiply 17m. 7fur. 36rds. 13ft. 7in. by 12.
23. Multiply 16deg. 39m. 3fur. 39rds. 5yds. 2ft. by 9.
24. Multiply 16deg. 29m. 7fur. 12rds. 8ft. 11in. 1bar. by 6.
25. Multiply 16A. 2R, 4p. 19yd. 7ft. 79in. by 11.
26. Multiply 7 cords, 116ft. 1629in. by 4.
27. Multiply 29hhd. 61gal. 3qt. 1pt. 3gi. by 7.
28. Multiply 8Tun. 8hhd. 56gal. 2qt. by 9.
29. Multiply 7hhd. 5gal. 2qt. 1pt. by 8.
30. Multiply 19bu. 2pk. 7qt. 1pt. by 6.
31. Multiply 36chald. 18bu. 3pk. 7qt. by 7.
32. Multiply 13y. 316da. 15h. 27m. 39sec. by 8.
33. If a man gives each of his 9 sons 23A. 3R. 19½p. what do they all receive?
34. If 12 men perform a piece of labor in 7h. 24m. 36s. how long would it take 1 man to perform the same task?
35. If 1 bag contain 3bu. 2pk. 4qt. what quantity does 8 bags contain?

CASE II.

When the quantity is more than 12 and is a composite number, that is, a number, which is the product of two or more numbers, the question is performed as in the following

EXAMPLE.

36. What will 42 yards of cloth cost at 6s. 9d. a yard?

£. s. d.
0   6   9
        6

2   0   6 = price of 6 yds.
        7

14 3   6 = price of 42 yds.

In this example, we find that 6 multiplied by 7 will produce the quantity 42 yards. We therefore multiply 6s. 9d. first by the 6, and then its product by 7; and the last product, 14£. 3s. 6d. is the answer or price of the 42 yards.

The pupil will now see the propriety of the following

RULE.

Multiply the price by one of the factors of the composite number and the product by the other.

37. What will 16 yards of velvet cost at 3s. 8d. per yard?
38. What will 72 yards of broadcloth cost, at 19s. 11d. per yard?
39. What will 84 yards of cotton cost, at 1s. 11d. per yard?
40. Bought 90 hogsheads of sugar, each weighing 12cwt. 2qr. 11lb.; what was the weight of the whole?
41. What cost 18 sheep at 5s. 9½d. a piece?
42. What cost 21 yards of cloth at 9s. 11d. per yard?
43. What cost 22 hats at 11s. 6d. each?

44. If 1 share in a certain stock be valued at 18£. 8s. 9¼d. what is the value of 96 shares?

45. If 1 spoon weigh 3oz. 5dwt. 15gr. what is the weight of 120 spoons?

46. If a man travel 24m. 7fur. 4rd. in 1 day, how far will he go in 1 month?

47. If the earth revolve 0°. 15'. per minute, how far per hour?

48. Multiply 39A. 3R. 17p. 30yd. 8ft. 100in. by 32.

49. If a man be 9da. 5h. 17m. 19sec. in walking 1 degree, how long would it take him to walk round the earth, allowing 365¼ days to a year?

CASE III.

When the quantity required is such a number, as cannot be produced by the product of two or more numbers, we should proceed as in the following

EXAMPLE.

50. What is the value of 58 tons of iron at 18£. 17s. 11d. a ton?

£.	s.	d.		£.	s.	d.	
18	17	11		18	17	11	
		5				3	
94	9	7	= price of 5 tons.	56	13	9	= price of 3 tons.
		10					
944	15	10	= price of 50 tons.				
56	13	9	= price of 3 tons.				
1001	9	7	= price of 53 tons.				

Because 53 is a *prime* number, that is, it cannot be produced by the product of any two numbers; we therefore find a convenient composite number less than the given number, viz. 50, which may be produced by multiplying 5 by 10. Having found the price of 50 tons by the last Case, we then find the price of the 3 remaining tons by Case I., and add it to the former, making the value of the whole quantity 1001£. 9s. 7d.

The pupil will hence perceive the propriety of the following

RULE.

Multiply the price by the continued product of such numbers, as will be nearest the given quantity, and then find the value of the remainder by Case I., and the two products will be the answer.

51. What will 57 gallons of wine cost at 8s. 3½d. per gallon?

52. Bought 29 lots of wild land, each containing 117A. 3R. 27p.; what were the contents of the whole?

53. Bought 89 pieces of cloth, each containing 37yd. 3qr 2na. 2in.; what was the whole quantity?

54. Bought 59 casks of wine, each containing 47gal. 3qt. 1pt. ; what was the whole quantity ?

55. If a man travel 17m. 3fur. 13rd. 14ft. in one day, how far will he travel in a year?

56. If a man drink 3gal. 1qt. 1pt. of beer in a week, how much will he drink in 52 weeks ?

57. There are 17 sticks of timber, each containing 37ft. 978in.; what is the whole quantity ?

58. There are 17 piles of wood, each containing 7 cords, 98 cubic feet ; what is the whole quantity ?

59. Multiply 2hhd. 19gal. 0qt. $1\frac{3}{13}$pt. by 39.

60. Multiply 3bu. 1pk. 4qt. 1pt. $0\frac{20}{136}$gi. by 53.

61. Multiply 16chal. 7bu. 1pk. 3qt. $0\frac{19}{171}$pt. by 17.

BILLS.

London, July 4, 1835.

Dow, Vance & Co., of Boston, U. S.,

Bought of Samuel Snow,

45 yds. Broadcloth,	at	8s.	4d.
50 " "	"	10s.	6d.
56 " "	"	3s.	7½d.
63 " "	"	12s.	11¾d.
72 " "	"	19s.	11d.
81 " "	"	9s.	3d.
35 " "	"	19s.	7½d.
99 " "	"	16s.	0½d.
66 " "	"	8s.	11d.
33 " "	"	16s.	11½d.

Received payment, 376£. 7s. 0¾d.

Samuel Snow.

Quebec, Jan. 8, 1835.

Mr. John Vose,

Bought of Vans & Conant,

46 Ivory Combs,	at	3s.	5¼d.
47 lb. Colored Thread,	"	6s.	9¼d.
51 yds. Durant,	"	1s.	8d.
52 Silk Vests,	"	6s.	7d.
53 Leghorns,	"	11s.	9½d.
57 ps. Nankin,	"	8s.	3¼d.
58 lbs. White Thread,	"	9s.	11¼d.

128£. 16s. 5¼d.

Received payment,

Vans & Conant.

Montreal, July 4, 1835.

Mr. James Savage,

Bought of Joseph Dowe,

83 galls.	Lisbon Wine,	at	6s.	7d.
85 "	Port do.	"	8s.	9½d.
86 "	Madeira do.	"	4s.	11½d.
87 "	Temperance do.	"	3s.	6½d.
89 "	Oil,	"	5s.	3d.
91	Leghorns,	"	19s.	10½d.
92 lbs.	Green Tea,	"	3s.	1½d.
93 pair	Thread Hose,	"	4s.	4½d.
94 "	Silk Gloves,	"	2s.	3½d.
95 "	Silk Hose,	"	6s.	6½d.
97 yds.	Linen,	"	5s.	5½d.
98 galls.	Winter Strained Oil,	"	7s.	7½d.

338£. 19s. 2½d.

Received payment,

Joseph Dowe.

Liverpool, June 2, 1835.

John Jones, of Philadelphia, U. S.,

Bought of Thomas Hogarth,

297 yds.	Black Broadcloth,	at	17s.	3½d.
473 "	Blue do.	"	9s.	11¼d.
512 "	Red do.	"	15s.	10d.
624 "	Green do.	"	12s.	6d.
765 "	White do.	"	19s.	9½d.
169 "	Black Velvet,	"	13s.	5¼d.
698 "	Green do.	"	15s.	6¾d.
315 "	Red do.	"	14s.	3½d.
713 "	White do.	"	11s.	7½d.
519 "	Carpet,	"	13s.	6¼d.
147 "	Black Kerseymere,	"	16s.	7¾d.
386 "	Blue do.	"	14s.	3¼d.
137 "	Green do.	"	19s.	9d.
999 "	Black Silk,	"	15s.	8d.

5012£. 0s. 11½d.

Received payment,

Thomas Hogarth.

Montreal, June 17, 1835.

Mr. Samuel Simpson,

Bought of Lackington, Grey & Co.

19	yds.	Cloth,	at	1s.	6d.
23	"	Worsted,	"	7s.	$8\frac{1}{4}$d.
26	"	Baize,	"	3s.	$11\frac{1}{2}$d.
29	"	Camlet,	"	6s.	$10\frac{1}{2}$d.
31	"	Bombazin,	"	1s.	$5\frac{1}{4}$d.
34	"	Linen,	"	3s.	7d.
37	"	Cotton,	"	11s.	9d.
38	"	Flannel,	"	6s.	11d.
39	"	Calico,	"	3s.	$10\frac{1}{4}$d.
41	"	Broadcloth,	"	6s.	$9\frac{1}{2}$d.
43	"	Nankin,	"	7s.	$5\frac{3}{4}$d.

106£. 1s. $11\frac{1}{2}$d

Received payment,

Lackington, Grey & Co.

London, May 11, 1835.

Messrs. Kimball & Fox, of Boston, U. S.

Bought of Benjamin Fowler,

297	yds.	Black Broadcloth,	at	17s.	$3\frac{1}{2}$d.
473	"	Blue do.	"	9s.	$11\frac{1}{4}$d.
512	"	Green do.	"	15s.	10d.
624	"	Black Silk,	"	12s.	8d.
765	"	Pongee do.	"	19s.	$9\frac{1}{2}$d.
169	"	Black Cassimere,	"	13s.	$5\frac{1}{4}$d.
698	"	Blue do.	"	15s.	$6\frac{3}{4}$d.
315	"	Durant,	"	14s.	$3\frac{1}{2}$d.
713	"	Silk Vesting,	"	11s.	$7\frac{1}{2}$d.
519	doz.	Silk Hose,	"	13s.	$6\frac{1}{4}$d.
147	ps.	Nankin,	"	16s.	$7\frac{3}{4}$d.
386	yds.	Irish Linen,	"	14s.	$3\frac{1}{4}$d.

4094£ 4s. $2\frac{1}{2}$d

Received payment,

Benjamin Fowler.

SECTION X.

COMPOUND DIVISION.

COMPOUND DIVISION is when the dividend consists of several denominations.

1. What is ½ of 3 pence? 4d.? 5d.? 7d.? 9d.? 11d.? 13d.?
2. What is ½ of 2 shillings? 3s.? 4s.? 5s.? 6s.? 7s.? 8s.?
3. What is ⅓ of 7£.? 8£.? 10£.? 11£.? 13£.? 14£.?
4. What is ¼ of 3cwt.? 5cwt.? 6cwt.? 7cwt.? 8cwt.? 9cwt.?
5. What is ¼ of 1yd.? 2yds.? 3yds.? 5yds.? 6yds.? 7yds.?
6. What is ⅓ of 1gal.? 2gal.? 3gal.? 4gal.? 5gal.? 6gal.?

EXAMPLES.

7 Divide 598£. 8s. 9d. equally among 5 persons.

```
       £.    s.   d.
  5)598    8    9
     119   13    9
```

Having divided the pounds by 5 we find 3£. remaining which are 60s.; to these we add the 8s. in the question, and again divide by 5 and find 3s. remaining, which are 36d.; to these we add the 11d. in the question, and divide their sum by 5. The several quotients we write under their respective denominations.

8. Divide 168£. 15s. 0d. equally among 36 men.

```
        £.    s.   d.
  36)168    15    0(4£.
     144
     ----
      24
      20
     ----
  36)495(13s.
     36
     ---
     135
     108
     ---
      27
      12
     ---
  36)324(9d.
     324
```

When the divisor is more than 12, we usually perform the operation by *long division*. In the present example, we first divide the pounds by 36, and obtain 4£. for the quotient and 24£. remaining, which we reduce to shillings and annex the 15s. and again divide by 36 and obtain 13s. for the quotient. The remainder we reduce to pence, and again divide and obtain 9d. for the quotient.

From the above examples and illustrations we deduce the following

RULE.

Divide the highest denomination, by the quantity, and if any thing remains, reduce it to the next lower denomination, and continue to divide until it is reduced to the lowest denomination.

NOTE. — When the operation is performed by *short division*, let the several quotients be placed under their respective denominations.

9. Divide 151£. 19s. 11¼d. by 9.   Ans. 16£. 17s. 9¼d.
10. Divide 350£. 17s. 8½d. by 519.   Ans. 13s. 6¼d.
11. Divide 225£. 1s. 10½d. by 63.   Ans. 3£. 11s. 5¼d.
12. Divide 159£. 4s. 9d. by 12.   Ans. 13£. 5s. 4¼d.
13. Divide 75£. 6s. 6d. by 4.   Ans. 18£. 16s. 7½d.
14. Divide 111£. 8s. 11d. by 7.
15. Divide 159£. 4s. 9d. by 12.
16. Divide 158lb. 9oz. 1dwt. 21gr. by 9.
17. Divide 111lb. 11oz. 17dwt. 9gr. by 7.
18. Divide 188 Ton, 1cwt. 2qr. 11lb. 4oz. by 11.
19. Divide 105 Ton, 7cwt. 0qr. 6lb. 8oz. by 8.
20. Divide 19lb. 9℥. 4ʒ. 2Э. 9gr. by 8.
21. Divide 335yd. 2qr. 0na. 0½in. by 7.
22. Divide 215m. 7fur. 1rd. 14ft. 6in. by 12.
23. Divide 149deg. 8m. 0fur. 0rd. 1yd. 1ft. 6in. by 9.
24. Divide 97deg. 55m. 7 fur. 35rd. 4ft. 2in. 1bar. by 6.
25. Divide 181A. 3R. 11p. 6yd. 4ft. 41in. by 11.
26. Divide 31 cords, 83ft. 1332in. by 4.
27. Divide 209hhd. 55gal. 3qt. 0pt. 1gi. by 7.
28. Divide 85 Tun, 3hhd. 4 gal. 2qt. by 9.
29. Divide 56hhd. 45gal. by 8.
30. Divide 118bu. 1pk. 5qt. by 6.
31. Divide 255chald. 24bu. 3pk. 1qt. by 7.
32. Divide 110y. 343d. 3h. 41m. 12sec. by 8.
33. A man divides his farm of 214A. 3R. 12p. equally among his 9 sons. How much does each receive ?
34. If one man perform a certain piece of labor in 3da. 16h. 54m., how long would it take 12 men to perform the same work?
35. A farmer has 29 bushels of rye which he wishes to put into 8 bags, how much must each bag contain ?

CASE II.

When the quantity is a composite number, proceed as in the following

EXAMPLE.

36. If 42 yards of cloth cost 14£. 3s. 6d., what is the value of 1 yard ?   Ans. 6s. 9d.

```
    £.  s.  d.
7)14   3   6      In this question, we find the component parts
 6)2   0   6    of 42 are 6 and 7; we therefore first divide the
   ———————       price by 7, and then divide the quotient by 6.
   0   6   9
```

From the above, we deduce the following

RULE.

Divide the dividend by one of the component parts, and the quotient thence arising, by the other, and the last quotient will be the answer to the question.

Note.—To find the true remainder; multiply the last remainder by the first divisor, and to the product add the first remainder.

37. If 16 yards of velvet cost 2£. 18s. 8d. what will 1 yard cost?

38. If 72 yards of broadcloth cost 71£. 14s. 0d. what is the value of 1 yard?

39. If 84 yards of cotton cost 8£. 1s. 0d. what will 1 yard cost?

40. If 90 hogsheads of sugar weigh 56T. 13cwt. 3qr. 10lb. what is the weight of 1 hogshead?

41. What will be the price of 1 sheep if 18 cost 5£. 4s. 3d.?

42. If 21 yards of cloth cost 10£. 8s. 3d. what is the price of 1 yard?

43. What is the value of 1 hat when 22 cost 12£. 13s. 0d.?

44. When 96 shares of a certain stock are valued at 1290£. 4s. 0d., what would be the cost of 1 share?

45. If 120 spoons weigh 32lb. 9oz. 15dwt., what does 1 weigh?

46. If a man in 1 month travel 746m. 5fur. 0rd. how far does he go in 1 day?

47. If the earth revolve 15° on its axis in 1 hour, how far does it revolve in 1 minute?

48. Divide 1275A. 2R. 16p. 22yd. 8ft. 32in. equally among 32 men.

49. If a man walk round the earth in 2y. 68da. 19h. 54m.; how long would it take him to walk 1 degree, allowing 365¼ days to a year?

The following questions are to be performed as the second example of this section.

50. If 53 tons of iron cost 1001£. 9s. 7d. what is the value of 1 ton?

51. If 57 gallons of wine cost 23£. 11s. 5¼d. what cost 1 gallon?

52. Divide 3419A. 2R. 23p. by 29.

53. If 89 pieces of cloth contain 3375yds. 3qr. 1na. 0¼in. how much does 1 piece contain?

54. If 59 casks contain 44hhd. 52gal. 2qt. 1pt. of wine, what are the contents of 1 cask?

55. If a man travel in 1 year (365 days) 6357m. 5fur. 14rd. 11½ft., how far is that per day?

56. When 175gal. 2qt. of beer are drunk in 52 weeks, how much is consumed in 1 week?

7

57. When 17 sticks of timber measure 15T. 38ft. 1074in., how many feet does 1 contain?

58. Divide 132 cords, 2ft. by 17.

59. Divide 89hhd. 54gal. by 39.

60. Divide 179bu. 3pk. 1qt. by 53.

61. Divide 275chald. 17bu. equally among 17 men.

CASE III.

When the value of a fractional part is required.

RULE.

Multiply the price by the numerator, and divide by the denominator, but if the price of a fractional part be given, and it is required to find the value or integer, multiply the price by the denominator and divide by the numerator.

EXAMPLE.

62. If the whole of a ship and cargo be valued at 2346£. 12s. 8d. what is ¾ worth?   Ans. 1759£. 19s. 6d.

	£.	s.	d.
	2346	12	8
			3
4)	7039	19	0
	1759	19	6

63. Bought ⅝ of a lottery ticket, which drew a prize of $1500; what was my share of the prize money?   Ans. $937.50.

64. If $\frac{7}{12}$ of an estate be valued at $1728.50, what is the value of the whole?   Ans. 2963.14¾.

65. A vessel and cargo is valued at $10.000, and ¾ of the cargo was sold for $5780; what was the value of the vessel?   Ans. $2775.00.

SECTION XI.

MENTAL OPERATIONS IN FRACTIONS.

NOTE. — The pupil is requested, before he commences the mental operations of Fractions, to examine carefully the definitions on page 78.

1. If $\frac{1}{3}$ of a barrel of cider cost 2 dollars, what will a barrel cost ?

2. Bought $\frac{1}{3}$ of a barrel of flour for 2 dollars, what would be the price of a barrel ?

3. If $\frac{1}{4}$ of a yard of cloth cost 3 dollars, what will $\frac{2}{4}$ cost ? What will $\frac{3}{4}$ cost ? What will a yard cost ?

4. If $\frac{1}{5}$ of a barrel of flour cost 2 dollars, what will $\frac{2}{5}$ cost ? $\frac{3}{5}$? $\frac{4}{5}$? $\frac{5}{5}$? $\frac{6}{5}$? $\frac{7}{5}$? $\frac{8}{5}$? $\frac{10}{5}$?

5. If $\frac{1}{8}$ of a pound of pepper cost 3 cents, what will $\frac{3}{8}$? $\frac{2}{8}$? $\frac{4}{8}$? $\frac{5}{8}$? $\frac{6}{8}$?

6. Bought $\frac{1}{8}$ of a barrel of sugar for 2 dollars, what was the price of $\frac{3}{8}$? $\frac{5}{8}$? $\frac{7}{8}$? $\frac{10}{8}$?

7. If a man bought $\frac{1}{8}$ of a firkin of butter for 7 cents, what did the whole firkin cost ?

8. Bought $\frac{1}{12}$ of an acre of land for 5 dollars, what was the price of an acre ? Of 2 acres ? Of 10 acres ? Of 20 acres ?

9. Bought $\frac{1}{7}$ of a pound of ginger for 4 cents, what did a pound cost ?

10. Bought $\frac{1}{10}$ of a chest of tea for 5 dollars, what did $\frac{2}{10}$ cost ? $\frac{3}{10}$? $\frac{4}{10}$? $\frac{5}{10}$? $\frac{10}{10}$?

11. Bought $\frac{1}{11}$ of a load of hay for 3 dollars, what did the load cost ? What did 2 loads cost ?

12. If $\frac{1}{7}$ of a cask of wine cost 9 dollars, what was that a cask ?

13. 2 is $\frac{1}{3}$ of what number? 4? 5? 6? 8? 9? 10? 12?

14. 4 is $\frac{1}{4}$ of what number ? 5? 6? 8? 9? 11? 15? 16?

15. 6 is $\frac{1}{5}$ of what number? 7? 8? 9? 11? 13? 18? 20?

16. 7 is $\frac{1}{8}$ of what number? 4? 5? 6? 8? 11? 15? 25?

17. 9 is $\frac{1}{4}$ of what number? 7? 2? 8? 11? 12? 13? 17?

18. 8 is $\frac{1}{7}$ of what number? 7? 11? 12? 15? 10? 11? 25?

19. 10 is $\frac{1}{6}$ of what number? 6? 7? 8? 9? 11? 15? 18?

20. 12 is $\frac{1}{9}$ of what number? 14? 15? 16? 17? 18? 19?

21. 11 is $\frac{1}{4}$ of what number? 6? 3? 2? 1? 7? 18? 14?

22. 14 is $\frac{1}{2}$ of what number? 11? 12? 13? 15? 16? 17?

23. 15 is $\frac{1}{3}$ of what number? 21? 18? 13? 14? 17? 19?

24. 7 is $\frac{1}{10}$ of what number? 10? 11? 12? 16? 18? 13?

25. 8 is $\frac{1}{4}$ of what number? 6? 7? 6? 5? 2? 7? 18?

26. If $\frac{2}{3}$ of a pound of sugar cost 4 cents, what will $\frac{1}{3}$ of a pound cost? What will a pound cost? 4 is $\frac{2}{3}$ of what number?

27. If $\frac{3}{4}$ of a pound of nutmeg cost 6 cents, what will $\frac{1}{4}$ of a pound cost? If $\frac{1}{4}$ of a pound cost 2 cents, what will a pound cost? 6 is $\frac{3}{4}$ of what number?

28. If $\frac{5}{6}$ of a cwt. of starch cost 10 dollars, what will $\frac{1}{6}$ of a cwt. cost? What will a cwt. cost? 10 is $\frac{5}{6}$ of what number?

29. If $\frac{2}{3}$ of a barrel of molasses cost 4 dollars, what is the value of $\frac{1}{3}$ of a barrel? If $\frac{1}{3}$ of a barrel cost 2 dollars, what will a barrel cost? 4 is $\frac{2}{3}$ of what number?

30. If $\frac{5}{6}$ of a pound of candles cost 15 cents, what will $\frac{1}{6}$ of a pound cost? If $\frac{1}{6}$ of a pound cost 3 cents, what will a pound cost? 15 is $\frac{5}{6}$ of what number?

31. Bought $\frac{5}{6}$ of a pound of mace for 20 cents, what was the cost of $\frac{1}{6}$ of a pound? If $\frac{1}{6}$ of a pound cost 4 cents, what will a pound cost? 20 is $\frac{5}{6}$ of what number?

32. If $\frac{11}{12}$ of a pound of opium cost 55 cents, what will $\frac{1}{12}$ of a pound cost? If $\frac{1}{12}$ of a pound cost 5 cents, what will a pound cost? 55 is $\frac{11}{12}$ of what number?

33. 6 is 3 times what number? 9? 12? 15? 18? 21? 24?

34. 15 is 3 times what number? 20? 25? 35? 65? 70?

35. 20 is 4 times what number? 4? 8? 12? 16? 40? 50?

36. 21 is 7 times what number? 14? 28? 56? 63? 70?

37. 24 is 2 times what number? 26? 30? 37? 39? 40?

38. 25 is 5 times what number? 80? 85? 95? 100? 200?

39. 3 is 6 times what number? 18? 21? 28? 31? 40? 60?

40. 18 is 9 times what number? 6? 7? 18? 30? 37? 28?

41. If 6 be $\frac{2}{3}$ of some number, what will be $\frac{1}{3}$ of the same number? If 3 be $\frac{1}{3}$ of a number, what is that number?

42. 3 is $\frac{1}{4}$ of what number?

43. If $\frac{6}{7}$ of a pound of sugar cost 12 cents, what will $\frac{1}{7}$ of a pound cost? If $\frac{1}{7}$ of a pound cost 2 cents, what will a pound cost?

44. If $\frac{4}{11}$ of a pound of nutmegs cost 20 cents, what will $\frac{1}{11}$ cost? If $\frac{1}{11}$ cost 5 cents, what will a pound cost? 20 is $\frac{4}{11}$ of what number?

45. If $\frac{8}{9}$ of a quarter of veal cost 64 cents, what will $\frac{1}{9}$ of a pound cost? If $\frac{1}{9}$ of a pound cost 8 cents, what will a pound cost? 64 is $\frac{8}{9}$ of what number?

46. Bought $\frac{7}{8}$ of a ton of potash for 84 dollars, what did $\frac{1}{8}$

of a ton cost?   If ½ of a ton cost 12 dollars, what did a ton cost?   84 is ⅞ of what number?

47. Bought ⅞ of a barrel of pork for 14 dollars, what was the cost of ⅛ of a barrel?   If ⅛ of a barrel cost 2 dollars, what was the price of a barrel?   14 is ⅞ of what number?

48. Bought ⅔ of a barrel of fish for 18 dollars, what did ⅓ of a barrel cost?   If ⅓ of a barrel cost 9 dollars, what was the value of a barrel?   18 is ⅔ of what number?

49. If ₇⁄₁₃ of an acre of land cost 42 dollars, what is the value of ₁⁄₁₃ of an acre?   If ₁⁄₁₃ of an acre cost 6 dollars, what is the value of an acre?   42 is ₇⁄₁₃ of what number?

50. If ₇⁄₁₀ of a bushel of apples cost 35 cents, what will ₁⁄₁₀ of a bushel cost?   If ₁⁄₁₀ of a bushel cost 5 cents, what is the value of a bushel?   35 is ₇⁄₁₀ of what number?

51. If ⅞ of a bushel of corn cost 63 cents, what is ⅛ of a bushel worth?   If ⅛ of a bushel cost 9 cents, what is the value of a bushel?   63 is ⅞ of what number?

52. If ₄⁄₁₁ of a pound of saleratus cost 8 cents, what is ₁⁄₁₁ of a pound worth?   If ₁⁄₁₁ of a pound cost 2 cents, what does a pound cost?

53. If ₇⁄₁₁ of a load of hay is sold for 14 dollars, what is ½ of a load worth?   If ½ of a load cost 2 dollars, what will 3 loads cost?   14 is ₇⁄₁₁ of what number?

54. If 10 cents will buy ⅔ of a pound of figs, how many cents will buy a pound?

55. If 12 dollars will buy ₄⁄₅ of a cwt. of sugar, how many dollars will it take to buy a cwt.?

56. 8 is ½ of what number? 7? 6? 11? 12? 15? 19? 20?
57. 9 is ⅓ of what number? 6? 7? 8? 15? 17? 18? 23?
58. 12 is ¾ of what number? 13? 16? 17? 19? 25? 27?
59. 4 is ⅖ of what number? 11? 7? 16? 17? 19? 21?
60. 3 is ₃⁄₁₁ of what number? 4? 5? 6? 8? 9? 11? 18?
61. 15 is ₃⁄₁₀ of what number? 16? 18? 20? 25? 26? 30?
62. 18 is ⅔ of what number? 3? 4? 5? 6? 7? 8? 11?
63. 20 is ⅘ of what number? 10? 8? 18? 21? 27? 28?
64. 21 is ₇⁄₁₁ of what number? 27? 28? 29? 33? 37? 38?
65. 25 is ₅⁄₁₂ of what number? 11? 16? 17? 18? 27? 36?
66. 24 is ₃⁄₁₁ of what number? 28? 29? 30? 37? 38? 61?
67. 27 is ₉⁄₁₀ of what number? 36? 42? 58? 59? 61? 72?
68. 30 is ⅔ of what number? 35? 40? 45? 55? 67? 85?
69. If ⅞ of a bushel of salt cost 42 cents, how many pounds of raisins at 6 cents a pound, would it take to purchase a bushel?

7*

70. If $\frac{6}{11}$ of a load of hay cost 18 dollars, how many barrels of cider at 3 dollars a barrel would it take to purchase a load? 18 is $\frac{6}{11}$ of how many times 3?

71. Bought $\frac{5}{9}$ of a yard of cloth for 60 cents; how many pounds of sugar at 9 cents would pay for one yard? 60 is $\frac{5}{9}$ of how many times 9?

72. A man sold a pair of oxen for 48 dollars, which was $\frac{8}{9}$ of what they cost him; he had paid for them in calves at 6 dollars a head. How many calves did it take to pay for the oxen?

73. 28 is $\frac{4}{7}$ of how many times 7? 8? 9? 10? 11? 12? 13?

74. 12 is $\frac{2}{11}$ of how many times 6? 7? 8? 12? 15? 16?

75. 32 is $\frac{2}{5}$ of how many times 10? 12? 13? 16? 17? 18?

76. 60 is $\frac{3}{7}$ of how many times 14? 7? 21? 24? 28? 30?

77. 72 is $\frac{4}{5}$ of how many times 9? 10? 11? 12? 15? 16?

78. 64 is $\frac{4}{7}$ of how many times 4? 5? 6? 7? 8? 9? 12?

79. 50 is $\frac{10}{3}$ of how many times 5? 6? 8? 9? 10? 12? 16?

80. 36 is $\frac{4}{7}$ of how many times 14? 16? 17? 18? 19? 20?

VULGAR FRACTIONS.

FRACTIONS are parts of an integer.

VULGAR FRACTIONS are expressed by two numbers, called the Numerator and Denominator; the former above, and the latter below a line.

Thus; $\left\{ \begin{array}{ll} \text{Numerator} & 7 \\ & — \\ \text{Denominator} & 11 \end{array} \right.$

The Denominator shows into how many parts the integer, or whole number, is divided.

The Numerator shows how many of these parts are taken.

1. A proper fraction is one, whose numerator is less than the denominator; as $\frac{3}{7}$.
2. An improper fraction is one, whose numerator exceeds or is equal to the denominator; as $\frac{17}{13}$ or $\frac{8}{8}$.
3. A simple fraction has a numerator and denominator only; as $\frac{5}{8}$, $\frac{11}{7}$.
4. A compound fraction is a fraction of a fraction, connected by the word *of*; as $\frac{2}{3}$ of $\frac{5}{6}$ of $\frac{3}{5}$ of $\frac{3}{4}$.
5. A mixed number is an integer with a fraction; as $7\frac{4}{11}$, $5\frac{2}{3}$.
6. A compound mixed fraction is one, whose numerator or denominator, or both, is a mixed number; as $\frac{7\frac{1}{2}}{11}$ or $\frac{4\frac{2}{3}}{7\frac{1}{2}}$

7. The greatest common measure of two or more numbers is the largest number, that will divide them without a remainder.

8. The least common multiple of two or more numbers is the least number, that may be divided by them without a remainder.

9. A fraction is in its lowest terms, when no number but a unit will measure both its terms.

10. A prime number is that, which can be measured only by itself or a unit; as 7, 11, and 19.

11. A perfect number is equal to the sum of all its aliquot parts ; as 6, 28, 496, &c.

12. A fraction is equal to the number of times the numerator will contain the denominator.

13. The value of a fraction depends on the proportion, which the numerator bears to the denominator.

CASE I.

To find the greatest common measure of two or more numbers, or to find the greatest number, that will divide two or more numbers.

RULE.

Divide the greater number by the less, and if there be a remainder, divide the last divisor by it, and so continue dividing the last divisor by the last remainder, until nothing remains, and the last divisor is the greatest common measure.

If there be more than two numbers, find the greatest common measure of two of them, and then of that common measure and the other numbers. If it should happen that 1 is the common measure, the numbers are prime to each other, and are incommensurable.

EXAMPLE.

1. What is the greatest common measure of 98 and 114?

$$98)114(1$$
$$98$$
$$\overline{}$$
$$16)98(6$$
$$96$$
$$\overline{}$$
Common measure 2)16(8
$$16$$

By this process, it is found that 2 is the greatest number, that will divide 98 and 114.

NOTE. —As 2 will divide 16, it will also divide 96, because it is a multiple of 16. It will also divide 98, because 98 is the sum of 96 and 2; and, as it will divide them *separate*, it will also *united*. Again 114 is equal to 98 + 16 and as 2 will divide both these numbers, it will also that of their sum. Therefore 2 will measure or divide 98 and 114 Q. E. D.

2. What is the greatest common measure of 56 and 168?
Ans. 56.

3. What is the greatest common measure of 96 and 128?
Ans. 32.

4. What is the greatest common measure of 57 and 285?
Ans. 57.

5. What is the greatest common measure of 169 and 175?
Ans. 1.

6. What is the greatest common measure of 175 and 455?
Ans. 35.

7. What is the greatest common measure of 169 and 866?
Ans. 1.

8. What is the greatest common measure of 47 and 478?
Ans. 1.

9. What is the greatest common measure of 84 and 1068?
Ans. 12.

10. What is the greatest common measure of 75 and 165?
Ans. 15.

11. What is the greatest common measure of 78, 234 and 468?
Ans. 78.

12. What is the greatest common measure of 144, 485 and 25?
Ans. 1.

13. What is the greatest common measure of 671, 2013 and 4026?
Ans. 671.

14. What is the greatest common measure of 16, 20 and 24?
Ans. 4.

15. What is the greatest common measure of 21, 27 and 81?
Ans. 3.

CASE II.

To reduce fractions to their lowest terms

1. Reduce $\frac{16}{48}$ to its lowest terms.

OPERATION.

$$4)\frac{16}{48}=4)\frac{4}{12}=\frac{1}{3} \text{ Ans.}$$

NOTE.—That $\frac{1}{3}$ is equal to $\frac{16}{48}$ may be demonstrated as follows :— 16 is the same multiple of 1, that 48 is of 3, therefore 16 has the same ratio to 48, that 1 has to 3; and as the value of a fraction depends on the ratio, which the numerator has to the denominator, it is evident when their ratios are the same that their values are equal; therefore $\frac{1}{3}$ is equal to $\frac{16}{48}$. Q. E. D.

Thus we see the propriety of the following

RULE.

Divide the numerator and denominator by any number, that will divide them both without a remainder; and so continue, until no number will divide them but unity. Or divide the numerator and denominator by their greatest common measure.

2. Reduce $\frac{16}{44}$ to its lowest terms. Ans. $\frac{4}{11}$.
3. Reduce $\frac{24}{32}$ to its lowest terms. Ans. $\frac{3}{4}$.
4. Reduce $\frac{36}{84}$ to its lowest terms. Ans. $\frac{3}{7}$.
5. Reduce $\frac{14}{108}$ to its lowest terms. Ans. $\frac{7}{54}$.
6. Reduce $\frac{15}{175}$ to its lowest terms. Ans. $\frac{3}{35}$.
7. Reduce $\frac{162}{244}$ to its lowest terms. Ans. $\frac{81}{122}$.
8. Reduce $\frac{617}{4319}$ to its lowest terms. Ans. $\frac{1}{7}$.
9. Reduce $\frac{416}{875}$ to its lowest terms. Ans. $\frac{416}{875}$.
10. Reduce $\frac{811}{1113}$ to its lowest terms. Ans. $\frac{811}{1113}$.
11. Reduce $\frac{615}{915}$ to its lowest terms. Ans. $\frac{41}{61}$.
12. Reduce $\frac{816}{1018}$ to its lowest terms. Ans. $\frac{408}{509}$.
13. Reduce $\frac{123}{456}$ to its lowest terms. Ans. $\frac{41}{152}$.

CASE III.

To reduce mixed numbers to improper fractions.

1. How many halves in 2 apples? In 3? In 4? In 5?
2. How many quarters in 7 oranges? In 9? In 10?
3. How many fifths in 3 bushels? In 4? In 5? In 6?
4. How many halves in $1\frac{1}{2}$? In $2\frac{1}{2}$? In $3\frac{1}{2}$? In $4\frac{1}{2}$?
5. How many quarters in $1\frac{1}{4}$? In $1\frac{3}{4}$? In $2\frac{1}{4}$? In $2\frac{1}{2}$?
6. How many sevenths in $1\frac{1}{7}$? In $1\frac{3}{7}$? In $2\frac{1}{7}$? In $2\frac{5}{7}$?

7. Reduce $17\frac{3}{5}$ to an improper fraction.

OPERATION.

$$\begin{array}{l} 17\frac{3}{5} \\ \underline{\ \ 5\ \ } \\ \underline{88} \\ \ \ 5 = \text{Ans.} \end{array}$$

The object in this question is to find how many fifths are contained in $17\frac{3}{5}$. This we obtain by multiplying 17 by 5, and adding 3 to the product. We may analyze this by saying, If 1 unit contain 5 fifths, 17 units will contain 17 times as much = 85 fifths; to which, if we add 3 fifths, the amount will be 88 fifths. Hence we deduce the following

RULE.

Multiply the whole number by the denominator of the fraction, and to the product add the numerator, and place their sum over the denominator of the fraction.

8. Reduce 16$\frac{2}{11}$ to an improper fraction.   Ans. $\frac{178}{11}$.
9. Reduce 14$\frac{3}{7}$ to an improper fraction.   Ans. $\frac{101}{7}$.
10. Reduce 126$\frac{11}{12}$ to an improper fraction.   Ans. $\frac{467}{12}$.
11. Reduce 149$\frac{13}{14}$ to an improper fraction.   Ans. $\frac{2089}{14}$.
12. Reduce 161$\frac{1}{17}$ to an improper fraction.   Ans. $\frac{347}{17}$.
13. Reduce 171$\frac{11}{12}$ to an improper fraction.   Ans. $\frac{10146}{12}$.
14. Reduce 98$\frac{3}{9}$ to an improper fraction.   Ans. $\frac{984}{9}$.
15. Reduce 116$\frac{31}{11}$ to an improper fraction.   Ans. $\frac{7197}{11}$.
16. Reduce 718$\frac{22}{11}$ to an improper fraction.   Ans. $\frac{62991}{11}$.
17. Reduce 100$\frac{100}{123}$ to an improper fraction.   Ans. $\frac{30000}{123}$.
18. Reduce 478$\frac{1}{111}$ to an improper fraction.   Ans $\frac{54453}{111}$.
19. Reduce 871$\frac{3}{116}$ to an improper fraction.   Ans. $\frac{101039}{116}$.
20. Reduce 167$\frac{100}{111}$ to an improper fraction.   Ans. $\frac{12639}{111}$.
21. Reduce 613$\frac{101}{307}$ to an improper fraction.   Ans. $\frac{188232}{307}$.
22. Reduce 159$\frac{100}{501}$ to an improper fraction.   Ans $\frac{79809}{501}$.
23. Reduce 999$\frac{222}{1000}$ to an improper fraction.   Ans. $\frac{999222}{1000}$.

CASE IV

To reduce improper fractions to integers or mixed numbers.

1. How many dollars in 2 half dollars?   In 4?   In 5?
2. How many dollars in 5 quarters?   In 6?   In 7?   In 8?
3. How many dollars in 16 eighths?   In 24?   In 30?
4. Reduce $\frac{117}{19}$ to a mixed fraction.

OPERATION.

19)117($6\frac{3}{19}$ Ans.
114
———
3

To apply this question, we may suppose a certain number of dollars to have been cut into 19 equal parts each, and we wish to know how many dollars 117 of these parts contain. To effect this, we must divide 117 by 19. Hence the following

RULE.

Divide the numerator by the denominator, and if there be a remainder, place it over the denominator at the right hand of the integer.

5. Reduce $\frac{177}{16}$ to a mixed number.   Ans. 11$\frac{1}{16}$.
6. Reduce $\frac{1621}{111}$ to a mixed number.   Ans. 14$\frac{7}{111}$.
7. Reduce $\frac{131}{17}$ to a mixed number.   Ans. 7$\frac{12}{17}$.
8. Reduce $\frac{517}{151}$ to a mixed number.   Ans. 3$\frac{64}{151}$.

9. Reduce $\frac{1161}{817}$ to a mixed number. Ans. $1\frac{344}{817}$

10. Reduce $\frac{1651}{17}$ to a mixed number. Ans. $97\frac{2}{17}$.

11. Reduce $\frac{1650}{11}$ to a mixed number. Ans. 150.

12. Reduce $\frac{679}{51}$ to a mixed number. Ans. $13\frac{16}{51}$.

13. Reduce $\frac{1000}{9}$ to a mixed number. Ans. $111\frac{1}{9}$.

14. Reduce $\frac{9999}{100}$ to a mixed number. Ans. $99\frac{99}{100}$.

15. Reduce $\frac{4123}{45}$ to a mixed number. Ans. $91\frac{28}{45}$.

16. Reduce $\frac{67856}{7}$ to a mixed number. Ans. $9693\frac{5}{7}$.

17. Reduce $\frac{1000000}{1}$ to a mixed number. Ans. 1000000.

CASE V.

To reduce compound fractions to simple fractions.

1. What part of an apple is $\frac{1}{2}$ of a half? $\frac{1}{3}$? $\frac{1}{4}$? $\frac{1}{5}$? $\frac{1}{6}$? $\frac{1}{7}$? $\frac{1}{8}$?

2. What part of an orange is $\frac{1}{2}$ of a third? $\frac{1}{3}$? $\frac{1}{4}$? $\frac{1}{5}$? $\frac{1}{6}$? $\frac{1}{7}$?

3. What part of a bushel is $\frac{1}{2}$ of a peck? $\frac{1}{3}$? $\frac{1}{4}$? $\frac{1}{5}$? $\frac{1}{6}$? $\frac{1}{7}$?

4. What part of a gallon is $\frac{1}{2}$ of a quart? $\frac{1}{3}$? $\frac{1}{4}$? $\frac{1}{5}$? $\frac{1}{8}$? $\frac{1}{10}$? $\frac{1}{12}$?

5. What is $\frac{3}{4}$ of $\frac{7}{8}$? $\frac{3}{4} \times \frac{7}{8} = \frac{21}{32}$ Ans.

This question may be analyzed as follows. $\frac{3}{4}$ of $\frac{7}{8}$ is = 3 times $\frac{1}{4}$ of $\frac{7}{8}$; and $\frac{1}{4}$ of $\frac{7}{8} = \frac{7}{32}$; and 3 times $\frac{7}{32} = \frac{21}{32}$ Ans. as before

Hence the propriety of the following

RULE.

Reduce the mixed numbers, if there be any, to improper fractions ; then multiply all the numerators together for a new numerator, and the denominators for a new denominator; the fraction should then be reduced to its lowest terms.

6. What is $\frac{2}{3}$ of $\frac{5}{6}$ of $\frac{7}{8}$ of $\frac{11}{12}$? Ans. $\frac{770}{1728} = \frac{385}{864}$.

7. What is $\frac{5}{6}$ of $\frac{11}{12}$ of of $\frac{1}{13}$? Ans. $\frac{11}{133}$.

8. What is $\frac{1}{2}$ of $\frac{5}{6}$ of $\frac{3}{4}$ of $\frac{11}{13}$? Ans. $\frac{165}{1456}$.

9. What is $\frac{7}{11}$ of $\frac{1}{2}$ of $\frac{3}{4}$ of $\frac{19}{50}$? Ans. $\frac{399}{4400}$.

10. What is $\frac{6}{11}$ of $\frac{1}{2}$ of $\frac{1}{2}$ of 21? Ans. $2\frac{5}{11}$.

11. What is $\frac{2}{5}$ of $\frac{9}{17}$ of 265? Ans. $84\frac{3}{17}$.

12. What is $\frac{4}{11}$ of $\frac{12}{13}$ of $\frac{6}{11}$ of 31? Ans. $5\frac{221}{2783}$.

13. What is $\frac{12}{13}$ of $\frac{13}{12}$ of 1728? Ans. $1067\frac{281}{437}$.

14. What is $\frac{2}{3}$ of $\frac{5}{9}$ of $11\frac{3}{4}$? Ans. $6\frac{33}{112}$.

15. What is $\frac{7}{11}$ of $15\frac{1}{7}$ of $5\frac{7}{10}$ of 100? Ans. $5758\frac{13}{44}$.

Note. — If there be numbers in the numerator similar to those in the denominator, they may be cancelled in the operation.

16. What is $\frac{3}{4}$ of $\frac{4}{7}$ of $\frac{7}{11}$ of $\frac{11}{14}$? 　　　Ans. $\frac{3}{14}=\frac{3}{14}$.
17. What is $\frac{2}{7}$ of $\frac{7}{15}$ of $\frac{15}{22}$ of $\frac{22}{33}$? 　　Ans. $\frac{2}{33}$.
18. What is $\frac{7}{11}$ of $\frac{11}{14}$ of $\frac{14}{17}$ of $\frac{17}{22}$? 　　Ans. $\frac{1}{2}$.
19. What is $\frac{1}{2}$ of $\frac{2}{3}$ of $\frac{3}{4}$ of $\frac{4}{5}$ of $\frac{5}{6}$ of 24? 　Ans. 2.

CASE VI.

To find the least common multiple of two or more numbers; that is, the least number, that may be divided by them without a remainder.

1. What is the least common multiple of 4, 6, 8, 16, 20?

OPERATION.

$$4)\overline{4\quad 6\quad 8\quad 16\quad 20}$$
$$2)\overline{1\quad 6\quad 2\quad 4\quad 5}$$
$$1\quad 3\quad 1\quad 2\quad 5$$

$4 \times 2 \times 3 \times 2 \times 5 = 240$ Ans.

We examine these numbers, and find that 4 will divide more of them than any other. Having divided them and written their quotients and undivided numbers in a line beneath, we again examine them as before, and find that 2 will divide more of the remaining numbers and quotients than any other number. After this division, we perceive, that no number but a unit will divide the quotients. We then multiply all the divisors and quotients together, and find the product to be 240, which is the least common multiple, or the least number, that can be divided by 4, 6, 8, 16, and 20, without a remainder. Hence the following

RULE.

Divide by such a number as will divide most of the given numbers without a remainder, and set the several quotients, with the several undivided numbers in a line beneath, and so continue to divide, until no number greater than unity will divide two or more of them. Then multiply all the divisors, quotients, and undivided numbers together, and the product is the least common multiple

2 What is the least common multiple of 6, 8, 10, 18, 20 and 24? 　　　　　　　　　　　　　　Ans. 360.

3. What is the least common multiple of 14, 19, 38 and 57? 　　　　　　　　　　　　　　　　Ans 798

4. What is the least common multiple of 20, 36, 48 and
50?   Ans. 3600.

5. What is the least common multiple of 15, 25, 35, 45
and 100?   Ans. 6300.

6. What is the least common multiple of 100, 200, 300,
400 and 575?   Ans. 27600.

7. I have four different measures; the first contains 4
quarts, the second 6 quarts, the third 10 quarts, and the
fourth 12 quarts.   How large is a vessel, that may be filled
by each one of these, taken any number of times full?
                                            Ans. 60 quarts.

CASE VII.

To reduce fractions to a common denominator.

1. Reduce $\frac{7}{8}$, $\frac{5}{12}$, and $\frac{11}{16}$ to other fractions of equal value,
having the same, or a common denominator.
                                          Ans. $\frac{42}{48}$, $\frac{20}{48}$, $\frac{33}{48}$.

First Method.

```
OPERATION.
4) 8  12  16      4×2×3×2 =    48 common denominator.
2) 2   3   4            8     ―――――――――――――――――――
   ―――――――――           12     6× 7 = 42 numerator for ⅞.
   1   3   2           16     4× 5 = 20 numerator for 5/12.
                              3×11 = 33 numerator for 11/16.
```

Having found the common denominator, 48, by the last
case, we divide it by the denominators 8, 12, and 16; and
the quotients 6, 4, and 3, we multiply by the numerators 7,
5, and 11, and the products 42, 20, and 33 are numerators
to be written over the common denominator; thus $\frac{42}{48}$, $\frac{20}{48}$, $\frac{33}{48}$.

Second Method.

```
OPERATION.
 7×12×16 = 1344 numerator for ⅞.
 5× 8×16 =  640 numerator for 5/12.
11× 8×12 = 1056 numerator for 11/16.
―――――――――――――――――――――――――――――――
 8×12×16 = 1536 common denominator.
```

The numerators are produced by multiplying the numerators of the given fraction by each of the other denominators,
and the common denominator is obtained by multiplying all
the denominators.   By this process, we obtain the following
                              Ans. $\frac{1344}{1536}$, $\frac{640}{1536}$, $\frac{1056}{1536}$.

The pupil will perceive, that this method does not express the fractions in so low terms as the other; although they both have the same value. From the above illustrations, we deduce the following

RULE.

Find the least common multiple of all the denominators by Case VI., and it will be the denominator required. Divide the common multiple by each of the denominators, and multiply the quotients by the respective numerators of the fractions and their products will be the numerators required.

Or, multiply each numerator into all the denominators except its own for a new numerator; and all the denominators into each other for a common denominator.

2. Reduce $\frac{3}{4}, \frac{1}{8}, \frac{7}{4}$ and $\frac{4}{13}$.   Ans. $\frac{54}{71}, \frac{19}{71}, \frac{42}{71}, \frac{30}{71}$.

3. Reduce $\frac{7}{11}, \frac{1}{15}, \frac{13}{30}$ and $\frac{3}{4}$.   Ans. $\frac{720}{1380}, \frac{704}{1380}, \frac{1364}{1380}, \frac{275}{1380}$.

4. Reduce $\frac{1}{7}, \frac{3}{14}, \frac{8}{21}$ and $\frac{4}{23}$.   Ans. $\frac{12}{84}, \frac{18}{84}, \frac{32}{84}, \frac{16}{84}$.

5. Reduce $\frac{16}{21}, \frac{1}{7}, \frac{11}{14}$ and $\frac{1}{3}$.   Ans. $\frac{32}{43}, \frac{24}{43}, \frac{33}{43}, \frac{21}{43}$.

6. Reduce $\frac{19}{23}, \frac{1}{11}, \frac{1}{7}$ and $\frac{3}{4}$.   Ans. $\frac{210}{508}, \frac{84}{508}, \frac{176}{508}, \frac{231}{508}$.

7. Reduce $\frac{1}{2}, \frac{3}{5}, \frac{3}{4}$ and $\frac{4}{5}$.   Ans. $\frac{30}{80}, \frac{48}{80}, \frac{48}{80}, \frac{48}{80}$.

8. Reduce $\frac{6}{7}, \frac{1}{4}, \frac{5}{11}$ and $\frac{7}{11}$.   Ans. $\frac{924}{1331}, \frac{308}{1331}, \frac{312}{1331}, \frac{441}{1331}$.

9. Reduce $\frac{3}{4}, \frac{7}{8}, \frac{1}{4}$ and $3\frac{1}{4}$.   Ans. $\frac{90}{120}, \frac{105}{120}, \frac{30}{120}, \frac{466}{120}$.

10. Reduce $\frac{5}{12}, \frac{3}{8}, \frac{1}{4}$ and $4\frac{3}{4}$.   Ans. $\frac{70}{168}, \frac{63}{168}, \frac{96}{168}, \frac{798}{168}$.

11. Reduce $\frac{1}{7}, \frac{5}{14}, \frac{11}{14}$, and $\frac{1}{3}$.   Ans. $\frac{24}{43}, \frac{15}{43}, \frac{33}{43}, \frac{21}{43}$.

12. Reduce $\frac{1}{5}, \frac{7}{13}, \frac{11}{17}$ and $\frac{3}{4}$.   Ans. $\frac{224}{1133}, \frac{117}{1133}, \frac{372}{1133}, \frac{44}{1133}$.

13. Reduce $\frac{19}{30}, \frac{7}{50}, \frac{11}{40}$ and $\frac{3}{20}$.   Ans. $\frac{570}{600}, \frac{140}{600}, \frac{165}{600}, \frac{12}{600}$.

14. Reduce $\frac{7}{8}, \frac{7}{12}, \frac{7}{16}$ and $\frac{7}{20}$.   Ans. $\frac{210}{240}, \frac{140}{240}, \frac{105}{240}, \frac{84}{240}$.

15. Reduce $\frac{1}{3}, 7, 8$, and $5\frac{1}{3}$.   Ans. $\frac{14}{63}, \frac{441}{63}, \frac{504}{63}, \frac{324}{63}$.

Let the following questions be performed by the second method.

16. Reduce $\frac{1}{2}, \frac{4}{5}$ and $\frac{7}{8}$ to fractions having a common denominator.   Ans. $\frac{80}{130}, \frac{96}{130}, \frac{105}{130}$.

17. Reduce $\frac{1}{7}, \frac{3}{8}$ and $\frac{7}{10}$.   Ans. $\frac{360}{830}, \frac{360}{830}, \frac{182}{830}$.

18. Reduce $\frac{7}{11}, \frac{1}{7}$ and $\frac{8}{13}$.   Ans. $\frac{546}{1001}, \frac{572}{1001}, \frac{616}{1001}$.

19. Reduce $\frac{7}{12}, \frac{4}{13}$ and $7\frac{1}{4}$.   Ans. $\frac{224}{624}, \frac{132}{624}, \frac{1836}{624}$.

20. Reduce $\frac{11}{13}, \frac{1}{3}$ and $\frac{1}{11}$.   Ans. $\frac{1433}{2283}, \frac{1029}{2283}, \frac{612}{2283}$.

21. Reduce $\frac{8}{5}, \frac{4}{17}$, and $11\frac{7}{11}$.   Ans. $\frac{2040}{2335}, \frac{840}{2335}, \frac{28315}{2335}$.

22. Reduce $\frac{1}{2}, \frac{3}{4}, \frac{1}{7}$ and 8.   Ans. $\frac{24}{43}, \frac{42}{43}, \frac{24}{43}, \frac{396}{43}$.

23. Reduce $\frac{1}{4}, \frac{7}{11}$ and $\frac{2}{3}$ of $7\frac{1}{2}$.   Ans. $\frac{178}{588}, \frac{252}{588}, \frac{3013}{588}$.

24. Reduce $\frac{1}{2}, \frac{3}{4}, \frac{1}{4}$ and 17.   Ans. $\frac{24}{48}, \frac{36}{48}, \frac{1}{48}, \frac{816}{48}$.

25. Reduce $\frac{11}{12}$, $\frac{2}{3}$ of 6 and $21\frac{1}{2}$. Ans. $\frac{110}{130}$, $\frac{475}{130}$, $\frac{2990}{130}$.
26. Reduce $\frac{5}{7}$, $\frac{3}{11}$, $\frac{4}{13}$, $\frac{4}{9}$ and $\frac{1}{2}$.

Ans. $\frac{12012}{14014}$, $\frac{4095}{14014}$, $\frac{4390}{14014}$, $\frac{8064}{14014}$, $\frac{7007}{14014}$.
27. Reduce $\frac{37}{45}$, $\frac{117}{?}$ and $\frac{7}{?}$.

Ans. $\frac{38504916}{17643153}$, $\frac{37088064}{17643153}$, $\frac{722313}{17643153}$.

CASE VIII.

To reduce fractions of a lower denomination to a higher.

1. What part of a pound is $\frac{3}{4}$ of a penny?

Now as pence are twelfths of a shilling and shillings are twentieths of a pound, it is evident, that the operation should be performed by Case V. of compound fractions;

$$\text{Thus, } \tfrac{3}{4}\times\tfrac{1}{12}\times\tfrac{1}{20}=\tfrac{3}{1920}=\tfrac{1}{640} \text{ Ans.}$$

From the above illustration, we see the propriety of the following

RULE.

Let the given fraction be reduced to a compound one by comparing it with all the denominations between the one given, and the one to which it is required to reduce it; then reduce this compound fraction to a simple one.

2. Reduce $\frac{3}{4}$ of a farthing to the fraction of a pound.
3. Reduce $\frac{2}{3}$ of a grain troy to the fraction of a pound.
4. Reduce $\frac{5}{8}$ of a scruple to the fraction of a pound.
5. Reduce $\frac{4}{11}$ of an ounce to the fraction of a hundred weight.
6. Reduce $\frac{3}{4}$ of a pound to the fraction of a ton.
7. Reduce $\frac{1}{2}$ of an inch to the fraction of an ell English.
8. Reduce $\frac{4}{5}$ of an inch to the fraction of a mile.
9. Reduce $\frac{1}{2}$ of a barleycorn to the fraction of a league.
10. Reduce $\frac{2}{3}$ of an inch to the fraction of an acre.
11. Reduce $\frac{1}{7}$ of a quart to the fraction of a tun, wine measure.
12. Reduce $\frac{2}{3}$ of a pint to the fraction of a bushel.
13. Reduce $\frac{1}{2}$ of a minute to the fraction of a year, (365$\frac{1}{4}$ days.)

CASE IX.

To reduce fractions of a higher denomination to a lower.

1. What part of a penny is $\frac{1}{240}$ of a pound?

As questions of this kind are the reverse of those in the last Case, it is evident they must be performed by multiplying the numerator, which is the same as dividing the fraction.

Thus, $\frac{1}{240} \times \frac{20}{1} \times \frac{12}{1} = \frac{240}{240} = \frac{1}{1} = 1$ Ans.

Hence the propriety of the following

RULE.

Let the given numerator be multiplied by all the denominations between it and the one to which it is to be reduced; then place the product over this denominator and reduce the fraction to its lowest terms.

2. Reduce $\frac{1}{1500}$ of a pound to the fraction of a farthing.
3. Reduce $\frac{1}{800}$ of a pound troy to the fraction of a grain.
4. Reduce $\frac{1}{2153}$ of a pound apothecaries' weight to the fraction of a scruple.
5. Reduce $\frac{3}{504}$ of a cwt. to the fraction of an ounce.
6. Reduce $\frac{1}{980}$ of a ton to the fraction of a pound.
7. Reduce $\frac{4}{225}$ of an ell English to the fraction of an inch.
8. Reduce $\frac{1}{110560}$ of a mile to the fraction of an inch.
9. Reduce $\frac{1}{114048}$ of a league to the fraction of a barleycorn.
10. Reduce $\frac{3}{3505056}$ of an acre to the fraction of an inch.
11. Reduce $\frac{1}{1153}$ of a tun of wine measure to the fraction of a quart.
12. Reduce $\frac{3}{550}$ of a bushel to the fraction of a pint.
13. Reduce $\frac{1}{2507680}$ of a year to the fraction of a minute.

CASE X.

To find the value of a fraction in the known parts of the integer.

1. What is the value of $\frac{3}{11}$ of a £.?

By Case IX. $\frac{3}{11}$ £. $= \frac{3}{11} \times \frac{20}{1} = \frac{60}{11} = 5\frac{5}{11}$s.; and $\frac{5}{11} = \frac{5}{11} \times \frac{12}{1} = \frac{60}{11}$d. $= 5\frac{5}{11}$d.; and $\frac{5}{11}$d. $= \frac{5}{11} \times \frac{4}{1} = \frac{20}{11}$qr. $= 1\frac{9}{11}$qr.

Ans. 5s. 5d. $1\frac{9}{11}$qr.

This question may be analyzed thus: — If 1£. is 20s., $\frac{3}{11}$ of a £. is $\frac{3}{11}$ of 20s. =5$\frac{5}{11}$s.; and if 1s. is 12d., $\frac{5}{11}$ of a shilling is $\frac{5}{11}$ of 12d. =5$\frac{5}{11}$d.; and if 1d. is 4qr., $\frac{5}{11}$ of a penny is $\frac{5}{11}$ of 4qr. = 1$\frac{9}{11}$qr. answer as before.

From these illustrations, we derive the following

RULE.

Multiply the numerator by the next lower denomination of the integer, and divide the product by the denominator; if anything remain, multiply it by the next less denomination, and divide as before, and so continue as far as may be required; and the several quotients will be the answer.

2. What is the value of $\frac{7}{24}$ of a shilling? -   Ans. 3$\frac{1}{2}$d.

-3. What is the value of $\frac{7}{8}$ of a guinea?

Ans. 21s. 9d. 1$\frac{1}{2}$qr.

4. What is the value of $\frac{7}{11}$ of a cwt.?

Ans. 2qr. 15lb. 4oz. 5$\frac{9}{11}$dr.

5. What is the value of $\frac{7}{15}$ of a lb. avoirdupois?

Ans. 7oz. 1$\frac{7}{15}$dr.

6 What is the value of $\frac{7}{8}$ of a lb. troy?

Ans. 10oz. 13dwt. 8gr.

7. What is the value of $\frac{4}{13}$ of a ℔. apothecaries' weight?

Ans. 3℥. 5ʒ. 1Ɖ. 12$\frac{4}{13}$gr.

8. What is the value of $\frac{7}{13}$ of a yard?

Ans. 2qr. 0na. 1$\frac{5}{13}$in.

9. What is the value of $\frac{7}{8}$ of an ell English?

Ans. 2qr. 3na. 0$\frac{1}{4}$in.

10. What is the value of $\frac{11}{13}$ of a mile?

Ans. 6fur. 30rd. 12ft. 8$\frac{4}{13}$in.

11. What is the value of $\frac{7}{8}$ of a furlong?

Ans. 35rd. 9ft. 2in.

12. What is the value of $\frac{7}{13}$ of an acre?

Ans. 2R. 6rd. 4yd. 5ft. 127$\frac{9}{13}$in.

13. What is the value of $\frac{9}{17}$ of a rod?

Ans. 144ft. 19$\frac{1}{17}$in

14. What is the value of $\frac{7}{13}$ of a cord?

Ans. 9ft. 1462$\frac{2}{13}$in.

15. What is the value of $\frac{3}{13}$ of a hhd. of wine?

Ans. 6gall. 2qt. 1pt. 0$\frac{4}{13}$gi.

16. What is the value of $\frac{7}{8}$ of a hhd. of beer?

Ans. 42gal.

17. What is the value of $\frac{11}{21}$ of a year (365¼ days)?

                      Ans. 174d. 16h. 26m. 5$\frac{5}{21}$sec.

18. What is the value of $7\frac{3\frac{1}{2}}{4\frac{1}{2}}$ of a dollar?

                            Ans. 7.74\frac{11}{18}$.

CASE XI.

To reduce any mixed quantity of weights, measures, &c. to the fraction of the integer.

1. What part of a shilling is 1d.? Is 2d.? Is 3d.? Is 4d.?
2. What part of a pound is 2s.? 3s.? 4s.? 5s.? 6s.? 9s.?
3. What part of a furlong is 3rd.? 4rd.? 5rd.? 8rd.? 9rd.?
4. What part of a hogshead is 5gal.? 8gal.? 10gal.?
5. What part of a foot is 2 inches? 3in.? 4in.? 5in.? 8in.?
6. What part of a £. is 5s. 5d. 1$\frac{9}{11}$qr.?

		s.	d.	qr.	
20		5	5	1$\frac{9}{11}$	In this question, the shillings,
12		12			pence, farthings, &c. are reduced
——		——			to elevenths of farthings for the
240		65			numerator of a fraction. A pound
4		4			is also reduced to the same de-
——		——			nomination, for a denominator.
960		261			The fraction is then reduced to
11		11			its lowest terms.
——		——			
10560		2880			

        $\frac{2880}{10560} = \frac{3}{11}$ Ans.

Hence the following

RULE.

Reduce the given number to the lowest denomination it contains for a numerator; and then reduce the integers to the same denomination, for the denominator of the fraction required.

7. Reduce 3¼d. to the fraction of a shilling.   Ans. $\frac{7}{24}$.
8. Reduce 21s. 9d. 1½qr. to the fraction of a guinea.

                            Ans. $\frac{7}{8}$.

9. Reduce 2qr. 15lb. 4oz. 5$\frac{9}{11}$dr. to the fraction of a cwt.

                            Ans. $\frac{7}{11}$.

10 What part of a pound are 7oz. 1½dr.?   Ans. $\frac{1}{2}$.

11. What part of a pound troy are 10oz. 13dwt. 8gr ? Ans. $\frac{8}{9}$.

12. What part of a pound apothecaries' weight are 3℥. 5ʒ. 1℈. 12$\frac{4}{13}$gr.? Ans. $\frac{4}{13}$.

13. What part of a yard are 2qr. 0na. 1$\frac{5}{13}$in.? Ans. $\frac{7}{13}$.

14. What part of an ell English are 2qr. 3na. 0$\frac{1}{4}$in.? Ans. $\frac{5}{9}$.

15. What part of a mile are 6fur. 30rd. 12ft. 8in. 0$\frac{12}{13}$br.? Ans. $\frac{11}{13}$.

16. Reduce 35rd. 9ft. 2in. to the fraction of a furlong. Ans. $\frac{8}{9}$.

17. What part of an acre are 2R. 6rd. 4yd. 5ft. 127$\frac{8}{13}$in.? Ans. $\frac{7}{13}$.

18. What part of a square rod are 144ft. 19$\frac{1}{17}$in.? Ans. $\frac{9}{17}$.

19. What part of a cord are 9ft. 1462$\frac{2}{13}$in.? Ans. $\frac{1}{13}$.

20. What part of a hhd. of wine are 6gal. 2qt. 1pt. 0$\frac{4}{19}$gi.? Ans. $\frac{2}{19}$

21. What part of a hhd. of beer are 42gal.? Ans. $\frac{7}{9}$.

22. What part of a year (365$\frac{1}{4}$) are 174da. 16h. 26m 5$\frac{5}{23}$sec.? Ans. $\frac{11}{23}$.

SECTION XII.

ADDITION OF VULGAR FRACTIONS.

CASE I.

To add fractions, that have a common denominator.

1. What part of an apple is $\frac{1}{4}$ and $\frac{3}{4}$? $\frac{1}{3}$ and $\frac{1}{3}$? $\frac{2}{5}$ and $\frac{1}{5}$?
2. What part of a dollar is $\frac{1}{5}$ and $\frac{3}{5}$? $\frac{2}{10}$ and $\frac{3}{10}$? $\frac{4}{10}$ and $\frac{6}{10}$?
3. What part of a shilling is $\frac{3}{12}$ and $\frac{4}{12}$? $\frac{5}{12}$, $\frac{4}{12}$, and $\frac{2}{12}$?
4. What part of an orange is $\frac{1}{3}$ and $\frac{2}{3}$? $\frac{1}{5}$ and $\frac{4}{5}$? $\frac{3}{5}$ and $\frac{2}{5}$?
5. Add $\frac{3}{12}$, $\frac{5}{12}$, $\frac{7}{12}$, and $\frac{11}{12}$ together.

OPERATION.

$$3+5+7+11=\frac{26}{12}=\frac{13}{6}=2\frac{1}{6}\text{ Ans.}$$

In this question, we add the numerators, and divide their sum by the denominator. Hence the following

RULE.

Write the sum of the numerators over the common denominator.

6. Add $\frac{5}{17}$, $\frac{6}{17}$, $\frac{7}{17}$, $\frac{11}{17}$, $\frac{14}{17}$, and $\frac{14}{17}$ together.   Ans. $3\frac{10}{17}$.
7. Add $\frac{4}{13}$, $\frac{5}{13}$, $\frac{11}{13}$, $\frac{13}{13}$, and $\frac{13}{13}$ together.   Ans. $2\frac{12}{13}$.
8. Add $\frac{5}{57}$, $\frac{17}{57}$, $\frac{37}{57}$, $\frac{47}{57}$, and $\frac{77}{57}$ together.   Ans. $2\frac{3}{57}$.
9. Add $\frac{11}{144}$, $\frac{12}{144}$, $\frac{22}{144}$, and $\frac{117}{144}$ together.   Ans. $1\frac{1}{8}$.
10. Add $\frac{212}{231}$, $\frac{221}{231}$, $\frac{267}{231}$, and $\frac{11}{231}$ together.   Ans. $2\frac{224}{231}$.
11. Add $\frac{421}{781}$, $\frac{741}{781}$, $\frac{882}{781}$, and $\frac{789}{781}$ together.   Ans. $3\frac{367}{781}$.
12. Add $\frac{1671}{8110}$, $\frac{8747}{8110}$, and $\frac{7188}{8110}$ together.   Ans. $1\frac{2141}{8110}$.
13. Add $\frac{3341}{3376}$, $\frac{3332}{3376}$, and $\frac{2000}{3376}$ together.   Ans. $2\frac{1477}{3376}$.
14. Add $\frac{999}{1000}$, $\frac{888}{1000}$, and $\frac{777}{1000}$ together.   Ans. $2\frac{83}{131}$.

CASE II.

To add fractions, that have not a common denominator.

1. What is the sum of $\frac{7}{8}$, $\frac{5}{12}$, $\frac{11}{16}$, and $\frac{13}{20}$?

First Method.

OPERATION.

$4)8\ \ 12\ \ 16\ \ 20 \qquad 4\times2\times3\times2\times5=240$ common denominator

$$2)\overline{2\quad 3\quad 4\quad 5}$$

$$\overline{1\quad 3\quad 2\quad 5}$$

Having found a common denominator by Case VII., we proceed as in the last Case.

$$\begin{array}{r|l}
8 & 30\times\ 7=210 \\
12 & 20\times\ 5=100 \\
16 & 15\times11=165 \\
20 & 12\times13=156 \\
\hline
 & \dfrac{631}{240}=2\frac{151}{240}\ \text{Ans.}
\end{array}$$

Second Method.

OPERATION.

$7\times12\times16\times20=26880$
$5\times\ \ 8\times16\times20=12800$
$11\times\ \ 8\times12\times20=21120$
$13\times\ \ 8\times12\times16=19968$

$$\frac{80768}{8\times12\times16\times20=30720}=2\frac{151}{240}\ \text{Ans.}$$

Let the pupil examine the second method of reducing fractions to a common denomiator in Case VII.

From the above illustrations, we deduce the following

RULE.

Reduce mixed numbers to improper fractions, and compound fractions to simple fractions; then reduce all the fractions to a common denominator, and the sum of their numerators written over the common denominator, will be the answer required.

2. Add $\frac{1}{4}$, $\frac{7}{8}$, $\frac{11}{12}$, and $\frac{3}{5}$ together.　　　　Ans. $2\frac{41}{60}$.

3. Add $\frac{7}{11}$, $\frac{2}{5}$, $\frac{1}{6}$, and $\frac{1}{4}$ together.　　　　Ans. $1\frac{331}{660}$.

4. Add $\frac{3}{10}$, $\frac{5}{11}$, $\frac{11}{12}$, and $\frac{7}{8}$ together.　　Ans. $2\frac{155}{660}$.

5. Add $\frac{5}{17}$, $\frac{3}{34}$, $\frac{12}{51}$, and $\frac{3}{8}$ together.　　Ans. 1.

6. Add $\frac{14}{15}$, $\frac{17}{30}$, $\frac{19}{45}$, and $\frac{7}{20}$ together.　Ans. $3\frac{13}{36}$.

7. Add $\frac{1}{2}$, $\frac{1}{3}$, $\frac{1}{4}$, $\frac{1}{5}$, $\frac{1}{6}$, and $\frac{1}{7}$ together.　Ans. $1\frac{79}{140}$.

8. Add $\frac{3}{8}$, $\frac{5}{7}$, and $5\frac{1}{2}$ together.　　Ans. $6\frac{33}{56}$.

9. Add $\frac{4}{11}$, $\frac{7}{22}$, and $9\frac{5}{11}$ together.　Ans. $9\frac{23}{22}$.

10. Add $\frac{2}{3}$, $\frac{1}{2}$, and $4\frac{4}{7}$ together.　　　Ans. $6\frac{1}{4}$.

11. Add $\frac{1}{2}$, $7\frac{1}{3}$, and $8\frac{3}{4}$ together.　　Ans. $17\frac{7}{12}$.

12. Add $\frac{7}{8}$, $3\frac{1}{4}$, and $5\frac{3}{7}$ together.　　Ans. $9\frac{111}{112}$.

13. Add $6\frac{3}{8}$, $7\frac{1}{3}$, and $4\frac{3}{4}$ together.　　Ans. $18\frac{11}{24}$.

NOTE.—If the quantity be a mixed number, the better way is to add their fractional parts separately, as in the following example.

14. What is the sum of $11\frac{3}{4}$, $15\frac{7}{8}$, $12\frac{5}{12}$, and $17\frac{5}{6}$?

$$
\begin{array}{r}
4)\overline{4 \quad 8 \quad 12 \quad 6} \\
3)\overline{1 \quad 2 \quad 3 \quad 6} \\
2)\overline{1 \quad 2 \quad 1 \quad 2} \\
\overline{1 \quad 1 \quad 1 \quad 1}
\end{array}
\qquad
\begin{array}{c}
4\times3\times2=24 \\
\begin{array}{c|c}
11\frac{3}{4} & 18 \\
15\frac{7}{8} & 21 \\
12\frac{5}{12} & 10 \\
17\frac{5}{6} & 20 \\
\hline
\text{Ans.}=57\frac{7}{8} & 69 \\
 & \overline{24} = 2\frac{7}{8}
\end{array}
\end{array}
$$

15. What is the sum of $11\frac{3}{8}$, $19\frac{5}{12}$ and $23\frac{5}{8}$?　Ans. $54\frac{10}{12}$.

16. What is the sum of $18\frac{1}{4}$, $27\frac{7}{10}$ and $49\frac{1}{2}$?　　Ans. 96.

17. What is the sum of $21\frac{1}{4}$, $18\frac{2}{3}$ and $26\frac{3}{8}$?　Ans. $66\frac{17}{24}$.

18. What is the sum of $17\frac{3}{4}$, $14\frac{1}{4}$ and $13\frac{2}{3}$?　Ans. $45\frac{1}{12}$.

19. What is the sum of $16\frac{3}{8}$, $8\frac{7}{8}$, $9\frac{3}{4}$, $3\frac{1}{4}$ and $1\frac{7}{8}$?

　　　　　　　　　　　　　　　Ans. $40\frac{1}{4}$.

20. What is the sum of $371\frac{11}{12}$, $614\frac{19}{20}$ and $81\frac{3}{4}$?

　　　　　　　　　　　　　　Ans. $1068\frac{37}{60}$.

21. Add $\frac{3}{4}$ of $18\frac{3}{11}$, $\frac{11}{12}$ of $\frac{3}{4}$ of $6\frac{3}{41}$ together.　Ans. $12\frac{331}{44}$.

22. Add $\frac{5}{8}$ of 18, and $\frac{4}{11}$ of $\frac{11}{12}$ of $7\frac{4}{11}$ together. Ans. $13\frac{31}{44}$.
23. Add $\frac{6}{7}$ of $15\frac{1}{4}$, $\frac{3}{4}$ of $107\frac{1}{7}$ together. Ans. $93\frac{3}{7}$.

24. Add $\frac{1}{3}$ of $\frac{3}{7}$ of $28\frac{3\frac{4}{7}}{4\frac{3}{7}}$ to $3\frac{39\frac{1}{2}}{105}$ Ans. $6\frac{3}{70}$.

CASE III.

To add any two fractions, whose numerators are a unit.

RULE.

Place the sum of the denominators over their product.

EXAMPLE.

1. Add $\frac{1}{4}$ to $\frac{1}{5}$.

$$\frac{4+5=9}{4\times5=20} \text{ Answer.}$$

2. Add $\frac{1}{3}$ to $\frac{1}{4}$, $\frac{1}{3}$ to $\frac{1}{5}$, $\frac{1}{7}$ to $\frac{1}{8}$, $\frac{1}{3}$ to $\frac{1}{8}$, $\frac{1}{3}$ to $\frac{1}{6}$, $\frac{1}{3}$ to $\frac{1}{9}$, $\frac{1}{3}$ to $\frac{1}{10}$.
3. Add $\frac{1}{12}$ to $\frac{1}{4}$, $\frac{1}{3}$ to $\frac{1}{8}$, $\frac{1}{5}$ to $\frac{1}{3}$, $\frac{1}{5}$ to $\frac{1}{4}$, $\frac{1}{3}$ to $\frac{1}{6}$, $\frac{1}{5}$ to $\frac{1}{6}$, $\frac{1}{5}$ to $\frac{1}{7}$.
4. Add $\frac{1}{11}$ to $\frac{1}{2}$, $\frac{1}{11}$ to $\frac{1}{3}$, $\frac{1}{11}$ to $\frac{1}{4}$, $\frac{1}{11}$ to $\frac{1}{5}$, $\frac{1}{11}$ to $\frac{1}{6}$, $\frac{1}{11}$ to $\frac{1}{7}$, $\frac{1}{11}$ to $\frac{1}{8}$.
5. Add $\frac{1}{10}$ to $\frac{1}{2}$, $\frac{1}{10}$ to $\frac{1}{3}$, $\frac{1}{10}$ to $\frac{1}{4}$, $\frac{1}{10}$ to $\frac{1}{5}$, $\frac{1}{10}$ to $\frac{1}{6}$, $\frac{1}{10}$ to $\frac{1}{7}$, $\frac{1}{10}$ to $\frac{1}{8}$.
6. Add $\frac{1}{4}$ to $\frac{1}{3}$, $\frac{1}{4}$ to $\frac{1}{5}$, $\frac{1}{4}$ to $\frac{1}{4}$, $\frac{1}{4}$ to $\frac{1}{5}$, $\frac{1}{4}$ to $\frac{1}{6}$, $\frac{1}{4}$ to $\frac{1}{7}$, $\frac{1}{5}$ to $\frac{1}{8}$.
7. Add $\frac{1}{8}$ to $\frac{1}{6}$, $\frac{1}{8}$ to $\frac{1}{7}$, $\frac{1}{8}$ to $\frac{1}{8}$, $\frac{1}{5}$ to $\frac{1}{9}$, $\frac{1}{8}$ to $\frac{1}{2}$, $\frac{1}{5}$ to $\frac{1}{3}$, $\frac{1}{8}$ to $\frac{1}{4}$.
8. Add $\frac{1}{7}$ to $\frac{1}{2}$, $\frac{1}{7}$ to $\frac{1}{3}$, $\frac{1}{7}$ to $\frac{1}{4}$, $\frac{1}{7}$ to $\frac{1}{5}$, $\frac{1}{7}$ to $\frac{1}{6}$, $\frac{1}{7}$ to $\frac{1}{8}$, $\frac{1}{7}$ to $\frac{1}{9}$.
9. Add $\frac{1}{9}$ to $\frac{1}{2}$, $\frac{1}{9}$ to $\frac{1}{3}$, $\frac{1}{9}$ to $\frac{1}{4}$, $\frac{1}{9}$ to $\frac{1}{5}$, $\frac{1}{9}$ to $\frac{1}{6}$, $\frac{1}{9}$ to $\frac{1}{7}$, $\frac{1}{7}$ to $\frac{1}{8}$.

NOTE.—The truth of this rule is evident from the fact, that this process reduces the fractions to a common denominator, and then adds the numerators.

If the numerators of the given fractions be alike, and more than a unit, multiply the sum of the denominators by one of the numerators for a new numerator, then multiply the denominators together for a new denominator.

10. Add $\frac{3}{4}$ to $\frac{3}{5}$.

$$\frac{4+5=9\times3=27}{4\times5=20}=1\frac{7}{20} \text{ Ans.}$$

11. Add $\frac{4}{5}$ to $\frac{4}{7}$, $\frac{4}{5}$ to $\frac{4}{9}$, $\frac{4}{5}$ to $\frac{4}{11}$, $\frac{3}{5}$ to $\frac{3}{7}$, $\frac{3}{5}$ to $\frac{3}{8}$, $\frac{3}{5}$ to $\frac{3}{11}$, $\frac{3}{8}$ to $\frac{3}{7}$.
12. Add $\frac{3}{5}$ to $\frac{3}{11}$, $\frac{3}{5}$ to $\frac{3}{8}$, $\frac{3}{5}$ to $\frac{3}{11}$, $\frac{5}{6}$ to $\frac{5}{7}$, $\frac{5}{6}$ to $\frac{5}{8}$, $\frac{5}{6}$ to $\frac{5}{9}$, $\frac{5}{6}$ to $\frac{5}{11}$.
13. Add $\frac{6}{7}$ to $\frac{6}{8}$, $\frac{6}{7}$ to $\frac{6}{11}$, $\frac{6}{7}$ to $\frac{6}{13}$, $\frac{6}{11}$ to $\frac{6}{13}$, $\frac{6}{11}$ to $\frac{6}{17}$, $\frac{6}{11}$ to $\frac{6}{15}$.
14. Add $\frac{8}{9}$ to $\frac{8}{11}$, $\frac{8}{11}$ to $\frac{8}{11}$, $\frac{8}{11}$ to $\frac{8}{13}$, $\frac{8}{13}$ to $\frac{8}{15}$, $\frac{8}{13}$ to $\frac{8}{17}$, $\frac{8}{13}$ to $\frac{8}{19}$.
15. Add $\frac{9}{10}$ to $\frac{9}{11}$, $\frac{9}{10}$ to [illegible], $\frac{9}{10}$ to $\frac{9}{14}$, $\frac{9}{10}$ to $\frac{9}{16}$, $\frac{9}{10}$ to $\frac{9}{17}$, $\frac{9}{10}$ to $\frac{9}{19}$.
16. Add $\frac{7}{8}$ to $\frac{7}{9}$, $\frac{7}{8}$ to $\frac{7}{10}$, $\frac{7}{8}$ to $\frac{7}{11}$, $\frac{7}{8}$ to $\frac{7}{12}$, $\frac{7}{8}$ to $\frac{7}{13}$, $\frac{7}{8}$ to $\frac{7}{14}$, $\frac{7}{8}$ to $\frac{7}{20}$.

17. Add $\frac{8}{9}$ to $\frac{8}{11}$, $\frac{8}{9}$ to $\frac{8}{13}$, $\frac{8}{9}$ to $\frac{8}{15}$, $\frac{8}{9}$ to $\frac{8}{17}$, $\frac{8}{9}$ to $\frac{8}{19}$, $\frac{8}{9}$ to $\frac{8}{21}$, $\frac{8}{9}$ to $\frac{8}{23}$.
18. Add $\frac{10}{11}$ to $\frac{10}{13}$, $\frac{10}{11}$ to $\frac{10}{17}$, $\frac{10}{11}$ to $\frac{10}{19}$, $\frac{10}{13}$ to $\frac{10}{13}$, $\frac{10}{13}$ to $\frac{10}{17}$, $\frac{10}{13}$ to $\frac{10}{19}$.
19. Add $\frac{11}{13}$ to $\frac{11}{15}$, $\frac{11}{13}$ to $\frac{11}{14}$, $\frac{11}{13}$ to $\frac{11}{15}$, $\frac{11}{13}$ to $\frac{11}{16}$, $\frac{11}{13}$ to $\frac{11}{17}$, $\frac{11}{13}$ to $\frac{11}{18}$.

NOTE.—The above rule may be found very useful, because all similar questions may be readily performed *mentally*.

CASE IV.

To add compound numbers..

1. Add $\frac{7}{13}$ of a £. to $\frac{9}{11}$ of a £.

Value of $\frac{7}{13}$ of a £. = 10s. 9d. $0\frac{12}{13}$ qr.
Value of $\frac{9}{11}$ of a £. = 16s. 4d. $1\frac{5}{11}$ qr.

1£. 7s. 1d. $2\frac{54}{143}$qr.

This question is performed, by finding the values of $\frac{7}{13}$ of a £. and $\frac{9}{11}$ of a £. by Case X. The fractions $\frac{12}{13}$ and $\frac{5}{11}$ are added by Case II, of Addition of Fractions. The following questions are performed in the same manner.

2. Add together $\frac{2}{3}$ of a £, $\frac{3}{7}$ of a £, and $\frac{3}{5}$ of a shilling.
Ans. 17s. $10\frac{24}{105}$d.

3. Add together $\frac{7}{11}$ of a ton, and $\frac{17}{22}$ of a cwt.
Ans. 13cwt. 2qr.

4. Add together $\frac{2}{3}$ of a yard, $\frac{1}{7}$ of an Ell English, and $\frac{5}{7}$ of a qr. Ans. 3qr. 3na. $1\frac{33}{140}$in.

5. Add together $\frac{7}{11}$ of a mile, $\frac{4}{13}$ of a furlong, and $\frac{2}{23}$ of a yard. Ans. 5fur. 16rd. 0ft. 3in. $1\frac{36}{143}$bar.

6. A. has three house lots; the first contains $\frac{5}{7}$ of an acre, the second $\frac{2}{3}$ of an acre, and the third $\frac{13}{14}$ of an acre. How many acres do they all contain?
Ans. 2A. 1R. 9p. 142ft. $87\frac{2}{7}$in.

7. A man travelled $18\frac{3}{7}$ miles the first day, $23\frac{11}{14}$ miles the second day, and $19\frac{1}{21}$ miles the third day. How far did he travel in the three days?
Ans. 61m. 2fur. 3rd. 13ft. $4\frac{3}{7}$in.

8. Add $\frac{11}{13}$ of a gallon of wine to $\frac{1}{12}$ of a hhd.
Ans. 6gal. 0qt. 1pt. $1\frac{1}{3}$gi.

9. Add $\frac{5}{16}$ of a week to $\frac{1}{5}$ of a day. Ans. 2d. 9h. 18m.

10. Add $\frac{3}{4}$ of a square foot to $\frac{1}{2}$ a foot square.
Ans. 1 foot.

11. Add 6 inches to 11rd. 16ft. 5in. Ans. 12rd 0ft. 5in.

SECTION XIII.

SUBTRACTION OF VULGAR FRACTIONS.

CASE 1.

To subtract fractions, that have a common denominator.

1. If $\frac{1}{4}$ be taken from $\frac{3}{4}$ what will be left?
2. If $\frac{3}{8}$ be taken from $\frac{5}{8}$ what will be left?
3. If $\frac{3}{10}$ be taken from $\frac{7}{10}$ what will be left?
4. What portion of a dollar will be left, if a $\frac{1}{4}$ be taken from $\frac{1}{2}$?
5. Subtract $\frac{5}{12}$ from $\frac{11}{12}$.

$$11 - 5 = 6. \quad \frac{6}{12} = \frac{1}{2} \text{ Ans.}$$

From the above we infer the following

RULE.

Subtract the less numerator from the greater, and under the remainder write the common denominator, and reduce the fraction if necessary.

6. Subtract $\frac{4}{17}$ from $\frac{11}{17}$. Ans. $\frac{7}{17}$.
7. Subtract $\frac{4}{15}$ from $\frac{11}{15}$. Ans. $\frac{7}{15}$.
8. Subtract $\frac{4}{27}$ from $\frac{23}{27}$. Ans. $\frac{19}{27}$.
9. Subtract $\frac{17}{35}$ from $\frac{19}{35}$. Ans. $\frac{2}{35}$.
10. Subtract $\frac{11}{15}$ from $\frac{8}{15}$. Ans. $\frac{13}{15}$.
11. Subtract $\frac{17}{27}$ from $\frac{29}{27}$. Ans. $\frac{12}{27}$.
12. Subtract $\frac{17}{42}$ from $\frac{31}{42}$. Ans. $\frac{14}{42}$.
13. Subtract $\frac{111}{137}$ from $\frac{148}{137}$. Ans. $\frac{37}{137}$.
14. Subtract $\frac{50}{100}$ from $\frac{75}{100}$. Ans. $\frac{1}{4}$.
15. Subtract $\frac{11}{32}$ from $\frac{9}{32}$. Ans. $\frac{1}{4}$.
16. Subtract $\frac{12}{44}$ from $\frac{35}{44}$. Ans. $\frac{5}{6}$.
17. Subtract $\frac{15}{70}$ from $\frac{47}{70}$. Ans. $\frac{7}{70}$.
18. Subtract $\frac{327}{1728}$ from $\frac{596}{1728}$. Ans. $\frac{269}{1728}$.
19. Subtract $\frac{98}{1000}$ from $\frac{100}{1000}$. Ans. $\frac{13}{500}$.

CASE II.

To subtract fractions whose denominators are unlike.

1. Subtract $\frac{4}{7}$ from $\frac{10}{11}$.

OPERATION.

Common denominator 77

$$
\begin{array}{c|l}
11 & 7 \times 10 = 70 \\
\;\;7 & 11 \times 4 = 44 \\
\hline
 & \quad\;\; 26 \\
 & \quad\;\; \overline{77} = \text{Ans.}
\end{array}
$$

In this question, we find the common denominator, 77, by multiplying the two denominators, 7 and 11; and then obtain the numerators, as in Case VII.; the difference of which we write over the common denominator.

Hence the propriety of the following

RULE.

Reduce compound fractions to simple ones, and mixed ones to improper fractions; then having found a common denominator, divide this by each of the denominators of the fraction, and multiply their quotients by their respective numerators. The difference of these products, placed over the common denominator, will give the answer required.

2. From $\frac{7}{13}$ take $\frac{2}{3}$. Ans. $\frac{13}{39}$.

3. From $\frac{5}{6}$ take $\frac{3}{7}$. Ans. $\frac{17}{42}$.

4. From $\frac{2}{5}$ take $\frac{1}{6}$. Ans. $\frac{1}{6}$.

5. From $\frac{13}{17}$ take $\frac{2}{5}$. Ans. $\frac{12}{11}$.

6. From $\frac{7}{10}$ take $\frac{5}{13}$. Ans. $\frac{17}{10}$.

7. From $\frac{14}{11}$ take $\frac{7}{13}$. Ans. $\frac{134}{11}$.

8. From $\frac{14}{13}$ take $\frac{3}{14}$. Ans. $\frac{611}{714}$.

9. From $\frac{14}{15}$ take $\frac{4}{11}$. Ans. $\frac{22}{11}$.

10. From $\frac{21}{50}$ take $\frac{3}{10}$. Ans. $\frac{3}{5}$.

11. From $\frac{17}{13}$ take $\frac{7}{50}$. Ans. $\frac{33}{100}$.

12. From $\frac{19}{7}$ take $\frac{5}{11}$. Ans. $\frac{4}{5}$.

13. From $\frac{5}{7}$ take $\frac{3}{14}$. Ans. $\frac{7}{5}$.

14. From $7\frac{3}{4}$ take $\frac{4}{5}$ of 9. Ans. $1\frac{3}{25}$.

15. From $\frac{2}{3}$ of $8\frac{3}{4}$ take $\frac{2}{3}$ of 5. Ans. $1\frac{11}{44}$.

16. From $\frac{1}{4}$ of 3 take $\frac{1}{3}$ of 2. Ans. $\frac{1}{15}$.

9

CASE III.

To subtract a proper fraction, or a mixed one, from a whole number.

1. From \$7. take \$$3\frac{5}{8}$.

OPERATION.

7
$3\frac{5}{8}$
\$$3\frac{3}{8}$ Ans.

It is evident that if $\frac{5}{8}$ of a dollar be taken from a whole dollar, that $\frac{3}{8}$ will remain. Having therefore added a unit to the minuend, we must also add one to the subtrahend as in Simple Subtraction. Hence the following

RULE.

Subtract the numerator from the denominator of the fraction, and under the remainder place the denominator, and carry one to the subtrahend to be subtracted from the minuend.

EXAMPLES.

2.	3.	4.	5.	6.	7.
32	16	671	385	16	18
$5\frac{7}{8}$	$4\frac{3}{4}$	$0\frac{41}{50}$	$16\frac{17}{35}$	$0\frac{117}{237}$	$1\frac{3}{4}$
$26\frac{1}{8}$	$11\frac{1}{4}$	$670\frac{9}{50}$	$368\frac{18}{35}$	$15\frac{120}{237}$	$16\frac{1}{4}$

8. From a hhd. of wine there leaked out $7\frac{8}{11}$ gallons; what quantity remained? Ans. $55\frac{3}{11}$gal.

9. A man engaged to labor 30 days, but was absent $5\frac{7}{13}$ days; how many days did he work? Ans. $24\frac{6}{13}$da.

10. From 144 pounds of sugar there was taken at one time $17\frac{5}{8}$ pounds, and at another $28\frac{7}{12}$ pounds; what quantity remains? Ans. $97\frac{19}{24}$lb.

11. A man sells $9\frac{7}{8}$ yards from a piece of cloth containing 34 yards; how many yards remain? Ans. $24\frac{1}{8}$yd.

12. The distance from Boston to Providence is 40 miles. A. having set out from Boston, has travelled $\frac{2}{17}$ of the distance; and B. having set out at the same time from Providence, has gone $\frac{2}{11}$ of the distance; how far is A. from B.? Ans. $28\frac{4}{187}$m.

13. From $\frac{2}{3}$ of a square yard take $\frac{2}{3}$ of a yard square. Ans. 2 square feet.

CASE IV.

To subtract one fraction from another, when both fractions have a unit for a numerator.

1. Take $\frac{1}{7}$ from $\frac{1}{3}$.

OPERATION.

$$\frac{7-3}{7 \times 3} = \frac{4}{21} \text{ Answer.}$$

The student will perceive, that this operation reduces the fractions to a common denominator. Hence the following

RULE.

Write the difference of the denominators over their product.

2. Take $\frac{1}{5}$ from $\frac{1}{2}$, $\frac{1}{3}$, $\frac{1}{3}$, $\frac{1}{4}$, $\frac{1}{4}$, $\frac{1}{4}$, $\frac{1}{4}$; $\frac{1}{20}$ from $\frac{1}{10}$, $\frac{1}{11}$, $\frac{1}{12}$, $\frac{1}{14}$.

3. Take $\frac{1}{8}$ from $\frac{1}{2}$, $\frac{1}{3}$, $\frac{1}{3}$, $\frac{1}{4}$, $\frac{1}{5}$, $\frac{1}{6}$, $\frac{1}{7}$; $\frac{1}{15}$ from $\frac{1}{13}$, $\frac{1}{14}$, $\frac{1}{14}$.

4. Take $\frac{1}{9}$ from $\frac{1}{2}$, $\frac{1}{3}$, $\frac{1}{4}$, $\frac{1}{4}$; $\frac{1}{13}$ from $\frac{1}{2}$, $\frac{1}{3}$, $\frac{1}{4}$, $\frac{1}{7}$, $\frac{1}{8}$.

5. Take $\frac{1}{6}$ from $\frac{1}{2}$, $\frac{1}{3}$, $\frac{1}{4}$; $\frac{1}{7}$ from $\frac{1}{2}$, $\frac{1}{3}$, $\frac{1}{4}$, $\frac{1}{6}$.

6. Take $\frac{1}{11}$ from $\frac{1}{10}$, $\frac{1}{2}$, $\frac{1}{3}$, $\frac{1}{4}$, $\frac{1}{5}$, $\frac{1}{6}$, $\frac{1}{7}$, $\frac{1}{8}$, $\frac{1}{9}$.

7. Take $\frac{1}{3}$ from $\frac{1}{2}$; $\frac{1}{4}$ from $\frac{1}{2}$, $\frac{1}{3}$; $\frac{1}{5}$ from $\frac{1}{2}$.

8. Take $\frac{1}{12}$ from $\frac{1}{2}$, $\frac{1}{3}$, $\frac{1}{4}$, $\frac{1}{5}$, $\frac{1}{6}$, $\frac{1}{7}$, $\frac{1}{8}$, $\frac{1}{9}$, $\frac{1}{10}$, $\frac{1}{11}$.

9. Take $\frac{1}{13}$ from $\frac{1}{2}$, $\frac{1}{3}$, $\frac{1}{4}$, $\frac{1}{5}$, $\frac{1}{6}$, $\frac{1}{7}$, $\frac{1}{8}$, $\frac{1}{9}$, $\frac{1}{10}$, $\frac{1}{11}$.

10. Take $\frac{1}{10}$ from $\frac{1}{2}$, $\frac{1}{3}$, $\frac{1}{4}$, $\frac{1}{4}$, $\frac{1}{5}$, $\frac{1}{6}$, $\frac{1}{7}$, $\frac{1}{8}$, $\frac{1}{9}$.

NOTE. — If the numerators of the given fractions be alike, and more than a unit, multiply the difference of the denominators by one of the numerators for a new numerator, then multiply the denominators together for a new denominator.

11. Take $\frac{2}{7}$ from $\frac{2}{3}$.

OPERATION.

$$7-3=4; \quad \frac{4 \times 2}{3 \times 7} = \frac{8}{21} \text{ Ans.}$$

12. Take $\frac{2}{3}$ from $\frac{2}{3}$, $\frac{3}{4}$ from $\frac{3}{7}$, $\frac{3}{10}$ from $\frac{3}{4}$, $\frac{3}{11}$ from $\frac{3}{5}$.

13. Take $\frac{3}{7}$ from $\frac{3}{5}$, $\frac{4}{11}$ from $\frac{4}{5}$, $\frac{4}{14}$ from $\frac{4}{7}$, $\frac{4}{11}$ from $\frac{4}{10}$.

14. Take $\frac{5}{7}$ from $\frac{5}{6}$, $\frac{5}{11}$ from $\frac{5}{6}$, $\frac{5}{14}$ from $\frac{5}{7}$, $\frac{5}{11}$ from $\frac{5}{10}$.

15. Take $\frac{5}{11}$ from $\frac{5}{7}$, $\frac{6}{11}$ from $\frac{6}{8}$, $\frac{6}{13}$ from $\frac{6}{7}$, $\frac{6}{13}$ from $\frac{6}{11}$.

16. Take $\frac{6}{13}$ from $\frac{6}{5}$, $\frac{6}{13}$ from $\frac{6}{7}$, $\frac{7}{14}$ from $\frac{7}{5}$, $\frac{7}{13}$ from $\frac{7}{11}$.

17. Take $\frac{7}{11}$ from $\frac{7}{8}$, $\frac{7}{11}$ from $\frac{7}{8}$, $\frac{7}{11}$ from $\frac{7}{8}$, $\frac{7}{11}$ from $\frac{7}{8}$.

18. Take $\frac{7}{11}$ from $\frac{7}{3}$, $\frac{8}{11}$ from $\frac{8}{3}$, $\frac{8}{14}$ from $\frac{8}{3}$, $\frac{8}{11}$ from $\frac{8}{7}$.

19 Take $\frac{10}{13}$ from $\frac{10}{9}$, $\frac{10}{15}$ from $\frac{10}{17}$, $\frac{10}{13}$ from $\frac{10}{9}$, $\frac{10}{17}$ from $\frac{10}{11}$.

20. Take $\frac{12}{7}$ from $\frac{12}{5}$, $\frac{12}{7}$ from $\frac{12}{5}$, $\frac{12}{5}$ from $\frac{12}{3}$, $\frac{12}{13}$ from $\frac{12}{7}$.

NOTE. — The above questions, and those of a similar kind, may readily be performed *mentally.*

CASE V.

To subtract compound numbers.

1. From $\frac{7}{11}$ of a £. take $\frac{1}{3}$ of a £.

Value of $\frac{7}{11}$ £. = 12s. 8$\frac{8}{11}$d. | 33 common denominator.
Value of $\frac{1}{3}$ £. = 4s. 5$\frac{1}{3}$d.
           Ans. 8s. 3$\frac{13}{33}$d.

| |
| 33 |
| 24 |
| 11 |
| 13 |
| 33 |

To perform this question, we find by Case X. Sect. XI. the value of $\frac{7}{11}$£. = 12s. 8$\frac{8}{11}$d.; and also of $\frac{1}{3}$£. = 4s. 5$\frac{1}{3}$d.; we then find a common denominator of the fractional part, by multiplying together their denominators, 11 and 3 = 33 We then proceed as in Case II., Sect. XIII. Hence the following

RULE.

Find the value of the fractions in integers; then subtract as in the foregoing rules.

2. From $\frac{3}{4}$ of a ton take $\frac{1}{3}$ of a cwt.
                        Ans. 1qr. 26lb. 14oz. 10$\frac{6}{11}$dr.
3. From $\frac{1}{3}$ of an ell English take $\frac{1}{2}$ of a yard.
                        Ans. 3qr. 0na. 2$\frac{1}{6}$in.
4. From $\frac{2}{3}$ of a mile take $\frac{7}{11}$ of a furlong.
                        Ans. 1fur. 5rd. 10ft 10in.
5. From $\frac{3}{7}$ of a degree take $\frac{2}{3}$ of a mile.
                        Ans. 49m. 0fur. 13rd. 11ft. 9in. 1$\frac{7}{9}$bar.
6. From $\frac{3}{11}$ of an acre take $\frac{1}{3}$ of a rod.
                        Ans. 1R. 17p. 22yd. 2ft. 108in.
7. From $\frac{3}{10}$ of a cord take $\frac{3}{11}$ of a cord.
                        Ans. 91ft. 1602$\frac{18}{55}$in.
8. From $\frac{7}{9}$ of a hhd. of wine there leaked out $\frac{1}{3}$ of it; what remained?   Ans. 6gal. 3qt. 0pt. 1$\frac{7}{9}$gi.
9. From Boston to Concord, N. H. the distance is 72 miles; having travelled $\frac{4}{9}$ of this distance, how much remains?
                        Ans. 30m. 6fur. 34rd. 4ft. 8in. 1$\frac{1}{9}$bar.
10. From $\frac{2}{7}$ of a year take $\frac{1}{3}$ of a week.
                        Ans. 101da. 5h. 54m. 17$\frac{1}{7}$sec.
11. From $\frac{3}{4}$ of an acre take $\frac{1}{2}$ of a foot.
                        Ans. 1R. 18p. 5yd. 4ft. 0in.

SECTION XIV.

MULTIPLICATION OF VULGAR FRACTIONS.

CASE I.

To multiply one fraction by another.

1. If 1 apple cost $\frac{1}{2}$ of a cent, what will 2 apples cost? 3?
2. If 1 orange cost $1\frac{1}{2}$ cents, what will 2 cost? 3? 4? 5?
3. If 1 pound of sugar cost $\frac{1}{8}$ of a dollar, what will 2lbs. cost? 3? 4? 5? 6? 7? 8? 9? 10?
4. If 1 pound of chalk cost $\frac{1}{3}$ of a cent, what cost 2lbs.?

5. Multiply $\frac{7}{8}$ by $\frac{3}{5}$.

$$\text{OPERATION.}$$
$$\frac{7\times3}{8\times5}=\frac{21}{40}\text{ Ans.}$$

When one number is multiplied by another, it is evident, that the multiplicand is repeated as many times as there are units in the multiplier. The multiplier in this case is $\frac{3}{5}$. The multiplicand, $\frac{7}{8}$, being less than a unit, is therefore to be repeated $\frac{3}{5}$ times. If, therefore, we multiply the denominator, 8, by 5, we make the fraction to be $\frac{7}{40}$, which is only $\frac{1}{5}$ of the value of $\frac{7}{8}$. If we now multiply the last numerator, 7, by 3, we shall increase its value 3 times; that is, $\frac{21}{40}$ is 3 times the value of $\frac{7}{40}$, and is $\frac{3}{5}$ of $\frac{7}{8}$. We may therefore analyze the question and say, $\frac{1}{5}$ of $\frac{7}{8}=\frac{7}{40}$; and $\frac{3}{5}$ of $\frac{7}{8}$ is 3 times $\frac{1}{5}=\frac{3}{5}\times\frac{7}{40}$ $=\frac{21}{40}$; therefore $\frac{21}{40}$ is $\frac{3}{5}$ of $\frac{7}{8}$. Hence the following

RULE.

Prepare the fractions as in Addition, and multiply the numerators for a new numerator, and the denominators for a new denominator; if the result be an improper fraction, reduce it to its equivalent whole number.

6. Multiply $\frac{5}{11}$ by $\frac{3}{7}$.   Ans. $\frac{15}{77}$.
7. Multiply $\frac{7}{9}$ by $\frac{13}{10}$.   Ans. $\frac{91}{90}$.
8. Multiply $\frac{3}{5}$ by $\frac{7}{11}$.   Ans. $\frac{21}{55}$.
9. Multiply $\frac{7}{13}$ by $\frac{11}{12}$.   Ans. $\frac{77}{156}$.
10. Multiply $\frac{5}{11}$ by $\frac{4}{7}$.   Ans. $\frac{20}{77}$.
11. Multiply $\frac{3}{4}$ by $\frac{11}{12}$.   Ans. $\frac{33}{48}$.

9*

12. Multiply $\frac{12}{17}$ by $\frac{11}{14}$.　　　　Ans. $\frac{11}{14}$.
13. Multiply $\frac{1}{4}$ by $8\frac{1}{2}$.　　　　Ans. $3\frac{1}{4}$.
14. Multiply $\frac{9}{10}$ by $17\frac{5}{11}$.　　　　Ans. $15\frac{8}{11}$.
15. Multiply $\frac{8}{9}$ by $71\frac{1}{4}$.　　　　Ans. $63\frac{37}{44}$.
16. Multiply $8\frac{7}{8}$ by $11\frac{2}{8}$.　　　　Ans. $100\frac{41}{44}$.
17. Multiply $161\frac{11}{14}$ by $19\frac{11}{14}$.　　　Ans. $3136\frac{33}{70}$.
18. Multiply $\frac{3}{4}$ of $9\frac{1}{4}$ by $\frac{1}{3}$ of 17.　　　Ans. $78\frac{5}{8}$.
19. Multiply $\frac{9}{10}$ of 7 by $\frac{11}{14}$ of $87\frac{3}{11}$.　　Ans. $403\frac{1}{4}$.
20. Multiply 7 by $3\frac{3}{4}$.　　　　Ans. $26\frac{1}{4}$.
21. Multiply 8 by $\frac{7}{8}$.　　　　Ans. $6\frac{1}{4}$.
22. Multiply 12 by $\frac{5}{7}$.　　　　Ans. $8\frac{4}{7}$.
23. Multiply 15 by $\frac{6}{11}$.　　　　Ans. $8\frac{2}{11}$.

24. A merchant owns $\frac{7}{8}$ of a ship; he sells $\frac{4}{11}$ of his share to A. What part is that of the whole?　　　Ans. $\frac{7}{22}$.

CASE II.

To multiply a mixed number by a whole number.

1. Multiply $7\frac{5}{8}$ by 9.

OPERATION.

$7\frac{5}{8}$
9
————
$68\frac{5}{8}$

In performing this question, we multiply the 5 of the numerator by the multiplier, 9, and find the product to be 45. This product we divide by 8, the denominator of the fraction, and find the quotient to be 5, which we carry to the product of 9 times 7. The remainder, 5, we write over the divisor, 8. Hence the following

RULE.

Multiply the numerator of the mixed number by the whole number, and divide the product by the denominator of the fraction, and as many times as it contains the denominator, so many units must be carried to the product of the integers. If, after division, anything remains, let it be a numerator and the divisor a denominator to a fraction to be affixed to the product.

2. Multiply $8\frac{3}{4}$ by 7.　　　　Ans. $60\frac{3}{4}$.
3. Multiply $9\frac{4}{7}$ by 8.　　　　Ans. $75\frac{4}{7}$.
4. Multiply $11\frac{7}{8}$ by 7.　　　　Ans. $82\frac{1}{4}$.
5. Multiply $15\frac{4}{11}$ by 7.　　　Ans. $108\frac{4}{11}$.
6. Multiply $14\frac{2}{3}$ by 9.　　　　Ans $131\frac{1}{3}$.

7. Multiply $23\frac{3}{8}$ by 11. Ans. $257\frac{1}{8}$.
8. Multiply $47\frac{3}{11}$ by 15. Ans. $709\frac{1}{11}$.
9. Multiply $37\frac{3}{4}$ by 18. Ans. $679\frac{1}{2}$.
10. Multiply $678\frac{11}{12}$ by 24. Ans. 16294.
11. What will $23\frac{7}{8}$ pounds of lead cost, at 8 cents a pound? Ans. $ 1.91.
12. What will $15\frac{11}{16}$ pounds of sugar cost, at 12 cents a pound? Ans. $ $1.88\frac{1}{4}$.
13. What will $29\frac{5}{12}$ cwt. of hay cost, at $ 1.12 per cwt.? Ans. $ $32.94\frac{2}{3}$.
14. What will $9\frac{7}{8}$ yards of broadcloth cost, at $ 8 per yard? Ans. $ 79.00.
15. What will $17\frac{9}{16}$ tons of potash cost, at $ 97 per ton? Ans. $ $1703.56\frac{1}{4}$.
16. What will $3\frac{17}{113}$ tons of plaster of Paris cost, at $ 12.75 per ton? Ans. $ $40.18\frac{12}{113}$.
17. What will $17\frac{5}{12}$ dozen candles cost, at 16 cents a dozen? Ans. $ $2.78\frac{2}{3}$.
18. What will $27\frac{3}{8}$ pounds of bar soap cost, at 7 cents a pound? Ans. $ $1.91\frac{5}{8}$.
19. What will $29\frac{7}{12}$ dozen of axes cost, at $ 11.62 per dozen? Ans. $ $343.75\frac{5}{6}$.
20. Bought 28 bales of cotton cloth, each bale containing $31\frac{3}{8}$ yards; what will be the cost, at 16 cents per yard? Ans. $ 140.56.

SECTION XV.

DIVISION OF VULGAR FRACTIONS.

CASE I.

To divide one fraction by another.

1. If 2 peaches cost $\frac{2}{3}$ of a cent, what will 1 peach cost?
2. If 1 orange cost 3 apples, what part of an orange will 1 apple buy?
3. If $\frac{3}{4}$ of a yard of cloth cost 1 dollar, what will $\frac{1}{4}$ of a yard cost? $\frac{3}{4}$ yard?
4. A certain teacher divided $\frac{1}{2}$ an orange among 8 of his pupils; what part did each receive?

5. A quarter of a dollar is divided equally among 4 persons, what part does each receive?

6. Among how many boys, must 8 apples be distributed to give each ⅛ of an apple?

OPERATION.

7. Divide $\frac{4}{7}$ by $\frac{1}{3}$. $\frac{4}{7} \times \frac{3}{1} = \frac{12}{7} = 1\frac{5}{7}$ Ans.

By multiplying the numerator of the dividend 4, by the denominator of the divisor 8, the dividend is reduced to eights; and then by multiplying the denominator of the dividend, 7, by 3, the numerator of the divisor, one third of the eights are taken.

Hence we deduce the following

RULE.

Prepare the fractions as in Addition; invert the divisor, and proceed as in Multiplication.

8. Divide $\frac{7}{15}$ by $\frac{5}{8}$.	$\frac{7}{15} \times \frac{8}{5} = \frac{56}{75} = \frac{56}{75}$ Ans.
9. Divide $\frac{11}{13}$ by $\frac{3}{4}$.	Ans. $1\frac{3}{5}$.
10. Divide $\frac{3}{4}$ by $\frac{8}{13}$.	Ans. $1\frac{7}{32}$
11. Divide $\frac{5}{8}$ by $\frac{3}{8}$.	Ans. $\frac{11}{15}$.
12. Divide $\frac{11}{15}$ by $\frac{7}{11}$.	Ans. $1\frac{16}{105}$.
13. Divide $\frac{8}{25}$ by $\frac{14}{7}$.	Ans. $\frac{63}{175}$.
14. Divide $\frac{15}{21}$ by $\frac{1}{15}$.	Ans. $11\frac{3}{7}$.
15. Divide $\frac{3}{25}$ by $7\frac{3}{4}$.	Ans. $\frac{36}{775}$.
16. Divide $\frac{8}{11}$ by $16\frac{3}{4}$.	Ans. $\frac{13}{407}$.
17. Divide $11\frac{1}{9}$ by $\frac{5}{9}$.	Ans. 20.
18. Divide $21\frac{3}{7}$ by $18\frac{3}{7}$.	Ans. $1\frac{113}{650}$.
19. Divide $17\frac{3}{11}$ by $28\frac{11}{14}$.	Ans. $\frac{242}{3154}$.
20. Divide $161\frac{3}{17}$ by $14\frac{3}{4}$.	Ans. $11\frac{49}{1547}$.
21. Divide $\frac{7}{11}$ of $\frac{2}{3}$ by $\frac{2}{3}$ of $\frac{3}{11}$.	Ans. $1\frac{1}{3}$.
22. Divide $\frac{2}{3}$ of $7\frac{3}{11}$ by $\frac{4}{11}$ of $17\frac{3}{4}$.	Ans. $\frac{3850}{6053}$.
23. Divide $\frac{6}{17}$ of 15 by $\frac{7}{13}$ of 22.	Ans. $\frac{675}{1309}$.

24. Bought $\frac{2}{7}$ of a coal mine for $\$3675$, and having sold $\frac{4}{7}$ of it, I gave $\frac{3}{5}$ of the remainder to a charitable society, and divided the residue among 7 poor persons; what was the share of each?   Ans. $\$50$ for each poor person.

25. Of an estate valued at $\$5000$, the widow receives $\frac{1}{3}$, the oldest son $\frac{2}{3}$ of the remainder; the residue is equally divided among 7 daughters; what is the share of each daughter?   Ans. $\$158\frac{12}{21}$.

CASE II.

To divide an integer or whole number by a fraction.

OPERATION.

1. Divide 17 by $\frac{3}{4}$. $17 \times 4 = \frac{68}{3} = 22\frac{2}{3}$ Ans.

By multiplying the dividend by 4, it is reduced to fourths; and by dividing the product by 3, the quotient will be three-fourths.   Hence the following

RULE.

Multiply the whole number by the denominator of the fraction, and divide the product by the numerator.

2. Divide 18 by $\frac{7}{11}$.	Ans. 28$\frac{4}{7}$.
3. Divide 28 by $\frac{13}{19}$.	Ans. 41$\frac{1}{13}$.
4. Divide 27 by $\frac{3}{17}$.	Ans. 459.
5. Divide 128 by $\frac{2}{15}$.	Ans. 960.
6. Divide 98 by $\frac{11}{17}$.	Ans. 151$\frac{7}{11}$.
7. Divide 19 by $\frac{13}{21}$.	Ans. 31$\frac{3}{13}$.
8. Divide 167 by $\frac{12}{15}$.	Ans. 200$\frac{2}{3}$.
9. Divide 49 by $\frac{11}{20}$.	Ans. 88$\frac{12}{11}$.
10. Divide 15 by $\frac{1}{15}$.	Ans. 225.

CASE III.

To divide a fraction by an integer.

OPERATION.

1. Divide $\frac{4}{5}$ by 17. $4 \times 17 = \frac{4}{85}$ Ans.

This case is the reverse of the last.   We therefore see the propriety of the following

RULE.

Multiply the integer by the denominator of the fraction, and write the product under the numerator.

2. Divide $\frac{7}{11}$ by 18.	Ans. $\frac{7}{198}$.
3. Divide $\frac{13}{19}$ by 28.	Ans. $\frac{13}{532}$.
4. Divide $\frac{1}{17}$ by 27.	Ans. $\frac{1}{459}$.
5. Divide $\frac{2}{15}$ by 128.	Ans. $\frac{1}{960}$.
6. Divide $\frac{11}{17}$ by 98.	Ans. $\frac{11}{1666}$.

7. Divide $\frac{14}{23}$ by 19. Ans. $\frac{14}{437}$.
8. Divide $\frac{5}{6}$ by 167. Ans. $\frac{5}{1002}$.
9. Divide $\frac{16}{29}$ by 49. Ans. $\frac{16}{1421}$.
10. Divide $\frac{1}{15}$ by 15. Ans. $\frac{1}{225}$.

CASE IV.

To divide a mixed fraction by an integer.

1. Divide $27\frac{3}{5}$ by 6.

OPERATION.

$6\,)\,\underline{27\frac{3}{5}}$

$4\frac{18}{30} = 4\frac{3}{5}$ Ans.

We divide the 27 by 6, and find it is contained 4 times, and that 3 remains, which we multiply by the denominator, 5, and to the product we add the numerator, 3, the sum of which is 18; this we write over the product of 6, the divisor, multiplied by the denominator, $5 = 30$. Hence the following

RULE.

Divide the integer as in whole numbers, and if any thing remains, multiply it by the denominator of the fraction, and to the product add the numerator of the fraction and write it over the product of the divisor, multiplied by the denominator.

2. Divide $29\frac{3}{8}$ by 9. Ans. $3\frac{19}{72}$.
3. Divide $14\frac{1}{2}$ by 7. Ans. $2\frac{1}{14}$.
4. Divide $13\frac{3}{8}$ by 8. Ans. $1\frac{43}{64}$.
5. Divide $14\frac{3}{5}$ by 6. Ans. $2\frac{13}{30}$
6. Divide \$$37\frac{3}{7}$ among 9 men. Ans. \$$4\frac{10}{63}$
7. Divide \$$96\frac{2}{3}$ among 11 persons. Ans. \$$8\frac{26}{33}$.
8. What is $\frac{1}{8}$ of $167\frac{7}{11}$ cwt. of iron? Ans. $20\frac{21}{22}$ cwt.
9. Divide $\frac{7}{8}$ of a prize, valued at \$1723, equally between 12 seamen. Ans. \$$125.63\frac{13}{24}$.
10. What will a barrel of flour cost, if 19 barrels can be purchased for \$$107\frac{3}{8}$? Ans. \$$5.65\frac{5}{38}$.
11. If 15 pounds of raisins can be obtained for \$$3\frac{2}{7}$, what will 1 pound cost? Ans. \$$0.21\frac{19}{21}$.
12. If 12 quarts of wine cost \$$3.75\frac{3}{4}$; what will a quart cost? Ans. \$$0.31\frac{5}{16}$.
13. If \$19 will buy $375\frac{11}{16}$ acres of land; how much can be bought for \$1. Ans. $19\frac{235}{304}$ acres.

MISCELLANEOUS QUESTIONS.

1. How far will a man walk in $17\frac{7}{11}$ hours, provided he goes at the rate of $4\frac{7}{8}$ miles an hour?
Ans. 82m. 4fur. 8rd. 1ft. 4in.

2. How much land is there in a field, which is $29\frac{7}{13}$ rods square?   Ans. 5A. 1R. 32p. 141ft. $109\frac{149}{169}$in.

3. How much wood in a pile, which is $17\frac{3}{4}$ feet long, $7\frac{1}{11}$ feet high, and $4\frac{3}{4}$ feet wide?   Ans. 4cd. $66\frac{97}{110}$ft.

4. What is the value of $19\frac{7}{8}$ barrels of flour, at $\$6\frac{3}{4}$ a barrel?   Ans. $\$134.15\frac{7}{8}$.

5. What is the value of $376\frac{11}{13}$ acres of land, at $\$75\frac{3}{8}$ per acre?   Ans. $\$28387.06\frac{1}{4}$.

6. What cost $17\frac{12}{13}$ quintals of fish, at $\$4.75$ per quintal?
Ans. $\$81.55\frac{40}{43}$.

7. What cost $1670\frac{7}{13}$ pounds of coffee, at $12\frac{3}{4}$ cents per pound?   Ans. $\$212.99\frac{11}{13}$.

8. What cost $28\frac{4}{11}$ tons of Lackawana coal, at $\$11\frac{3}{4}$ a ton?
Ans. $\$333.27\frac{3}{11}$.

9. Bought $37\frac{11}{13}$ hogsheads of molasses, at $\$17.62\frac{1}{4}$ a hhd.; what was the whole cost?   Ans. $\$655.20\frac{5}{11}$.

10. What cost $\frac{7}{8}$ of a cord of wood, at $\$5.75$ a cord?
Ans. $\$5.03\frac{1}{8}$.

11. What are the contents of a field, which is $139\frac{3}{4}$ rods long, and $38\frac{3}{4}$ rods wide?   Ans. 33A. 3R. $15\frac{13}{16}$p.

12. Bought 15 loads of wood, each containing $11\frac{3}{4}$ feet, cord measure. I divide it equally between 9 persons; what does each receive?   Ans. $19\frac{7}{12}$ ft.

13. If the transportation of $18\frac{3}{4}$ tons of iron cost $\$48.15\frac{3}{4}$, what is it per ton?   Ans. $\$2.62\frac{4}{15}$.

14. If a hhd. of wine cost $\$98\frac{7}{8}$, what is the price of one gallon?   Ans. $\$1.56\frac{115}{128}$.

15. If 5 bushels of wheat cost $\$8\frac{2}{3}$; what will a bushel be worth?   Ans. $\$1.64\frac{1}{3}$.

16. What will 11 hogsheads and $17\frac{1}{4}$ gallons of wine cost, at $19\frac{3}{4}$ cents a gallon?   Ans. $\$140.32\frac{3}{4}$.

17. How many bottles, each containing $1\frac{3}{4}$ pints, are sufficient for bottling a hhd. of cider?   Ans. 288 bottles.

18. I have a shed, which is $18\frac{7}{13}$ feet long, $10\frac{6}{13}$ feet wide, and $7\frac{11}{13}$ feet high; how many cords of wood will it contain?
Ans. 11cd. $124\frac{332}{1718}$ft.

19. What will $6\frac{7}{8}$ pounds of tea cost, at $65\frac{3}{4}$ cents per lb.? Ans. \$4.52$\frac{1}{32}$.

20. How many cubic feet does a box contain, that is $8\frac{3}{4}$ feet long, $5\frac{7}{12}$ feet wide, and 3 feet high? Ans. $146\frac{9}{16}$ft.

21. How many feet of boards will it take to cover a side of a house, which is $46\frac{5}{12}$ feet long, and $17\frac{1}{2}$ feet high? Ans. $812\frac{7}{24}$ft.

22. How many feet of boards will it take to make 7 boxes, that shall be $5\frac{1}{2}$ feet long, $2\frac{5}{12}$ feet high, $3\frac{1}{12}$ feet wide; and how many cubic feet will they contain? Ans. $527\frac{59}{72}$ft. $286\frac{253}{288}$ cub. ft.

23. A certain room is 12 feet long, $11\frac{1}{2}$ feet wide, and $7\frac{1}{2}$ feet high; how much will it cost to plaster it, at $2\frac{3}{4}$ cents per square foot? Ans. \$13.48$\frac{7}{8}$.

24. A man has a garden that is $14\frac{1}{2}$ rods long, and $10\frac{1}{4}$ rods wide; he wishes to have a ditch dug around it, that shall be 3 feet wide and $4\frac{1}{2}$ feet deep; what will be the expense if he gives 2 cents per cubic foot? Ans. \$223.76$\frac{1}{4}$.

25. How many bushels of grain will a box contain, which is $14\frac{7}{12}$ feet long, $5\frac{11}{12}$ feet deep, and $4\frac{1}{4}$ feet wide, there being $2150\frac{2}{5}$ cubic inches in a bushel? Ans. $294\frac{2429}{3584}$bush.

26. Which will contain the most, and by how much,—a box that is 10 feet long, 8 feet wide, and 6 feet deep; or a cubical one, whose each side measures 8 feet? Ans. The last contains 32 cubic feet most.

27. Divide \$1112$\frac{5}{8}$ equally among 129 men. Ans. \$8$\frac{5}{8}$.

28. Bought 68 barrels of flour at \$7$\frac{11}{12}$ per barrel, wha was the amount of the whole? Ans. \$538$\frac{1}{3}$.

29. What cost $8\frac{2}{3}$ acres of land at \$42$\frac{3}{5}$ per acre? Ans. \$369.20.

30. How shall four 3's be arranged, that their value shall be nothing?

31. From $\frac{7}{0}$ take 11.

SECTION XVI.

DECIMAL FRACTIONS.

A DECIMAL FRACTION is that whose integer is always divided into 10, 100, 1000, &c. equal parts. Its denominator is always an unit, with as many ciphers annexed, as there are places in the given decimal. There is therefore no need of having the denominator *expressed ;* for the value of the fraction is always known by placing a point before it, at the left hand, called the separatrix. Thus, .5 is $\frac{5}{10}$, .37 is $\frac{37}{100}$, .348 is $\frac{348}{1000}$.

Ciphers annexed to the right hand of decimals do not increase their value ; for .4 or .40 or .400 are decimals having the same value, each being equal to $\frac{4}{10}$ or $\frac{2}{5}$; but when ciphers are placed on the left hand of a decimal, they decrease the value in a tenfold proportion. Thus .4 is $\frac{4}{10}$, or four tenths ; but 04 is $\frac{4}{100}$, or four hundredths ; and .004 is $\frac{4}{1000}$, or four thousandths. The figure next the separatrix is reckoned so many tenths ; the next at the right, so many hundredths ; the third is so many thousandths ; and so on, as may be seen by the following

TABLE. ·

Millions.	Hundreds of Thousands.	Tens of Thousands.	Thousands.	Hundreds.	Tens.	Units.	Tenths.	Hundredths.	Thousandths.	Ten Thousandths.	Hundred Thousandths	Millionths.
7	6	5	4	3	2	1.	2	3	4	5	6	7

From this table it is evident, that in decimals, as well as in whole numbers, each figure takes its value by its distance from the place of units

NOTE. — If there be one figure in the decimal, it is so many tenths ; if there be two figures, they express so many hundredths ; if there be three figures, they are so many thousandths, &c.

10

NUMERATION OF DECIMAL FRACTIONS.

Let the pupil write the following numbers.

1. Three hundred twenty-five, and seven tenths.
2. Four hundred sixty-five, and fourteen hundredths.
3. Ninety-three, and seven hundredths.
4. Twenty-four, and nine millionths.
5. Two hundred twenty-one, and nine hundred thousandths.
6. Forty nine thousand, and forty-nine thousandths.
7. Seventy-nine million two thousand, and one hundred five thousandths.
8. Sixty-nine thousand fifteen, and fifteen hundred thousandths.
9. Eighty thousand, and eighty-three ten thousandths.
10. Nine billion, nineteen thousand nineteen, and nineteen hundredths.
11. Twenty-seven, and nine hundred twenty-seven thousandths.
12. Forty-nine trillion, and one trillionth.
13. Twenty-one, and one ten thousandth.
14. Eighty-seven thousand, and eighty-seven millionths.
15. Ninety-nine thousand ninety-nine, nine thousand nine billionths.
16. Seventeen, and one hundred seventeen ten thousandths.
17. Thirty-three, and thirty-three hundredths.
18. Forty-seven thousand, and twenty-nine ten millionths.
19. Fifteen, and four thousand seven hundred thousandths.
20. Eleven thousand, and eleven hundredths.
21. Seventeen, and eighty-one quatrillionths.
22. Nine, and fifty-seven trillionths.
23. Sixty-nine thousand, and three hundred forty-nine thousandths.

Let the following numbers be written in words.

27.86	86.0007	1.000007	16.300000007
48.07	5.6001	5.101016	1.315
15.716	34.1063	6.716678	0.0000001
161.3	15.0016	1.631	10.10101
87.006	16.1004	3.760701	1.000327

SECTION XVII.

ADDITION OF DECIMALS.

1. Add 23.61 and 161.5 and 2.6789 and 61.111 and 27.0076 and 116.71 and 6151.7671 together.

OPERATION.
```
  23.61
 161.5
   2.6789
  61.111
  27.0076
 116.71
6151.7671
---------
6544.3846
```

In this question, it will be perceived, that tenths are written under tenths, hundredths under hundredths, &c.; and that the operation of addition is performed as in addition of whole numbers. We therefore induce the following

RULE.

Write the numbers under each other according to their value, add as in whole numbers, and point off from the right hand as many places for decimals, as there are in that number, which contains the greatest number of decimals.

2. Add together the following numbers — 31.61356, 6716.31, 413.1678956, 35.14671, 3.1671, 314.6.   Ans. 7564.0052656.

3. What is the sum of the following numbers — 1121.6116, 61.87, 46.67, 165.13, 676.167895.   Ans. 2071.449495.

4. Add 7.61, 637.1, 6516.14, 67.1234, 6.1234, together.
Ans. 7234.0968.

5. Add 21.611, 6888.32, 3.6167, together.   Ans. 6913.5477.

6. Add seventy-three and twenty-nine hundredths, eighty-seven and forty-seven thousandths, three thousand and five and one hundred six ten thousandths, twenty-eight and three hundredths, twenty-nine thousand and five thousandths together.
Ans. 32193.3826.

7. Add two hundred nine thousand and forty-six millionths, ninety-eight thousand two hundred and seven and fifteen ten thousandths, fifteen and eight hundredths, and forty-nine ten thousandths, together.   Ans. 307222.086446.

8. What is the sum of twenty-three million and ten, and one thousand and five hundred thousandths, twenty-seven and nineteen millionths, seven and five tenths ?
Ans. 23001044.500069.

9. Add the following numbers ; fifty-nine and fifty-nine thousandths, twenty-five thousand and twenty-five ten thousandths, five and five millionths, two hundred five and five hundredths.
Ans. 25269.111505.

10. What is the sum of the following numbers ; twenty-five and seven millionths, one hundred forty-five, and six hundred forty-three thousandths, one hundred seventy-five and eighty-nine hundredths, seventeen and three hundred forty-eight hundred thousandths ?   **Ans. 363.556487.**

SECTION XVIII.

SUBTRACTION OF DECIMALS.

RULE.

Let the numbers be so written, that the separatrix of the subtrahend be directly under that of the minuend, subtract as in whole numbers, and point off so many places for decimals, as there are in that number, which contains the greatest number of decimals.

OPERATIONS.

1.	2.	3.	4.
61.9634	39.3	5.	6.1
9.182	1.6789	1.678	1.99999
52.7814	37.6211	3.322	4.10001

5. From 41.7   take 21.9767.   **Ans. 19.7233.**
6. From 29.167 take 19.66711.   **Ans. 9.49989.**
7. From 91.61 take 2.6671.   **Ans. 88.9429.**
8. From 96.71 take 96.709.   **Ans. .001.**
9. Take twenty-seven and twenty-eight thousandths from ninety-seven and seven tenths.   **Ans. 70.672.**
10. Take one hundred fifteen and seven hundredths from three hundred fifteen and twenty-seven ten thousandths.
          Ans. 199.9327.
11. From twenty-nine million four thousand and five, take twenty-nine thousand, and three hundred forty-nine thousand two hundred, and twenty-four hundred thousandths.
          Ans. 28625804.99976.
12. From one million take one millionth.
          Ans. 999999.999999.

SECTION XIX.

MULTIPLICATION OF DECIMALS.

	EXAMPLES.	OPERATIONS.	VULG. FRACT.	DECIMALS.
1.	Multiply $\frac{3}{10}$ by $\frac{6}{10}$.	$\frac{3}{10}\times\frac{6}{10}$ = $\frac{18}{100}$ =	.18
2.	Multiply $\frac{7}{10}$ by $\frac{8}{10}$.	$\frac{7}{10}\times\frac{8}{10}$ = $\frac{56}{100}$ =	.56
3.	Multiply $\frac{9}{10}$ by $\frac{9}{10}$.	$\frac{9}{10}\times\frac{9}{10}$ = $\frac{81}{100}$ =	.81
4.	Multiply $\frac{7}{10}$ by $\frac{7}{10}$.	$\frac{7}{10}\times\frac{7}{10}$ = $\frac{49}{100}$ =	.49
5.	Multiply $\frac{9}{100}$ by $\frac{4}{10}$.	$\frac{9}{100}\times\frac{4}{10}$ = $\frac{36}{1000}$ =	.036
6.	Multiply $\frac{2}{100}$ by $\frac{4}{100}$.	$\frac{2}{100}\times\frac{4}{100}$ = $\frac{8}{10000}$ =	.0008
7.	Multiply $\frac{7}{1000}$ by $\frac{9}{100}$.	$\frac{7}{1000}\times\frac{9}{100}$ = $\frac{63}{100000}$ =	.00063
8.	Multiply $\frac{789}{1000}$ by $\frac{426}{1000}$.	$\frac{789}{1000}\times\frac{426}{1000}$ = $\frac{335234}{1000000}$ =	.335234
9.	Multiply $\frac{671}{1000}$ by 32.	$\frac{671}{1000}\times\frac{32}{1}$ = $\frac{21472}{1000}$ =21.472
10.	Multiply $\frac{7}{100}$ by 46.	$\frac{7}{100}\times\frac{46}{1}$ = $\frac{322}{100}$ = 3.22

11. Multiply 76.81 by 3.2.   12. Multiply .1234 by .0046.

OPERATION.
```
  76.81
   3.2
 ─────
 15362
 23043
 ─────
245.792
```

OPERATION.
```
 .1234
 .0046
 ─────
  7404
  4936
 ─────
.00056764
```

From the above examples, we deduce the following

RULE.

Multiply as in whole numbers, and point off as many figures for decimals in the product, as there are decimals in the multiplicand and multiplier; but, if there should not be so many figures in the product as in the multiplicand and multiplier, supply the defect by prefixing ciphers. See Example 12th.

13. Multiply 61.76   by   .0071.   Ans. .438496.
14. Multiply   .0716 by 1.326.   Ans. .0949416.
15. Multiply   .61001 by   .061.   Ans. .03721061.
16. Multiply 71.61   by 365.   Ans. 26137.65.
17. Multiply   .1234 by 1234.   Ans. 152.2756.
18. Multiply 6.711   by 6543.   Ans. 43910.073.
10*

19. Multiply .0009 by .0009.   Ans. .00000081.

20. Multiply forty-nine thousand by forty-nine thousandths.
        Ans. 2401.

21. What is the product of one thousand and twenty-five, multiplied by three hundred and twenty-seven ten thousandths ?
        Ans. 33.5175.

22. What is the product of seventy-eight million two hundred and five thousand and two by fifty-three hundredths ?
        Ans. 41448651.06.

23. Multiply one hundred and fifty-three thousandths by one hundred twenty-nine millionths.   Ans. .000019737.

24. What is the product of fifteen thousand, multiplied by fifteen thousandths ?   Ans. 225.

25. What will 26.7 yards of cloth cost, at $ 5.75 a yard ?
        Ans. $ 153.52.5.

26. What will 14.75 bushels of wheat cost, at $ 1.25 a bushel ?
        Ans. $ 18.43.7.5.

27. What will 375.6 pounds of sugar cost, at $0.125 per lb. ?
        Ans. $ 46.95.

28. What will 26.58 cords of wood cost, at $ 5.625 a cord ?
        Ans. $ 149.51.2¼.

29. What will 28.75 tons of potash cost, at $ 125.78 per ton ?
        Ans. $ 3616.17.5.

30. What will 369 gallons of molasses cost, at $ 0.375 a gallon ?   Ans. $ 138.37.5.

31. What will 97.48cwt. of hay cost, at $ 1.125 per cwt. ?
        Ans. $ 109.66.5.

32. What will 63.5 bushels of corn be worth, at $ 0.78 per bushel ?   Ans. $ 49.53.

SECTION XX.

DIVISION OF DECIMALS.

	EXAMPLES.			OPERATIONS.	VULG. FRAC.	DECIMALS
1. Divide	$\frac{2}{10}$	by	$\frac{5}{10}$.	$\frac{2}{10} \times \frac{10}{5} =$	$\frac{20}{50} = \frac{4}{10} =$	.4
2. Divide	$\frac{3}{10}$	by	$\frac{6}{10}$.	$\frac{3}{10} \times \frac{10}{6} =$	$\frac{30}{60} = \frac{5}{10} =$	.5
3. Divide	$\frac{36}{100}$	by	$\frac{4}{10}$.	$\frac{36}{100} \times \frac{10}{4} =$	$\frac{360}{400} = \frac{9}{10} =$	.9
4. Divide	$\frac{48}{10}$	by	4.	$\frac{48}{10} \times \frac{1}{4} =$	$\frac{48}{40} = \frac{12}{10} =$	1.2
5. Divide	$\frac{96}{100}$	by	$\frac{12}{100}$.	$\frac{96}{100} \times \frac{100}{12} =$	$\frac{9600}{1200} = \frac{80}{10} =$	8.
6. Divide	$\frac{9}{100}$	by	$\frac{9}{1000}$.	$\frac{9}{100} \times \frac{1000}{9} =$	$\frac{9000}{900} = \frac{10}{1} =$	10.

7. Divide 1.728 by 1.2.　　8. Divide $\frac{1728}{1000}$ by $\frac{12}{10}$.

OPERATION BY DECIMALS.

1.2)1.728(1.44 Ans.

1 2
――
52
48
――
48
48
――

OPERATION BY VULGAR FRACTIONS.

$\frac{1728}{1000}\times\frac{10}{12}=\frac{17280}{12000}=\frac{1728}{1200}=\frac{144}{100}=1\frac{44}{100}$ Ans.

Hence the propriety of the following

RULE.

Divide as in whole numbers, and point off as many decimals in the quotient as the number of decimals in the dividend exceed those of the divisor; but if the number of those in the divisor exceed that of the dividend, supply the defect by annexing ciphers to the dividend. And if the number of decimals in the quotient and divisor together are not equal to the number in the dividend, supply the defect by prefixing ciphers to the quotient.

9. Divide 780.516 by 2.43.　　　　Ans. 321.2.
10. Divide 7.25406 by 9.57.　　　　Ans. .758.
11. Divide .21318 by .38.　　　　Ans. .561.
12. Divide 7.2091365 by .5201.　　Ans. 13.861+.
13. Divide 56.8554756 by .0759.　Ans. 749.084.
14. Divide 30614.4 by .9567.　　　Ans. 32000.
15. Divide .306144 by 9567.　　　Ans. .000032.

16. Divide four thousand three hundred twenty-two and four thousand five hundred seventy-three ten thousandths, by eight thousand and nine thousandths.　　Ans. .5403+.

17. Divide thirty-six and six thousand nine hundred forty-seven ten thousandths, by five hundred and eighty-nine.

Ans. .0623.

18. Divide three hundred twenty-three thousand seven hundred sixty-five, by five millionths.　　Ans. 64753000000.

SECTION XXI.

REDUCTION OF DECIMALS.

CASE I.

To reduce a vulgar fraction to its decimal.

1. Reduce $\frac{3}{8}$ to its decimal.

OPERATION.

8)3.000
.375 Ans.

That the decimal .375 is equal to $\frac{3}{8}$ may be shown by writing it in a vulgar fraction and reducing it; thus, $\frac{375}{1000}=\frac{75}{200}=\frac{15}{40}=\frac{3}{8}$ Ans.

Hence the following

RULE.

Divide the numerator by the denominator, annexing one or more ciphers to the numerator, and the quotient will be the decimal required.

NOTE. — It is not usually necessary, that decimals should be carried to more than six places.

2. Reduce $\frac{5}{8}$ to a decimal. Ans. .625.
3. Reduce $\frac{1}{2}$ to a decimal. Ans. .5.
4. Reduce $\frac{2}{3}$, $\frac{3}{4}$, $\frac{5}{6}$, $\frac{11}{12}$, $\frac{3}{16}$, $\frac{1}{25}$ and $\frac{1}{8}$ to decimals.
   Ans. .666+, .75, .833+, 91666+, .1875, .04.125.
5. Reduce $\frac{1}{17}$, $\frac{2}{27}$, $\frac{5}{37}$, $\frac{3}{431}$, $\frac{8}{111}$ and $\frac{1}{1234}$ to decimals.
   Ans. .05882+, .07407+, .1351+, .00696+, .07207+, .0008103+.

CASE II.

To reduce compound numbers to decimals.

1. Reduce 15s. 9¾d. to the decimal of a £. Ans. .790625.

OPERATION.

4	3.00
12	9.75000
20	15.81250
	.790625 Ans.

The 3 farthings are ¾ of a penny, and these reduced to a decimal are .75 of a penny, which we annex to the pence and proceed in the same manner with the other terms. Hence the following

RULE.

Write the given numbers perpendicularly under each other for divisors, proceeding orderly from the least to the greatest: opposite to

each dividend, on the left hand, place such a number for a divisor, as will bring it to the next superior name, and draw a line between them. Begin at the highest, and write the quotient of each division, as decimal parts, on the right of the dividend next below it, and so on, till they are all divided ; and the last quotient will be the decimal required.

2. Reduce 9s. to the fraction of a pound. Ans. .45.

3. Reduce 15cwt. 3qr. 14lb. to the decimal of a ton.
Ans. .79375.

4. Reduce 2qr. 21lb. 8oz. 12dr. to the decimal of a cwt.
Ans. .6923828125.

5. Reduce 1qr. 3na. to the decimal of a yard. Ans. .4375.

6. Reduce 5fur. 35rd. 2yd. 2ft. 9in. to the decima' of a mile.
Ans. .73603219 +.

7. Reduce 3gal. 2qt. 1pt. of wine to the decimal of a hogshead.
Ans. .575396 +.

8. Reduce 1pt. to the decimal of a bushel. Ans. .015625.

9. Reduce 2R. 16p. to the decimal of an acre. Ans. .6.

CASE III.

To find the decimal of any number of shillings, pence, and farthings by *inspection.*

NOTE. — A demonstration of this case has been given, page 57.

RULE.

Write half of the greatest number of shillings for the first decimal figure, and if there be an odd shilling, annex a 5 to the half number of shillings, let the farthings in the given pence and farthings, occupy the second and third places, observing to increase their number by 1, if they exceed 12, and by 2, if they exceed 36.

EXAMPLES.

1. Find the decimal of 15s. 9¾d. by inspection.

.7 = ½ of 14s.
.05 = for odd shilling.
39 = farthings in 9¾d.
2 = for excess of 36.
———
.791

2. Find the value of 13s. 6¾d. by inspection. Ans. .678.

3. Find the value of 19s. 8¼d. by inspection. Ans. .984.

4. Value the following sums by inspection, and find their total —19s. 11¾d., 16s. 9½d., 1s. 11d., 3s. 0¾d., 17s. 5½d., 13s. 4½d., 18s. 8½d., 19s. 11¾d., 13s. 3¼d., 16s. 0¼d., 17s. 7¾d. Ans. 7.91£

CASE IV.

To find the value of any given decimal in the terms of the integer.

1. What is the value of .790625 £. ?   Ans. 15s. 9¼d.

OPERATION.

```
.790625
      20
15.812500
      12
 9.750000
       4
 3.000000
```

As a lower denomination consists of more units than the same value in a higher one, therefore to bring pounds to farthings, we must multiply by the same numbers as in common Reduction.

Hence we deduce the following

RULE.

Multiply the given decimal by that number which it takes of the next denomination to make one of that greater, and cut off as many places for a REMAINDER *on the right, as there are places in the given decimal. Multiply the* REMAINDER *by the next lower denomination, and cut off for a remainder as before, and so proceed, until the decimal is reduced to the denomination required; the several denominations, standing at the left hand, are the answers required.*

2. What is the value of .625 of a shilling ?   Ans. 7½d.
3. What is the value of .6725 of a cwt. ?
        Ans. 2qr. 19lb. 5⅖oz.
4. What is the value of .9375 of a yard ?   Ans. 3qr. 3na.
5. What is the value of .7895 of a mile ?
        Ans. 6fur. 12rd. 10ft. 6¹³⁄₁₆in.
6. What is the value of .9378 of an acre ?
        Ans. 3R. 30p. 13ft. 9⁴⁄₁₃ in.
7. Reduce .5615 of a hogshead of wine to its value in gallons, &c.   Ans. 35 gal. 1qt. 0pt. 3¹¹³⁄₂₅₃gi.
8. Reduce .367 of a year to its value in days, &c.
        Ans. 134da. 1h. 7m. 19½sec.
9. What is the value of .6923828125 of a cwt. ?
        Ans. 2qr. 21lb. 8oz. 12dr.
10. What is the value of .015625 of a bushel ?
        Ans. 1 pint.
11. What is the value of .55 of an ell English ?
        Ans. 2qr. 3na.
12. What is the value of .6 of an acre ?   Ans. 2R. 16p.

SECTION XXII.

MISCELLANEOUS EXAMPLES.

1. What is the value of 7cwt. 2qr. 18lb. of sugar, at $11.75 per cwt. ?   Ans. $90.01.3$\frac{11}{14}$.

2. What cost 19cwt. 3qr. 14lb. of iron, at $9.25 per cwt. ?
                                         Ans. $183.84.3$\frac{3}{4}$.

3. What cost 39A. 2R. 15p. of land, at $87.37.5 per acre ?
                                         Ans. $3459.50.3$\frac{99}{33}$.

4. What would be the expense of making a turnpike 87m. 3fur. 15rds., at $578.75 per mile ?   Ans. 50595.41\frac{1}{4}$.

5. What is the cost of a board 18ft. 9in. long, and 2ft. 3$\frac{1}{4}$in. wide, at $.05.3 per foot ?   Ans. $2.27.7$\frac{11}{14}$.

6. Goliah of Gath was 6$\frac{1}{2}$ cubits high ; what was his height in feet, the cubit being 1ft. 7.168in. ?   Ans. 10ft. 4.592in.

7. If a man travel 4.316 miles in an hour ; how long would he be in travelling from Bradford to Boston, the distance being 29$\frac{1}{2}$ miles ?   Ans. 6h. 50m. 6sec.+

8. What is the cost of 5yd. 1qr. 2na. of broadcloth, at 5.62\frac{1}{2}$ per yard ?   Ans. $30.23.4$\frac{3}{4}$.

9. Bought 17 bags of hops, each weighing 4cwt. 3qr. 7lb., at 5.87\frac{1}{2}$ per cwt. ; what was the cost ?   Ans. $480.64.8$\frac{7}{16}$.

10. Purchased a farm, containing 176A. 3R. 25rds., at 75.37\frac{1}{2}$ per acre : what did it cost ?   Ans. $13334.30.8$\frac{11}{14}$.

11. What cost 17625 feet of boards, at $12.75 per thousand ?
                                         Ans. $224.71.8$\frac{3}{4}$.

12. How many square feet in a floor 19ft. 3in. long, and 15ft. 9in. wide ?   Ans. 303ft. 27in.

13. How many square yards of paper will it take to cover a room 14ft. 6in. long, 12ft. 6in. wide, and 8ft. 9in. high ?
                                         Ans. 52$\frac{1}{2}$yd.

14. How many solid feet in a pile of wood 10ft. 7in. long; 4ft. wide, and 5ft. 10in. high ?   Ans. 246$\frac{17}{17}$ft.

15. How many garments, each containing 4yd. 2qr. 3na., can be made from 112yd. 2qr. of cloth ?   Ans. 24.

16. Bought 1gal. 2qt. 1pt. of wine for $1.32 ; what would be the price of a hogshead ?   Ans. $70.56.

17. Bought 125$\frac{1}{2}$yd. of lace for $15.06 ; what was the price of 1 yard ?   Ans. $0.12.

18. What cost 17cwt. 3qr. of wool, at $35.75 per cwt. ?
                                         Ans. $634.56.2$\frac{1}{4}$.

19. What cost 7hhd. 47gal. of wine, at $87.25 per hhd. ?
                                         Ans. 675.84\frac{3}{4}$.

20. How many solid feet in a stick of timber 34ft. 9in. long, 1ft. 3in. wide, and 1ft. 6in. deep ?   Ans. 65.15625ft.

21. How many cwt. of coffee in 17¾ bags, each bag containing 2cwt. 1qr. 7lb. ? Ans. 41cwt. 0qr. 5¼lb.

22. If 18yd. 1qr. of cloth cost $36.50, what is the price of 1 yard ? Ans. $2.00.

23. If $477.72 be equally divided among 9 men, what will be each man's share ? Ans. $53.08.

24. A man bought a barrel of flour for $5.37.5, 7gal. of molasses for $1.78, 9 gal. of vinegar for $1.1875, 1 gal. of wine for $1.125, 14lb. of sugar for $1.275, and 5lb. of tea, for $2.625 ; what did the whole amount to ? Ans. $13.36.7¼.

25. A man purchased 3 loads of hay; the first contained 2¾ tons, the second 3¼ tons, and the third 1 1/14 tons; what was the value of the whole, at $17.625 a ton ? Ans. $128.88.2¹³.

26. At $13.625 per cwt., what cost 3cwt. 2qr. 7lb. of sugar? Ans. $48.53.9¼.

27. At $125.75 per acre, what cost 37A. 3R. 35rds. ? Ans. $4774.57.0⁴₇.

28. At $11.25 per cwt. what cost 17cwt. 2qr. 21lb. of rice ? Ans. $198.98.4½.

29. What cost 7¼ bales of cotton, each weighing 3.57cwt., at $9.37½ per cwt. ? Ans. $244.85.1⁹₁₆.

30. What cost 7hhd. 49gal. of wine, at $97.625 per hhd. ? Ans. $759.30.5³⁴₆₃.

31. What cost 7yd. 3qr. 3na. of cloth, at $4.75 per yard ? Ans. $37.70.3⅛.

32. What cost 27T. 15cwt. 1qr. 3¼lb. of hemp, at $183.62 per ton ? Ans. $5098.03.7⁴₃₂.

33. What is the cost of constructing a railroad 17m. 3fur. 15rd., at $1725.87.5 per mile ? Ans. $30067.97.3¹³₁₂.

SECTION XXIII.

EXCHANGE OF CURRENCIES.

PREVIOUS to the year 1776, all accounts in this country were kept in pounds, shillings, pence, and farthings ; but owing to the depreciation of the currency, a dollar was estimated differently in different countries.

In New England, Virginia, Kentucky, Tennessee, and Ohio, the dollar is valued at	6s. 0d.
New York. North Carolina, and New Jersey,	8s. 0d.
Pennsylvania, Delaware, and Maryland,	7s. 6d.
South Carolina and Georgia,	4s. 8d.
Canada and Nova Scotia,	5s. 0d.
England and Newfoundland (sterling),	4s. 6d.

In order, therefore, to change any of the above currencies to federal money, the shillings, pence and farthings, if there be any, must first be reduced to decimals of a pound, and annexed to the pounds. We then adopt this general

RULE.

Divide the pounds by the value of a dollar in the given currency, EXPRESSED BY A FRACTION OF A POUND ; *that is, to change the old New England currency to federal money, divide by* $\frac{3}{10}$; *because 6 shillings is* $\frac{3}{10}$ *of a pound.*

To change the old currency of New York, &c., to federal money, divide by $\frac{4}{10}$; *because 8 shillings is* $\frac{4}{10}$ *of a pound.*

To change the old currency of Pennsylvania, &c., to federa! money, divide by $\frac{3}{8}$; *because 7 shillings and 6 pence is* $\frac{3}{8}$ *of a pound.*

To change the old currency of South Carolina and Georgia to federal money, divide by $\frac{7}{30}$; *because 4 shillings and 8 pence is* $\frac{7}{30}$ *of a pound.*

To change Canada and Nova Scotia currency to federal money, divide by $\frac{1}{4}$; *because 5 shillings is* $\frac{1}{4}$ *of a pound.*

To change English (sterling) money to federal money, divide by $\frac{9}{40}$; *because 4 shillings and 6 pence is* $\frac{9}{40}$ *of a pound.*

To change sterling money to lawful money, add $\frac{1}{3}$ *to the sterling, and the sum will be the lawful; because 4 shillings and 6 pence sterling is 6 shillings lawful; and, for the same reason, take* $\frac{1}{4}$ *from the lawful, and the remainder will be sterling.*

To reduce federal money to any of the above currencies, the federal money must be MULTIPLIED *by the above fractions.*

EXAMPLES.

1. Change 18£. 4s. 6d. of the old New England currency to federal money.

$$18.225£. \div \tfrac{3}{10} = \$60.75 \text{ Answer.}$$

In this example we reduce the 4 shillings and 6 pence to a decimal of a pound, which we find to be .225. This decimal we annex to the pounds, and multiply the 18.225£. by 10, and divide by 3, and it produces the answer $ 60.75. The reason for this process has already been shown.

2. Change $ 60.75 to the old currency of New England.

$$\$60.75 \times \tfrac{3}{10} = 18.225 = 18£. \text{ 4s. 6d.} = \text{Answer.}$$

The decimal .225 is reduced to shillings and pence by Case IV. of Decimal Fractions.

11

3. Change 78£. 7s. 6d. of the old currency of New England to federal money.   Ans. $ 261.25.

4. Change $ 261.25 to the old currency of New England.
                                                Ans. 78£. 7s. 6d.

5. Change 46£. 16s. 6d. of the old currency of New York to federal money.   Ans. $ 117.06¼.

6. Change $ 117.06¼ to the old currency of New York.
                                                Ans. 46£. 16s. 6d.

7. Change 387£. of the old currency of New Jersey and Pennsylvania to federal money.   Ans. $ 1032.

8. Change $ 1032 to the old currency of New Jersey and Pennsylvania.   Ans. 387£.

9. Change 12£. 12s. of the old currency of South Carolina and Georgia to federal money.   Ans. $ 54.

10. Change $ 54 to the old currency of South Carolina and Georgia.   ,   Ans. 12£. 12s.

11. Change 128£. 18s. 6d. of Canada and Nova Scotia to federal money.   Ans. $ 515.70.

12. Change $ 515.70 to Canada and Nova Scotia currency.
                                                Ans. 128£. 18s. 6d.

13. Change 162£. 18s. English money (sterling) to federal money.   Ans. $ 724.

14. Change $ 724 to sterling money.   Ans. 162£. 18s.

15. Change 347£. sterling to lawful money.
                                                Ans. 462£. 13s. 4d.

16. Change 462£. 13s. 4d. lawful to sterling.   Ans. 347£.

SECTION XXIV.

CIRCULATING DECIMALS.*

DEFINITIONS.

1. Those decimals that are produced from Vulgar Fractions, whose denominators do not measure their numerators, and are distinguished by the continual repetition of the same figure or figures, are called *infinite decimals.*

2. The circulating figures, that is, those that continually repeat, are called *repetends ;* and, if the same figure only repeats, it is called a *single repetend ;* as .11111 or .5555, and is expressed by writing the circulating figure with a point over it ; thus, .11111, and is denoted by .1̇, and .5555 by .5̇.

* As Circulating Decimals are not so much of a practical nature as many other rules, and as they are somewhat difficult in their operation, the student can omit them until he reviews arithmetic.

3. If the *same* figures circulate alternately, it is called a *compound repetend* ; as .475475475, and is distinguished by putting a point over the first and last repeating figures ; thus, .475475475 is written .4̇7̇5̇.

4. When other figures arise before those which circulate, it is called a *mixed repetend* ; as .124̇6̇, or .17̇8̇3̇5̇.

5. *Similar repetends* begin at the same place ; as .3̇ and .6̇; or 5.1̇2̇3̇ and 3.4̇7̇8̇.

6. *Dissimilar repetends* begin at different places ; as .98̇6̇ and .46̇2̇5̇.

7. *Conterminous repetends* end at the same place ; as .63̇1̇ and .46̇5̇.

8. *Similar and conterminous repetends* begin and end at the same place ; as .1̇72̇8̇ and .4̇98̇7̇.

REDUCTION OF CIRCULATING DECIMALS.

CASE I.

To reduce a simple repetend to its equivalent vulgar fraction.

If a unit with ciphers annexed to it be divided by 9 ad infinitum, the quotient will be 1 continually ; that is, if ⅑ be reduced to a decimal, it will produce the circulate .1̇ ; and since .1̇ is the decimal equivalent to ⅑, .2̇ will be equivalent to ²⁄₉, .3̇ to ³⁄₉, and so on, till .9̇ is equal to ⁹⁄₉ or 1. Therefore every single repetend is equal to a vulgar fraction, whose numerator is the repeating figure, and denominator 9. Again, ¹⁄₉₉, or ¹⁄₉₉₉, being reduced to decimals, makes .01010101, and .001001001 ad infinitum = .0̇1̇ and .0̇01̇ ; that is ¹⁄₉₉ = .0̇1̇, and ¹⁄₉₉₉ = .0̇01̇ ; consequently ²⁄₉₉ = .0̇2̇, and ²⁄₉₉₉ = .0̇02̇ ; and, as the same will hold universally, we deduce the following

RULE.

Make the given decimal the numerator, and let the denominator be a number consisting of as many nines, as there are recurring places in the repetend.

If there be integral figures in the circulate, as many ciphers must be annexed to the numerator, as the highest place of the repetend is distant from the decimal point.

EXAMPLES.

1. Required the least vulgar fraction equal to $.\dot{6}$ and $.1\dot{2}\dot{3}$.

$\dot{6} = \frac{6}{9} = \frac{2}{3}$ Ans. $.1\dot{2}\dot{3} = \frac{123}{999} = \frac{41}{333}$ Ans.

2. Reduce $.\dot{3}$ to its equivalent vulgar fraction.   Ans. $\frac{1}{3}$.

3. Reduce $\dot{1}.6\dot{2}$ to its equivalent vulgar fraction.   Ans. $1\frac{62}{99}$.

4. Reduce $.\dot{7}6923\dot{0}$ to its equivalent vulgar fraction.

                              Ans. $\frac{10}{13}$.

CASE II.

To reduce a mixed repetend to its equivalent vulgar fraction.

1. What vulgar fraction is equivalent to $.13\dot{8}$?

OPERATION.

$$.13\dot{8} = \frac{13}{100} + \frac{8}{900} = \frac{117}{900} + \frac{8}{900} = \frac{125}{900} = \frac{5}{36} \text{ Ans.}$$

As this is a mixed circulate, we divide it into its finite and circulating parts; thus $.13\dot{8} = .13$ the finite part, and $.008$ the repetend or circulating part; but $.13 = \frac{13}{100}$; and $.00\dot{8}$ would be equal to $\frac{8}{9}$, if the circulation began immediately after the place of units; but, as it begins after the place of hundreds, it is $\frac{8}{9}$ of $\frac{1}{100} = \frac{8}{900}$. Therefore $.13\dot{8} = \frac{13}{100} + \frac{8}{900} = \frac{117}{900} + \frac{8}{900} = \frac{125}{900} = \frac{5}{36}$ Ans.   Q. E. D.

From the above demonstration we deduce the following

RULE.

To as many nines as there are figures in the repetend, annex as many ciphers as there are finite places for a denominator; multiply the nines in the denominator by the finite part, and add the repeating decimal to the product for the numerator. If the repetend begins in some integral place, the finite value of the circulating, must be added to the finite part.

2. What is the least vulgar fraction equivalent to $.5\dot{3}$?

                              Ans. $\frac{8}{15}$.

3. What is the least vulgar fraction equivalent to $.5\dot{9}2\dot{5}$?

                              Ans. $\frac{16}{27}$.

4. What is the least vulgar fraction equivalent to $.008497133$?

                              Ans. $\frac{83}{9768}$.

5. What is the finite number equivalent to $3\dot{1}.6\dot{2}$?

                              Ans. $31\frac{62}{99}$.

CASE III.

To make any number of dissimilar repetends, similar and conterminous.

1. Dissimilar made similar and conterminous.

OPERATION.

9.1$\dot{6}\dot{7}$= 9.16767676	Any given repetend whatever, whether
14.6 =14.60000000	single, compound, pure, or mixed, may be
3.16$\dot{5}$= 3.16555555	transformed into another repetend, that
12.4$\dot{3}\dot{2}$=12.43243243	shall consist of an equal or greater number
8.1$\dot{8}\dot{1}$= 8.18181818	of figures at pleasure ; thus .$\dot{4}$ may be
1.$\dot{3}0\dot{7}$= 1.30730730	changed into $\ddot{4}4$ or .$\dot{4}\dot{4}\dot{4}$; and .$\ddot{2}\ddot{9}$ into

.$\dot{2}92\dot{9}$ or .$\dot{2}9\dot{2}\dot{9}$. And as some of the circu-lates in this question consist of 1, some of 2, and others of 3 places ; and, as the least common multiple of 1, 2, and 3, is 6, we know, that the *new* repetend will consist of 6 places, and will begin just so far from unity, as is the farthest among the dissimilar repetends, which, in the present example, is the third place.

Hence the following

RULE.

Change the given repetends into other repetends, which shall consist of as many figures, as the least common multiple of the several number of places, found in all the repetends, contains units.

2. Make 3.6$\dot{7}\dot{1}$, 1.00$\dot{7}\dot{1}$, 8.5$\ddot{2}$ and 7.$\dot{6}1632\dot{5}$ similar and conterminous.

3. Make 1.5$\ddot{2}$, 8.$\dot{7}15\dot{6}$, 3.56$\dot{7}$ and 1.$\dot{3}7\dot{8}$ similar and conterminous.

4. Make .00$\dot{0}\dot{7}$, .14$\dot{1}41\dot{4}$ and 887.$\dot{1}$ similar and conterminous.

CASE IV.

To find whether the decimal fraction, equal to a given vulgar fraction, be finite or infinite, and of how many places the repetend will consist.

11*

RULE.

Reduce the given fraction to its least terms, and divide the denominator by 2, 5 or 10, as often as possible. If the whole denominator vanish in dividing by 2, 5 or 10, the decimal will be finite, and will consist of so many places, as you perform division. If it do not vanish, divide 9999, &c., by the result till nothing remain, and the number of 9's used will show the number of places in the repetend; which will begin after so many places of figures, as there are 10's, 2's or 5's used in dividing.

NOTE.—In dividing 1.0000, &c. by any prime number whatever except 2 or 5, the quotient will begin to repeat as soon as the remainder is 1. And since 9999, &c. is less than 10000, &c. by 1, therefore 9999, &c. divided by any number whatever, will leave a 0 for a remainder, when the repeating figures are at their period. Now whatever number of repeating figures we have, when the dividend is 1, there will be exactly the same number, when the dividend is any other number whatever. For the product of any circulating number, by any other given number, will consist of the same number of repeating figures as before. Thus, let .378137813781, &c. be a circulate, whose repeating part is 3781. Now every repetend (3781) being equally multiplied, must produce the same product. For these products will consist of more places, yet the overplus in each being alike, will be carried to the next, by which means, each product will be equally increased, and consequently every four places will continue alike. And the same will hold for any other number whatever. Hence it appears, that the dividend may be altered at pleasure, and the number of places in the repetend will be still the same; thus, $\frac{1}{11}=.\dot{0}\dot{9}$, and $\frac{3}{11}=.\dot{2}\dot{7}$, where the number of places in each are alike, and same will be true in all cases.

EXAMPLES.

1. Required to find whether the decimal equal to $\frac{210}{1150}$ be finite or infinite; and, if infinite, of how many places the repetend will consist.

$\frac{210}{1150}=_{2)}\frac{3}{16}=\overset{(2)}{8}=\overset{(2)}{4}=\overset{(2)}{2}=1$; therefore, because the denominator vanishes in dividing, the decimal is finite, and consists of four places; thus, $16)3.\underset{.1875}{0000}$.

2. Required to find whether the decimal equal to $\frac{475}{3800}$ be finite or infinite; and, if infinite, of how many places that repetend will consist.

$\frac{475}{3800}=\frac{19}{112}$ $2)112=\overset{(2)}{56}=\overset{(2)}{28}=\overset{(2)}{14}=7$. Thus, $7)\frac{999999}{142857}$; therefore, because the denominator, 112, did not vanish in dividing by 2, the decimal is infinite; and as six 9's were used, the circulate consists of six places, beginning at the fifth place, because four 2's were used in dividing.

3. Let $\frac{1}{11}$ be the fraction proposed.

4. Let $\frac{107}{253}$ be the fraction proposed.

SECTION XXV.

ADDITION OF CIRCULATING DECIMALS.

1. Let $3.\dot{5}+7.6\dot{5}\dot{1}+1.\dot{7}6\dot{5}+6.1\dot{7}\dot{3}+51.\dot{7}+3.7+27.\dot{6}3\dot{1}$ and $1.\dot{0}0\dot{3}$ be added together.

EXAMPLE.

OPERATION.

Dissimilar. Similar and Conterminous.

$3.\dot{5}\ \ =3.555555\dot{5}$
$7.6\dot{5}\dot{1}=7.651651\dot{6}$
$1.\dot{7}6\dot{5}=1.765765\dot{7}$
$6.1\dot{7}\dot{3}=6.173737\dot{3}$
$51.\dot{7}\ \ =51.717777\dot{7}$
$3.7\ \ =3.700000\dot{0}$
$27.\dot{6}3\dot{1}=27.631631\dot{6}$
$1.\dot{0}0\dot{3}=1.003003\dot{0}$

103.2591227

Having made all the numbers similar and conterminous by Sect. XXIV. Case III., we add the first six columns, as in Simple Addition, and find the sum to be $3591224=\frac{3591224}{9999999}=3.591227$. The repeating decimals $.591227$ we write in its proper place, and carry 3 to the next column, and then proceed as in whole numbers.

Hence the following

RULE.

Make the repetends similar and conterminous, and find their sum, as in common Addition. Divide this sum by as many 9's as there are places in the repetend, and the remainder is the repetend of the sum, which must be set under the figures added, with ciphers on the left, when it has not so many places as the repetends. Carry the quotient of this division to the next column, and proceed with the rest as with finite decimals.

2. Add $27.5\dot{6}+5.6\dot{3}\dot{2}+6.\dot{7}+16.3\dot{5}\dot{6}+.\dot{7}\dot{1}$ and $6.\dot{1}23\dot{4}$ together.
        Ans. $63.1\dot{6}9067086888\dot{8}$.

3. Add $2.7\dot{6}\dot{5}+7.1\dot{6}67\dot{4}+3.67\dot{1}+.\dot{7}$ and $.\dot{1}72\dot{8}$ together.
        Ans. $14.5\dot{5}43\dot{6}$.

4. Add $5.\dot{1}6345\dot{5}+8.\dot{6}381\dot{1}+3.\dot{7}\dot{5}$ together.
        Ans. $17.\dot{5}5919120847374090303\dot{2}$.

5. Reduce the following numbers to decimals and find their sum ; $\frac{1}{3}$, $\frac{1}{7}$, and $\frac{1}{9}$.
        Ans. $.\dot{5}8730\dot{1}$.

SECTION XXVI.

SUBTRACTION OF CIRCULATING DECIMALS.

EXAMPLE.

1. From 87.164̇5̇ take 19.47̇9̇16̇7̇.

OPERATION.

87.164̇5̇   =87.164545

19.47̇9̇16̇7̇=19.479167

———————

67.685̇3̇7̇7̇

Having made the numbers similar and conterminous, we subtract as in whole numbers, and find the remainder of the circulate to be 5378, from which we subtract 1, and write the remainder in its place, and proceed with the other part of the question as in whole numbers. The reason why 1 should be added to the repetend may be shown as follows. The minuend may be considered 16⁴⁰⁴⁵⁄₅₅₅₅, and the subtrahend 7²¹⁶⁷⁄₅₅₅₅; we then proceed with these numbers as in Case II. Subtraction of vulgar fractions; and the numerator 5377 will be the repeating decimal. Q. E. D. Hence the following

OPERATION.

16⁴⁰⁴⁵⁄₅₅₅₅

7²¹⁶⁷⁄₅₅₅₅

————

8⁵³⁷⁷⁄₅₅₅₅

RULE.

Make the repetends similar and conterminous, and subtract as usual; observing, that if the repetend of the subtrahend be greater than the repetend of the minuend, then the remainder on the right must be less by unity, than it would be, if the expressions were finite.

2. From 7.1̇ take 5.0̇2̇.		Ans. 2.0̇8̇.
3. From 315.87̇ take 78.03̇7̇8̇.		Ans. 237.83̇8̇072095497̇.
4. Subtract ⅐ from ⅑.		Ans. .07̇9̇36̇5̇.
5. From 16.134̇7̇ take 11.0884̇.		Ans. 5.046̇2̇.
6. From 18.167̇8̇ take 3.27̇.		Ans. 14.895̇1̇.
7. From 3.12̇3̇ take 0.71̇.		Ans. 2.405̇9̇51̇.
8. From ⅞ take ₁₁⁄₁₃.		Ans. .2̇46755̇3̇.
9. From ⅛ take ⅞.		Ans. .15̇8̇73̇0̇.
10. From ₁⁄₁₇ take ₄⁄₁₇.		Ans. .1̇76470588235294̇1̇.

From 5.1̇2345̇ take 2.35̇2̇3456̇.

Ans. 2.771̇10553̇2̇1̇666927777̇9̇88888599̇9̇94̇.

SECTION XXVII.

MULTIPLICATION OF CIRCULATING DECIMALS.

1. Multiply $.3\dot{6}$ by $.2\dot{5}$.

First Method.
OPERATION.

$.3\dot{6}=\frac{33}{90}=\frac{4}{11}$, $.2\dot{5}=\frac{2}{10}+\frac{5}{90}=\frac{23}{90}$.

$\frac{4}{11}\times\frac{23}{90}=\frac{92}{990}=.09\dot{2}\dot{9}$ Answer.

In the first method we reduce the numbers to vulgar fractions, and then multiply and reduce them.

2. Multiply $582.3\dot{4}\dot{7}$ by $.08$.

Second Method.
OPERATION.

$582.3\dot{4}\dot{7}\times.08=46.58\dot{7}\dot{7}\dot{8}$ Answer.

In the second method we multiply as in whole numbers, but we add two units to the product; for $8\times347=2776=\frac{2776}{999}=2\frac{778}{999}$. Thus we see the repeating number is $77\dot{8}$. Hence the following

RULE.

Turn both the terms into their equivalent vulgar fractions, and find the product of those fractions as usual. Then change the vulgar fraction, expressing the product, into an equivalent decimal, and it will be the product required. But, if the multiplicand ONLY *has a repetend, multiply as in whole numbers, and add to the right hand place of the product as many units as there are tens in the product of the left hand place of the repetend. The product will then contain a repetend, whose places are equal to those in the multiplicand.*

3. Multiply $87.3\dot{2}5\dot{8}\dot{6}$ by 4.37.   Ans. $381.6140\dot{3}3\dot{8}$.

4. Multiply $3.14\dot{5}$ by $4.2\dot{9}\dot{7}$.   Ans. $13.5\dot{1}6953\dot{3}$

5. What is the value of $.\dot{2}8571\dot{4}$ of a guinea ?   Ans. 8s.

6. What is the value of $.461607\dot{1}42857$ of a ton ?

Ans. 9cwt. 0qr. 26lb

7. What is the value of $.2\dot{8}4931506$ of a year ?   Ans. 104da.

SECTION XXVIII.

DIVISION OF CIRCULATING DECIMALS.

1. Divide $.\dot{5}\dot{4}$ by $.1\dot{5}$.

OPERATION.

$.\dot{5}\dot{4} = \frac{54}{99} = \frac{6}{11}.$

$.1\dot{5} = \frac{1}{10} + \frac{5}{90} = \frac{14}{90} = \frac{7}{45}.$

$\frac{6}{11} \div \frac{7}{45} = \frac{6}{11} \times \frac{45}{7} = \frac{270}{77}.$

$\frac{270}{77} = 3\frac{39}{77} = 3.\dot{5}0649\dot{3}$ Ans.

Having reduced the numbers to vulgar fractions, we divide one by the other, and change the quotient to a decimal. Hence the following

RULE.

Change both the divisor and the dividend into their equivalent vulgar fractions, and find their quotient as usual. Change the vulgar fraction, expressing the quotient, into its equivalent decimal, and it will be the quotient required.

2. Divide $345.\dot{8}$ by $.\dot{6}$.   Ans. $518.8\dot{3}$.
3. Divide $234.\dot{6}$ by $.\dot{7}$.   Ans. $301.\dot{7}1428\dot{5}$.
4. Divide $.\dot{3}\dot{6}$ by $.2\dot{5}$.   Ans. $1.\dot{4}229249011857707509881$.

SECTION XXIX.

QUESTIONS TO BE PERFORMED BY ANALYSIS.

1. If a man travel 48 miles in 12 hours; how far will he travel in 17 hours ?   Ans. 68 miles.

[The following is the most obvious solution of this question. If he travel 48 miles in 12 hours, in 1 hour he will travel $\frac{1}{12}$ of 48 miles, which is 4 miles. Then if he travel 4 miles in 1 hour, he will in 17 hours travel 17 times as far, which is 68 miles, the answer.]

2. If 72 pounds of beef cost $6.48, what will 1 pound cost ? hat will 675 pounds cost ?   Ans. $60.75.

3. If $\frac{1}{8}$ of a dollar buy 1 pound of sugar, how much may be ught for $1 ? how much for 29\frac{7}{8}$?   Ans. 239lb.

4. If a hogshead of wine cost $73.50, what is the value of 1
llon ? what cost 17hhd. 45 gallons ? Ans. $1302.00.

5. Bought 11 bushels of rye for $9.00 ; what cost 25 bushels ?
Ans. $20.45⁵₁₁.

6. If a crew of 15 hands consume in 3 months 1620 pounds of
)eef, how much would be sufficient for 27 hands for the same
:ime ? Ans. 2916lb.

7. Bought 9 yards of flannel for $7.00; what would be the
value of 37¹¹₁₅ of a yard ? Ans. $29.31½.

8. If a certain field will pasture 8 horses 9 weeks, how long
will it pasture 23 horses ? Ans. 3³₂₃ weeks.

9. If 7₁₅ dozen of hats cost $318.50, what will 19¹¹₁₂ dozen cost?
Ans. $874.95¹¹₁₂.

10. If 1 ton of hay cost $25 00, what will 17 tons 13 cwt. 19
pounds cost ? Ans. $441.46+.

11. What part of 9¼ is 25 ? Ans. 2¹²₃₇.

12. What part of 25 is 9¼ ? Ans. ³⁷₁₀₀.

13. If $25 will pay for 7¾ yards of broadcloth, what would be
the price of 97 yards ? Ans. $328.81¹¹₃₁.

14. If $47.25 will pay for the keeping of 7 horses 2 months,
what would it cost to keep 43 horses for the same time ?
Ans. $290.25.

15. If 7½ yards of cloth a yard wide is sufficient to make a
cloak, how many yards would it take if the cloth was 1½ yards
wide ? Ans. 4⁴₇ yards.

16. If a barrel of beer will last 10 men a week, how long
would it last 1 man ? how long 37 men ? Ans. 1³³₃₇ days.

17. If 5 calves are worth 9 sheep, how many calves will purchase 108 sheep ? Ans. 60 calves.

18. If 11 yards of cotton 3 quarters wide, are sufficient to line
a garment, how many yards would it require that were 5 quarters wide ? Ans. 6¾ yds.

19. If 7 pair of shoes will purchase 2 pair of boots, how many
pair of shoes would it take to buy 18 pair of boots ?
Ans. 63 pair.

20. If 4 gallons of vinegar be worth 7 gallons of cider, what
quantity of vinegar would it take to buy 47 gallons of cider ?
Ans. 26⁴₇ gallons.

21. If a man travel 377 miles in 15 days, how far would he
travel in 1 day ; how far in 100 days ? Ans. 2513⅓ miles.

22. If 12 men can dig a ditch 50 feet long, and 4 feet wide, 3
feet deep, in 30 days, how long would it take 1 man ? how long
47 men ? Ans. 7³¹₄₇ days.

23. If 5 barrels of flour cost $25.75, what will 39 barrels cost?
Ans. $200.85.

24. Bought 17 acres of land for $791.01 ; what would 99
acres, 3 roods, and 14 perches cost ? Ans. $4598.90.8¼

25. Bought $97\frac{3}{8}$ yards of cloth for \$275.20; what is the price of 7 yards? Ans. \$19.78+.

26. If a pole 7 feet long cast a shadow of 5 feet, how high is that steeple, whose shadow is 97 feet? Ans. $135\frac{4}{5}$ feet.

27. Gave $3\frac{3}{7}$ cwt. of sugar, at \$9 for $\frac{11}{17}$ of an acre of land; how much sugar would it have required to purchase an acre? Ans. $5\frac{23}{77}$ cwt.

28. If a vessel sail $47\frac{3}{7}$ miles in $3\frac{2}{7}$ hours, how far would it sail in 1 week? Ans. $2425\frac{1}{23}$ m.

29. James can mow a field in 7 days, by laboring 10 hours a day; how many days would it take him to perform the work by laboring 12 hours a day? Ans. $5\frac{5}{6}$ days.

30. If $\frac{7}{11}$ of a lot of land is worth \$42.12, what is the value of $\frac{4}{9}$ of it? Ans. \$$29.41\frac{5}{7}$.

31. If a man can earn \$10.27 in $\frac{4}{7}$ of a week, how much would he earn in a month? Ans. \$71.89.

32. If 18 men can reap 72 acres in 5 days, how long would it take 6 men to perform the labor? Ans. 15 days.

33. If 19 gallons of wine can be bought for \$25, how many gallons will \$71.25 buy? Ans. $54\frac{3}{20}$.

34. If a penny loaf weighs 8 ounces, when wheat is \$1 a bushel, what should it weigh, when wheat is sold for \$1.25 a bushel? Ans. $6\frac{2}{5}$ ounces.

35. If a basket, which contains $1\frac{1}{4}$ bushels, must be filled with apples 7 times to make 1 barrel of cider, how many barrels may be made by its being filled 126 times? Ans. 18 barrels.

36. If a cloak can be made of $4\frac{1}{8}$ yards of cloth, that is $1\frac{3}{8}$yds. wide, how many yards would it take, of cloth that is $\frac{7}{8}$ of a yard wide? Ans. 6yds. 1qr. $3\frac{5}{7}$na.

37. If a box 4 feet long, 2 feet wide, $1\frac{1}{2}$ feet high, contains 300 pounds of sugar, how much will a box that is 8 feet long, 4 feet wide, and 3 feet high contain? Ans. 2400lbs.

38. How much in length, that is $12\frac{4}{7}$ rods in breadth, will make an acre? Ans. $12\frac{8}{11}$rd.

39. Sound, uninterrupted, moves 1142 feet a second; how long, after a cannon's being discharged at Boston, is the time before it is heard at Bradford, the nearest distance being $25\frac{21}{22}$ miles? Ans. 2 minutes.

40. Bought $\frac{7}{11}$ of a ton of potash, and sold $\frac{11}{14}$ of it for \$46.70; what was the value of a ton? Ans. \$93.40.

41. If $\frac{4}{11}$ of a lot of land be worth \$97, what would the whole lot be worth? Ans. \$266.75.

42. If 19 pounds of salmon be worth 50 pounds of beef, how much salmon would buy 77 pounds of beef? Ans. $29\frac{13}{50}$lbs.

43. What part of $17\frac{3}{11}$ is $4\frac{1}{4}$? Ans. $\frac{187}{760}$.

44. What part of 11s. 3d. is 15s. Ans. $\frac{4}{3}$.

45. What part of $7\frac{1}{4}$ yards is $3\frac{3}{5}$ ells English? Ans. $\frac{18}{29}$.

SECTION XXX.

SIMPLE INTEREST.

INTEREST is the compensation, which the *borrower* of money makes to the *lender*.

PRINCIPAL, is the sum lent.

AMOUNT, is the interest added to the principal.

PER CENT., a contraction of per centum, is the rate establish-ed by law; or that which is agreed on by the parties, and is so much for a hundred dollars for one 1 year.

GENERAL RULE *

Multiply the principal by the rate per cent. and divide by 100, and the quotient is the answer for one year; or, multiply by the per cent. and cut off two figures from the right hand, and those at the left will be the interest in dollars, and those at the right will be decimals of a dollar.

SECOND METHOD. — *Let the per cent. be considered a deci-mal of a hundred dollars, and multiply the principal by it. Thus, .06 is the decimal for six per cent. and .08 is the decimal for eight per cent.*

NOTE. — When no particular per cent. is named, 6 per cent. is to be under-stood; as it is the legal interest in the New England States generally. In New York, the legal interest is 7 per cent.

EXAMPLES.

1. What is the interest of $ 144 for 1 year ?

First Method.	Second Method.
144	144
6	.06
100)864	$ 8.64 Answer.
$ 8.64 Answer.	

2. What is the interest of $78 for 1 year ?   Ans. $ 4.68.
3. What is the interest of $ 675 for 1 year ?   Ans. $ 40.50.

* This rule is obvious, from the fact, that the *rate per cent.* is such a part of every *hundred*. Thus, 6 per cent. is $\frac{6}{100}$ of the *principal*.

4. What is the interest of $1728 for 1 year ?

Ans. $103.68.

5. What is the interest of $19.64 for 2 years ?

Ans. $2.35.6.

6. What is the interest of $896.28 for 3 years ?

Ans. $161.33.

7. What is the interest of $349.25 for 10 years ?

Ans. $209.55.

8. What is the interest of $3967.87 for 2 years ?

Ans. $476.14.4.

9. What is the interest of $123.45 for 6 years ?

Ans. $44.44.2.

10. What is the interest of $89.25 for 50 years ?

Ans. $267.75.

11. What is the interest of $17.25 for 7 years ?

Ans. $7.24.5.

12. What is the interest of $29.19 for 9 years ?

Ans. $15.76.2.

13. What is the interest of $617.56 for 25 years ?

Ans. $926.34.

14. What is the amount of $31.75 for 100 years ?

Ans. $222.25.

15. What is the amount of $76.47 for 7 years ?

Ans. $108.58.7.

16. What is the amount of $716.57 for 4 years ?

Ans. $888.54.6.

17. What is the amount of $178.56.5 for 30 years ?

Ans. $499.98.2.

18. What is the interest of $97.06 for 9 years ?

Ans. $52.41.2.

19. What is the interest of $0.75 for 75 years ?

Ans. $3.37.5.

20. What is the interest of $750 for 12 years ?

Ans. $540.

CASE I.

To find the interest for months, at six per cent.

RULE.*

Multiply the principal by half the number of months, and proceed as in the general rule; or, let half the number of months be considered as a decimal.

* It is obvious, that if the rate per cent. were 12, it would be 1 per cent. a month. If, therefore, it be 6 per cent. it will be half per cent. a month; that is, half the months will be the per cent.

EXAMPLES.

1. What is the interest of $ 368 for 8 months ?

$$368$$
$$.04 = \text{half the months.}$$
$$\overline{\$ 14.72} = \text{Answer.}$$

2. What is the interest of $ 637 for 10 months ?
Ans. $ 31.85.

3. What is the interest of $ 1671.32 for 14 months ?
Ans. $ 116.99.

4. What is the interest of $ 891.24 for 9 months ?
Ans. $ 40.10.

5. What is the interest of $ 819.75 for 11 months ?
Ans. $ 45.08.6.

6. What is the interest of $ 3671.25 for 13 months ?
Ans. $ 238.63.

7. What is the interest of $ 61.18 for 15 months ?
Ans. $ 4.58.

8. What is the interest of $ 3181.29 for 18 months ?
Ans. $ 286.31.

9. What is the interest of $ 11.39 for 19 months ?
Ans. $ 1.08.

10. What is the interest of $ 9.98 for 23 months ?
Ans. $ 1.14.

11. What is the interest of $87.19 for 27 months ?
Ans. $ 11.77.

12. What is the interest of $ 32.18 for 36 months ?
Ans. $ 5.79.

13. What is the interest of $ 167.18 for 50 months ?
Ans. $ 41.79.

14. What is the interest of $ 336.19 for 100 months ?
Ans. $ 193.09.

CASE II.

To find the interest of any sum for months and days, at 6 per cent.

RULE.*

Multiply by half the number of months and one-sixth of the days; and if the principal be dollars, cut off three figures from the right hand, and those at the left will be the interest in dol

* By taking one-sixth of the days, and annexing them to half of the months, they become decimals of a month; for as days are thirtieths of a month, they will therefore be sixtieths of half months; and if one-sixth of the days be taken they become tenths, or decimals of a month.

lars, and those at the right will be cents and mills. But if the principal be dollars and cents, five figures must be cut off from the right hand, and those at the left will be the interest, as before, &c.

NOTE. — If any other per cent. is wanted, proceed as above, and then multiply by the given rate per cent. and divide by 6, and the quotient is the interest.

EXAMPLES.

1. What is the interest of $ 68.25 for 8 months and 24 days ?
                                        Ans. $ 3.00.3.

```
 68.25
 .0.44
 ─────
 27300
 27300
 ─────
3.00.300
```

The first 4 in the multiplier is half of the 8 months ; the second 4 is one-sixth of the 24 days.

2. What is the interest of $ 637.28 for 17 months and 19 days, at 8 per cent. ?   Ans. $ 74.91.5.
3. What is the interest of $ 396.15 for 13 months and 9 days ?
                                        Ans. $ 26.34.3.
4. What is the interest of $ 16.75 for 7 months and 17 days, at 7 per cent. ?   Ans. $ 0.73.9.
5. What is the interest of $ 976.18 for 29 months and 23 days, at 9 per cent. ?   Ans. $ 217.93.2.
6. What is the interest of $ 36.18 for 3 months and 7 days ?
                                        Ans. $ 0.58.4.
7. What is the interest of $51.17 for 9 months and 29 days, at 4 per cent. ?   Ans. $ 1.69.9.
8. What is the interest of $ 365.19 for 33 months and 4 days, at 2 per cent. ?   Ans. $ 20.16.6.
9. What is the interest of $ 125.75 for 5 months and 4 days ?
                                        Ans. $ 3.22.7.
10. What is the interest of $ 35.49 for 1 month and 2 days, at 7½ per cent. ?   Ans. $ 0.23.6.
11. What is the interest of $ 112.50 for 3 months and 1 day, at 9½ per cent. ?   Ans. $ 2.70.1.
12. What is the interest of $ 97.15 for 35 months and 27 days ?
                                        Ans. $ 17.43.8.
13. What is the interest of $ 47.15 for 1 month and 19 days, at 13½ per cent. ?   Ans. $ 0.86.6.
14. What is the interest of $ 678.75 for 67 months and 20 days ?   Ans. $ 257.51.8.
15. What is the interest of $ 86 for 99 months and 29 days, at 25 per cent. ?   Ans. $ 179.10.6.
16. What is the interest of $ 33.35.8 for 15 months and 17 days ?   Ans. $ 2.59.6.
17. What is the interest of $ 144 for 5 days ?
                                        Ans. $ 0.12.0

CASE III.

When the interest is required on any sum, from a certain day of the month in a year, to a particular day of a month in the same, or in another year.

RULE.

Find the time, by placing the latest date in an upper line, and the earliest date under it. Let the year be placed first; the number of months that have elapsed since the year commenced, at its right hand, and the day of the month next; then subtract the earlier from the latest date, and the remainder is the time for which the interest is required. Then proceed as in the last rule. Or, the months may be considered as fractional parts of a year, and the days as fractional parts of a month.

EXAMPLES.

1. What is the interest of $84.97, from Sept. 25, 1833, to March 8, 1835 ?

Yrs.	m.	d.
1835	2	8
1833	8	25
1	5	13

First Method.

```
   84.97
  .0.87½
  ───────
   59479
   67976
    1416
  ───────
 7.40.6.55       Ans. $7.40.6.
```

Second Method.

```
          84.97
            .06
          ──────
         5.0982
4 months, ⅓ = 1.6994
1 month,  ¼ =  .4248
10 days,  ⅓ =  .1416
2 days,   ⅕ =   283
1 day,    ½ =   141
```

It is evident, that 4 months' interest is ⅓ of a year's interest; and, for the same reason, 1 month's interest is ¼ of 4 months' interest; and as 10 days is ⅓ of a month, the interest for that time is ⅓ of a month's interest; and if the interest of 10 days be divided by 5, the quotient will be 2 day's interest, and the half of this will be 1 day's interest.

$7.40.64 Interest as before.

2. What is the interest of $786.75, from Dec. 9, 1831, to May 11, 1833 ?   Ans. $67.13.6.

3. What is the interest of $98.25, from July 4, 1826, to Oct. 19, 1829 ?   Ans. $19.40.4.

4. What is the interest of $76.89.5, from Jan. 11, 1822, to July 27, 1833 ?   Ans. $53.26.2.

12*

5. What is the interest of $ 22.76.3, from Feb. 19, 1806, to July 18, 1830 ?   Ans. $ 33.34.4.

6. What is the interest of $76.35, from August 17, 1830, to May 5, 1832 ?   Ans. $7.86.4.

7. What is the interest of $ 97.86, from May 17, 1821, to Dec. 19, 1828 ?   Ans. $ 44.55.3.

8. What is the interest of $ 1728.75, from Nov. 19, 1823, to June 18, 1826 ?   Ans. $ 267.66.8.

9. What is the interest of $ 99.99.9, from Jan. 1, 1800, to Feb. 29, 1832 ?   Ans. $ 192.96.4.

10. What is the interest of $ 16.76, from Dec. 17, 1811, to June 17, 1822 ?   Ans. $ 10.55.8.

11. What is the interest of $ 35.61, from Nov. 11, 1831, to Dec. 15, 1833 ?   Ans. $ 4.47.4.

12. What is the interest of $ 786.97, from Oct. 19, 1827, to August 17, 1831, at 7½ per cent. ?   Ans. $ 225.92.5.

13. What is the interest of $ 96.84, from Nov. 27, 1829, to July 3, 1832, at 7½ per cent. ?   Ans. $ 18.88.3.

14. What is the interest of $ 11.10.5, from April 17, 1832, to Dec. 7, 1832, at 7 per cent. ?   Ans. $ 0.49.6.

15. What is the interest of $ 117.21, from June 19, 1806, to June 17, 1819, at 8½ per cent. ?   Ans. $ 129.46.1.

16. What is the interest of $ 17869.75, from Feb. 7, 1830, to Jan. 11, 1832, at 5 per cent. ?   Ans. $ 1722.44.5.

17. What is the interest of $ 71.09.1, from July 29, 1823, to June 19, 1827, at 12 per cent. ?   Ans. $ 33.17.5.

18. What is the interest of $ 83.47, from Nov. 8, 1830, to July 11, 1833, at 8¼ per cent. ?   Ans. $ 19.53.7.

19. What is the interest of $ 79.25, from Dec. 8, 1831, to July 17, 1833 ?   Ans. $7.64.7.

20. What is the interest of $ 175.07, from Jan. 7, 1825, to Oct. 12, 1829 ?   Ans. $ 50.04.0.

21. What is the interest of $ 12.75, from June 16, 1831, to August 20, 1833 ?   Ans. $ 1.66.6.

22. What is the interest of $ 197.28.5, from Dec. 6, 1832, to Jan. 11, 1834 ?   Ans. $ 12.98.7.

23. What is the interest of $ 12.69, from Jan. 2, 1833, to Aug. 30, 1834, at 7 per cent. ?   Ans. $ 1.47.5.

24. What is the interest of $79.15, from Feb. 11, 1831, to June 10, 1833, at 7½ per cent. ?   Ans. $ 13.57.3.

25. What is the amount of $ 88.33, from March 11, 1831, to Jan. 1, 1833, at 7½ per cent. ?   Ans. $ 94.61.4.

26. What is the amount of $ 100.25, from March 2, 1831, to June 1, 1831, at 4 per cent. ?   Ans. $ 101.24.1.

27. What is the amount of $ 369.29, from April 30, 1830, to July 31, 1832, at 9 per cent. ?   Ans. $ 444.16.3.

28. What is the interest of $ 769.87, from Jan. 1, 1830, to June 17, 1835, at 9½ per cent. ?   − Ans. $ 399.41.2.

29. What is the interest of $ 69.75, from Jan. 11, 1833, to June 29, 1833, at 17½ per cent. ?   Ans. $ 5.69.6.

30. What is the interest of $ 368.18, from April 2, 1816, to June 19, 1835, at 2 per cent. ?   Ans. $ 141.48.3.

31. What is the interest of $ 16.16, from March 3, 1831, to Dec. 6, 1833, at 1 per cent. ?   Ans. $ 0.44.5.

32. What is the interest of $ 1728.19, from May 7, 1824, to July 17, 1830, at ¼ per cent. ?   Ans. $ 26.76.2.

33. What is the interest of $ 897.16, from Dec. 29, 1831, to June 30, 1833, at 5½ per cent. ?   Ans. $ 82.82.6.

34. What is the amount of $ 1760.07, from Feb. 17, 1831, to Dec. 19, 1832, at ½ per cent. ?   Ans. $ 1776.25.2.

CASE IV.

To find the interest of any sum for days.

RULE.

If the rate per cent. be 6, multiply the principal by the number of days, and divide by 6083, and the quotient is the interest; but if the rate per cent. be 5, divide by 7300.

NOTE. — The reason for dividing by these numbers is, that the interest of any sum at 6 per cent. will amount to the principal in 6083 days; and at 5 per cent. in 7300 days.

EXAMPLES.

1. What is the interest of $ 1835, for 35 days ?   Ans. $ 10.55.8.

2. What is the interest of $ 165.37, for 165 days ?   Ans. $ 4.48.5.

3. What is the interest of $ 16.87, for 79 days, at 5 per cent. ?   Ans. $ 0.18.2.

4. What is the interest of $ 167, for 87 days, at 5 per cent. ?   Ans. $ 1.99.

5. What is the interest of $ 761.81, for 165 days ?   Ans. $ 20.66.8.

6. What is the interest of $ 76.18.5, for 315 days, at 5 per cent. ?   Ans. $ 3.28.7.

7. What is the interest of $ 178.69.7, for 271 days ?   Ans. $ 7.96.1.

8. What is the interest of $ 1728.79, for 318 days, at 5 per cent. ?   Ans. $ 75.30.8.

9. What is the interest of $ 73 for 73 days, at 5 per cent. ?   Ans. $ 0.73.

SECTION XXXI.

PARTIAL PAYMENTS.

When notes are paid within one year from the time they become due, it has been the usual custom to find the amount of the
principal from the time it became due, until the time of payment;
and then to find the amount of each endorsement, from the time
it was paid until settlement, and to subtract their sum from the
amount of the principal.

EXAMPLES.

(1.)   Baltimore, January 1, 1833.
For value received, I promise Riggs, Peabody, & Co. to pay
them, or order, on demand, one thousand seven hundred and
twenty-eight dollars, with interest.   John Paywell, Jr.

On this note are the following endorsements.   March 1, 1833, received three
hundred dollars.   May 16, 1833, received one hundred and fifty dollars.   Sept.
1, 1833, received two hundred and seventy dollars.   December 11, 1833, received one hundred and thirty-five dollars.

What was due at the time of payment, which was December
16, 1833 ?   Ans. $ 948.03.

OPERATION.

Principal, - - - - - - - -	$ 1728.00
Interest for 11 months and 15 days, - - - -	99.36
	$ 1827.36

First payment, - - - -	300.00
Interest for 9 months and 15 days, -	14.25
Second payment, - - - -	150.00
Interest for 7 months, - - - -	5.25
Third payment, - - - -	270.00
Interest for 3 months and 15 days, -	4.72
Fourth payment, - - - -	135.00
Interest for 5 days, - - - -	.11
	$ 879.33

Balance remaining due, Dec. 16, 1833, $ 948.03

(2.)   Concord, Feb. 4, 1834.
For value received, we jointly and severally promise James
Thomas, to pay him, or order, on demand, seven hundred dol-
lars, with interest.   Sampson Phillips,
Attest, Henry Dix.   Quarls Hoopoo.

On this note are the following payments.   March 18, 1834, received one
hundred and sixty dollars.   June 24, 1834, received two hundred dollars.
September 11, 1834, received one hundred and twenty dollars.   October 5,
1834, received sixty dollars.

What was due on this note, Nov. 28, 1834 ?   Ans.$180.43.

(3.)   Portland, May 16, 1834.
For value received, we, Luke A. Homer, as principal, and
Daniel D. Snow and Ichabod Frost, as sureties, promise John
Webster, to pay him, or order, in one year, six hundred dollars,
with interest after three months.   Luke A. Homer,
                                           Daniel D. Snow,
Attest, M. Peters.   Ichabod Frost.

On this note are the following endorsements. Sept. 18, 1834, received one
hundred and thirty-six dollars. December 5, 1834, received one hundred and
ninety-seven dollars. February 11, 1835, received two hundred dollars. April
19, 1835, received forty dollars.

What was due August 1, 1835 ?   Ans. $ 40.31.2.

In the United States' Court, and in most of the Courts of the
several States, the following rule is adopted for estimating interest on notes and bonds, when partial payments have been
made.

RULE.

*Compute the interest on the principal sum, from the time when
the interest commenced, to the first time when a payment was
made, which exceeds, either alone, or in conjunction with the
preceding payments, if any, the interest at that time due; add
that interest to the principal, and from the sum subtract the payment made at that time, together with the preceding payments,
if any, and the remainder forms a new principal; on which
compute and subtract the interest, as upon the first principal,
and proceed in the same manner to the time of judgment.*

The following example will illustrate the above rule.

(4.)   Boston, June 17, 1827.
For value received, I promise James E. Snow, to pay him,
or order, on demand, one hundred and sixty-five dollars and
eighteen cents, with interest.   James Y. Frye.
   ($ 165.18.)
      Attest, John True.

On this note are the following endorsements. December 7, 1827, received
eighteen dollars and thirteen cents of the within note. October 19, 1828, received twenty-eight dollars and sixteen cents. September 25, 1829, received
thirty-six dollars and twelve cents. July 10, 1830, received three dollars and
eighteen cents. June 6, 1831, received thirty-six dollars and twenty-eight cents.
December 28, 1832, received thirty-one dollars and seventeen cents. May 5,
1833, received three dollars and eighteen cents.' September 1, 1833, received
twenty-five dollars and eighteen cents. October 18, 1834, received ten dollars.

How much remains due, September 27, 1835 ?
                                           Ans. $ 15.41.7.

Principal, carrying interest from June 17, 1827, $ 165.18.0
Interest from June 17, 1827, to Dec. 7, 1827, 5mo. 20 days,   4.68.0

                                                 Amount, 169.86.0

First payment, Dec. 7, 1827,   -   -   -   -   -   18.18.0

Balance for new principal, -   -   -   -   -   - 151.73.0
Interest from Dec. 7, 1827, to Oct. 19, 1828, 10mo. 12da.   7.38.9

                                                ,Amount, 159.61.9

Second payment, Oct. 19, 1828,   -   -   -   28.16.0

Balance for new principal, -   -   -   -   -   - 131.45.9
Interest from Oct. 19, 1828, to Sept. 25, 1829, 11mo. 6da.   7.36.1

                                                 Amount, 138.82.0

Third payment, Sept. 25, 1829,   -   -   -   -   36.12.0

Balance for new principal, -   -   -   -   - 102.70.0
Interest from Sept. 25, 1829, to June 6, 1831, 20mo. 11da. 10.45.8

                                                 Amount, 113.15.8

Fourth pay't, July 10, 1830, a sum less than int'st, 3.18
Fifth pay't, June 6, 1831, a sum greater than int'st. 36.28
                                                        _____
                                                         39.46.0

Balance for new principal,   -   -   -   -   -   73.69.8
Interest from June 6, 1831, to Dec. 28, 1832, 18mo. 22da.   6.90.3

                                                 Amount, 80.60.1

Sixth payment, Dec. 28, 1832,   -   -   -   -   -   31.17.0

Balance for now principal,   -   -   -   -   -   49.43.1
Interest from Dec. 28, 1832, to May 5, 1833, 4mo. 7da.   1.04.6

                                                 Amount, 50.47.7

Seventh payment, May 5, 1833,   -   -   -   -   -   3.18.0

Balance for new principal,   -   -   -   -   -   47.29.7
Interest from May 5, 1833, to Sept. 1, 1833, 3mo. 26 days,   91.4

                                                 Amount, 48.21.1

Amount brought forward, 48.21.1

Eighth payment, Sept. 1, 1833, - - - - - 25.18.0

Balance for new principal, - - - - - - 23.03.1
Interest from Sept. 1, 1833, to Oct. 18, 1834, 13mo. 17da. 1.56.2

Amount, 24.59.3

Ninth payment, - - - - - - 10.00.0

Balance for new principal, - - - - - - 14.59.3
Interest from Oct. 18, 1834, to Sept. 27, 1835, 11mo. 9da. 82.4

Balance due at the time of payment, - - - $15.41.7

(5.)   Salem, June 17, 1829.

For value received, I promise L. Swan, to pay him, or order, on demand, seven hundred and sixty-nine dollars and eighty-seven cents, with interest.   Samuel Q. Peters.
Attest, Moses Haynes.

On this note are the following payments. March 1, 1830, received seventy-five dollars and fifty cents. June 11, 1831, received one hundred and sixty-five dollars. September 15, 1831, received one hundred and sixty-one dollars. Jan. 21, 1832, received forty-seven dollars and twenty-five cents. March 5, 1833, received twelve dollars and seventeen cents. December 6, 1833, received ninety-eight dollars. July 7, 1834, received one hundred and sixty-nine dollars.

What remains due Sept. 25, 1835 ?   Ans. $226.29.7.

(6.)   Haverhill, April 30, 1831.

For value received, I promise Kimball & Hammond, to pay them, or order, on demand, three hundred dollars, with interest.
Attest, James Quintire.   Simpson W. Leavet.

The following partial payments were made on this note. June 27, 1832, received one hundred and fifty dollars. December 9, 1832, received one hundred and fifty dollars.

What was due Oct. 9, 1833 ?   Ans. $26.73.5

(7.)   New York, Feb. 11, 1832.

For value received, I promise John Trow, to pay him, or order, on demand, fifty-four dollars, and eighteen cents, with interest.   Luke M. Sampson.

On this note are the following payments. July 11, 1833, received twelve dollars and twenty-five cents. August 15, 1834, received two dollars and ten cents. July 9, 1835, received three dollars and twelve cents. August 21, 1835, received thirty-seven dollars and eighteen cents.

What was due Dec. 17, 1835.   Ans. $10.22.2.

(8.)   Boston, Jan. 7, 1831.

For value received, we jointly and severally promise Jones, Oliver, & Co. to pay them, or order, on demand, one thousand seven hundred and twenty-eight dollars, with interest.

                            John Bountiful,
  Attest, Timothy True.   James Trusty.

On this note are the following payments. February 9, 1832, received seven hundred and sixty dollars and twenty-eight cents and five mills. March 5, 1833, received sixty-eight dollars and fifty cents. December 28, 1833, received eight hundred and seventy-six dollars and twenty-eight cents. July 17, 1834, received sixty dollars.

What was due at the time of payment, which was Oct. 1, 1834?   Ans. $ 209.22.9.

(9.)   Philadelphia, May 7, 1829.

For value received, I promise John Jordan, to pay him, or order, on demand, five hundred dollars, with interest.
                           Thomas C. True.

The following partial payments were endorsed on this note. June 29, 1830, received one hundred dollars. December 5, 1831, received one hundred dollars. March 12, 1832, received five dollars. July 4, 1833, received ninety-five dollars. December 1, 1834, received two hundred dollars.

What was due Jan. 1, 1836?   Ans. $ 141.50.4.

(10.)   Newburyport, March 19, 1831.

For value received, we, jointly and severally, promise John Frost, to pay him, or order, on demand, eighty-nine dollars and seventy-five cents, with interest.   Henry Augustus,
  Attest, James Snow.   Marcus T. Cicero.

On this note are the following endorsements. December 6, 1831, received twelve dollars and twelve cents. February 17, 1832, received twelve dollars and twelve cents. March 19, 1833, received three dollars and sixteen cents. December 28, 1834, received two dollars and eighteen cents. January 1, 1835, received twenty-five dollars and twenty-five cents. March 11, 1835, received thirty-one dollars and eighteen cents. July 17, 1835, received five dollars and eighteen cents. September 1, 1835, received six dollars and twenty-nine cents.

What was due Dec. 29, 1835?   Ans. $ 10.57.

The following rule is established by the Supreme Court of the State of Connecticut.

RULE.

Compute the interest to the time of the first payment; if that be one year or more from the time the interest commenced, add it to the principal, and deduct the payment from the sum total. If there be after payments made, compute the interest on the *balance due to the next payment, and then deduct the payment as above; and, in like manner, from one payment to another, till all the payments are absorbed; provided the time between one payment and another be one year or more. But if any payments be made before one year's interest hath accrued, then compute the interest on the principal sum due on the obligation, for one year, add it to the principal, and compute the interest on the sum paid, from the time it was paid, up to the end of the year; add it to the sum paid, and deduct that sum from the principal and interest, added together.**

If any payments be made of a less sum than the interest arisen at the time of such payment, no interest is to be computed, but only on the principal sum for any period.

(11.)   Wethersfield, June 1, 1828.

For value received, I promise J. D. to pay him, or order, nine hundred dollars, on demand, with interest.

James L. Emerson.

On this note are the following endorsements.   June 16, 1829, received two hundred dollars of the within note.   August 1, 1830, received one hundred and sixty dollars.   November 16, 1830, received seventy-five dollars.   February 1, 1832, received two hundred and twenty dollars.

What was due August 1, 1832 ?   Ans. $444.98.2.

OPERATION.

Principal, ...	$900.00
Interest from June 1, 1828, to June 16, 1829, 12½ mo.	56.25
	956.25
First payment, ...	200.00
	756.25
Interest from June 16, 1829, to Aug. 1, 1830, 13½ mo.	51.04.6
	807.29.6
Second payment, ...	160.
	647.29.6
Interest for one year, ...	38.83.7
	686.13.3

* If the year extends beyond the time of payment, find the amount of the remaining principal to the time of payment; find, also, the amount of endorsement, or endorsements, and subtract their sum from the amount of the principal

13

Brought forward, $686.13.3
Am't of 3d pay't, fr. Nov. 16, 1830, to Aug. 1, 1831, 8½ m.   78.18.7

607.94.6
Interest from Nov. 16, 1830, to Feb. 1, 1832, 14¼ mo.   44.07.6

652.02.2
Fourth payment.   .   .   .   .   .   220.

432.02.2
Interest from Feb. 1, 1832, to Aug. 1, 1832, 6 mo.   12.96.0

Balance due August 1, 1832, ... $444.98.2

SECTION - XXXII.

MISCELLANEOUS PROBLEMS IN INTEREST.

Principal, interest, and time, given, to find the rate per cent.

1. At what rate per cent. must $500 be put on interest to gain $120, in 4 years ?

OPERATION.

```
500
.01
─────
5.00
  4
─────────
20.00)120.00(6 per cent. Ans.
      120.00
```

Hence the following

BY ANALYSIS.

The interest of $1.00 for the given time at one per cent is 4 cents. $500.00 will be 500 times as much. = $500 × .04 = $20,00. Then, if $20, give 1 per cent. $120 will give ¹²⁰⁄₂₀ = 6 per cent.

RULE.

Divide the given interest by the interest of the given sum at 1 per cent. for the given time; the quotient will be the rate per cent. required.

2. At what rate per cent. must $120 be on interest, to amount $22.80 in 16 months?   Ans. 8¼ per cent.
3. At what rate per cent. must $280 be on interest to amount $35 in 6¼ years ?   Ans. 7¼ per cent.

Principal, interest, and rate per cent. given to find the time.

4. How long must $500 be on interest, at 6 per cent. to gain $120?

<table>
<tr><td>OPERATION.</td><td>BY ANALYSIS.</td></tr>
</table>

5 00
.06
——
30.00)120.00 (4 years Ans.
   120.00
——

We find the interest of $1.00 at the given rate for one year is 6 cents. $500 will, therefore, be 500 times as much=$500×.06=$30.00. Now, if it take 1 year to gain $30, it will require to gain $120, $\frac{120}{30}=$ 4 years Ans.

Hence the following

RULE.

Divide the given interest, by the interest of the principal, for 1 year, and the quotient is the time.

5. How long must $120 be on interest at 8¼ per cent. to amount to $133.20?   Ans. 16 months.
6. How long must $280 be on interest at 7¼ per cent. to amount to $411.95?   Ans. 6¼ years.

Interest, time, and rate per cent. given to find the principal.

7. What principal at 6 per cent. is sufficient, in 4 years to gain $120.

<table>
<tr><td>OPERATION.</td><td>BY ANALYSIS.</td></tr>
</table>

$1
.06
——
.06
 4
——
.24)120.00($500. Ans.
   120.00
——

The interest of $1 for the given rate and time, is 24 cents. If 24 cents then give $1 for principal, $120.00 will give $\frac{12000}{24}$ times as much = $500 Ans.

Hence the following

RULE.

Divide the given interest, by the interest of $1, for the given rate and time, and the quotient is the principal.

8. What principal at 8¼ per cent. is sufficient in 16 months to gain $13.20?   Ans. $120.00.
9. What principal at 7¼ per cent. is sufficient in 6¼ years, to amount to $411.95?   Ans. $280.00.

MISCELLANEOUS QUESTIONS.

1. A merchant, having $1729 in the Union Bank, wishes to withdraw 15 per cent. ; how much will remain ?

Ans. $1468.90.

2. A gentleman has an estate of $15.000. He has given 12½ per cent. to his wife, and 25 per cent. of the remainder to his children ; how much will he have left ? Ans. $9843.75.

3. What is the difference between 12½ per cent. of $560, and 18 per cent. of $400 ? Ans. $2.00.

4. A merchant expended $1890 as follows. He laid out 12 per cent. of his money for calicoes ; 20 per cent. for hard ware ; 40 per cent. for broadcloths ; and with the remainder he bought 2646 pounds of sugar ; what did it cost him a pound ?

Ans. $0.20.

SECTION XXXIII.

DISCOUNT.

THE object of Discount is to show us what allowance should be made, when any sum of money is paid before it becomes due.

The *present worth* of any sum is the principal, that must be put at interest, to amount to that sum in the given time. That is, $100 is the *present worth* of $106, due one year hence ; because $100 at 6 per cent. will amount to $106, and $6 is the discount.

1. What is the present worth of $12.72, due one year hence ?

First Method.	*Second Method.*
$12.72	$ cts.
100	1.06)12.72($12 Ans.
106)1272.00($12 Ans.	10.6
106	2.12
212	2.12
212	

As $100 will amount to $106 in one year at 6 per cent., it is evident, that if $\frac{100}{106}$ of any sum be taken, it will be its present worth for one year ; and that $\frac{6}{106}$ will be the discount. And, as $1 is the present worth of $1.06 due one year hence, it is evident, that the present worth of $12.72 must be equal to the number of times $12.72 will contain $1.06. From the above illustrations we deduce the following

RULE.

Divide the given sum by the amount of $1. for the given rate and time; and the quotient will be the present worth. If the present worth be subtracted from the given sum, the remainder will be the discount.

2. What is the present worth of $117.60, due one year hence, at 12 per cent. ? 　　　　　　　　　　Ans. $105.00.

3. What is the discount of $802.50, at 7 per cent. due one year hence ? 　　　　　　-　　　　Ans. $52.50.

4. What is the present worth of $769.60, due 3 years and 5 months hence ? 　　　　　　　Ans. $ 638.67.2.$\frac{48}{341}$.

5. What is the present worth of $986.40, due 7 years, 9 months and 20 days hence ? 　　　　Ans. $ 671.78.2.$\frac{58}{881}$.

6. How much grain must be sent to the miller, that a bushel of meal may be returned ; the miller taking $\frac{1}{15}$ part for toll ?
　　　　　　　　　　　　　　Ans. 34$\frac{2}{15}$ quarts.

7. What is the present worth of $678.75, due 3 years, 7 months hence, at 7½ per cent. ? 　.　　Ans. $ 534.97.5.$\frac{75}{203}$.

8. I have given my note for $1000, to be paid Dec. 18, 1835. What is the note worth June 7, 1834 ? 　Ans. $ 915.89.$\frac{461}{5551}$.

9. James has a note against Samuel for $715.50, dated Aug. 17, 1834, which becomes due, January 11, 1835. What ready money will pay the note Sept. 25, 1834 ? 　Ans. $ 703.07.8+

10. A has B's note, which becomes due Nov. 25, 1835, for $914.75. What is this note worth January 1, 1835 ?
　　　　　　　　　　　　　　Ans. $ 867.88.4+

11. A merchant has given two notes ; the first for $79.87, to be paid January 21, 1836; the second $87.75, to be paid Dec. 17, 1836. What ready money will discharge both notes Feb. 10, 1835 ? 　　　　　　　　　Ans. $ 154.54.4+

12. It being now Oct. 14, 1833, and A owing me $1728, to be paid Dec. 17, 1837 ; what ought I now to receive as an equivalent ? 　　　　　　　　Ans. $ 1381.84.7.$\frac{683}{2501}$.

13. Bought cloth in Boston at $5.00 per yard. What must be my "asking price," in order that I may *fall* on it 10 per cent. and still make 10 per cent. on my purchase ? 　Ans. $6.11¼.

14. James Ober owes Samuel Hall as follows: $365.87, to be paid Dec. 19, 1835 ; $161.15, to be paid July 16, 1836 ; $112.50, to be paid June 23, 1834 ; $96.81, to be paid April 19, 1838. What should he receive as an equivalent, Jan. 1, 1834 ?
　　　　　　　　　　　　　　Ans. $653.40+

13*

15. A B, of Bradford, has sent shoes at several different times
to Spofford and Tileston, New York, as follows.
January 16, 1834, were received shoes to the value of $865.00.
Feb. 17, 1834,   "   " $86.27.
March 29, 1834,   "   "   769.25.
May 25, 1834,   "   "   183.75.
June 19, 1834,   "   " $96.81.
The above were sold on 6 months' credit.
A B has received of Spofford and Tileston, as follows: Sep-
tember 1, 1834, $1000; Oct. 19, 1834, $375.25; Nov. 15, 1834,
$681,29; Dec. 8, 1834, $100; March 12, 1835, $275.28. Re-
quired the balance at the time of settlement, Sept. 9, 1835.
Ans. Spofford and Tileston owe to A B, $195.51+

SECTION XXXIV.

COMMISSION AND BROKERAGE.

COMMISSION AND BROKERAGE are compensations made to factors, brokers, and other agents, for their services, either for
buying or selling goods.

NOTE.—A factor is an agent, employed by merchants, *residing in other
places*, to buy and sell, and to transact business on their account. A Broker is
an agent employed by merchants to transact business.

RULE.

The method of operation is the same as in interest.

EXAMPLES.

1. My agent in New Orleans, has purchased cotton, on my
account, to the amount of $18768; what is his commission, at
1¾ per cent. ?   Ans. $328.44.
2. If a broker sells goods for me to the amount of $896, what
is his commission, at 2 per cent. ?   Ans. $17.92.
3. My factor, in London, writes that he has purchased for me
to the amount of 395£. 15s. 5d.; what is his commission, at 2¼
per cent. ?
Ans. 8£. 18s. 1$\frac{69}{100}$d.

NOTE.—As his commission is to be taken out of the sum *remitted*, he will
not receive 4 dollars on every 100, but 4 on every 104; that is, he will receive $\frac{4}{104}$.

4. A factor receives $1976, which he is to lay out for goods ; having deducted his commission of 4 per cent. how much will remain to be laid out ?   Ans. $1900.

5. I have remitted to my correspondent a certain sum of money, which he is to lay out for me in iron, and having reserved to himself 2½ per cent. on the purchase, which amounted to $90, he buys for me the iron at $95, per ton. Required the sum remitted, and the quantity of iron purchased.

Ans. $\begin{cases} \text{Sum remitted, \$3690.} \\ \text{Iron purchased, 37 T. 17 cwt. 3 qr. 16} \frac{4}{11}. \end{cases}$

SECTION XXXV

STOCKS.

STOCKS is a general name used for funds, established by government, or individuals, in their corporate capacity, the value of which is often variable.

When stocks will bring more in the market than their original cost, their value is said to be *above par*, and when, from any circumstance, their value is less than the original cost, they are said to be *below par*.

The method for computation is the same as in interest.

EXAMPLES.

1. What is the value of $24360 of the National Bank stock, at 135 per cent. ?   24360 × 1.35 = $32886 Answer.

2. Sold 15 shares, $100 each, of the Boston Bank, at 13 per cent. advance. To what did they amount ?   Ans. $1695.

3. What must I give for 12 shares in the Haverhill Bank, at 15 per cent. advance, shares being $100 each ?
                              Ans. $1380.

4. What is the purchase of 1058£. 12s. bank stock, at 115¾ per cent. ?   Ans. 1225£. 6s. $7\frac{2}{35}$d.

5. Sold 30 shares, $100 each, in the Boston and Providence Railroad, at 8¾ per cent. advance. To what did they amount ?
                          Ans. $3262.50.

6. What is the value of 10 shares in the Philadelphia and Trenton Railroad stock, at 85 per cent., original shares being $100 ?   Ans. $850.

7. What must be given for 5 shares of the stock in the Ocean Insurance Company, at 7 per cent. advance ; the original shares being $100 ?   Ans. $535.

SECTION XXXVI.

INSURANCE AND POLICIES.

INSURANCE is a security, by paying a certain sum, to indemnify the secured against such losses as shall be specified in the policy.

Policy is the name of the writ, or instrument, by which the contract or indemnity is effected between the parties.

NOTE.—If the average loss does not exceed 5 per cent. the *underwriters* are free, and the insured bears the loss himself.

RULE.

The same as in interest and discount.

EXAMPLES.

1. What is the premium on $896, at 12 per cent. ?
Ans. $ 107.52.
2. What is the premium on $850, at 18½ per cent. ?
Ans. $ 157.25.
3. What is the premium of insuring $9870, at 7 per cent. ?
Ans. $ 690.90.
4. What sum will be secured for a policy of $1728, deducting 15 per cent. ?   Ans. $ 1468.80.
5. What sum must a policy be taken out for, to cover $ 2475, when the premium is 10 per cent ?   Ans. $ 2750.

NOTE.—As 10 per cent. is already taken out, the sum covered must be $\frac{90}{100}$ of the policy.

6. A certain company own a cotton factory, valued at $ 26250. For what sum must a policy be taken out, to cover the whole property, at 12½ per cent. ?   Ans. $ 30.000.
7. If a policy be taken out, for $ 3600, at 40 per cent. what is the sum covered ?   Ans. $2160.

NOTE.—It is evident, that, if 40 per cent. be taken from any sum, 60 per cent. will be left.

8. If a policy be taken out for $ 600, at 10 per cent., what is the sum covered ?   Ans. $ 540.
9. A merchant adventured $ 1000 from Boston to New Orleans, at 3 per cent. ; from thence to Chili, at 5 per cent. ; and thence to Canton, at 6 per cent. ; and from thence to Boston, at 7 per cent. For what sum must he take out a policy, to cover his adventure the voyage round ?   Ans. $ 1241.34.8+.

SECTION XXXVII.

BANKING.

A BANK is a place of deposite for money, which is usually divided into shares, and owned by persons called *stockholders*. Its concerns are managed by a Board of Directors. It issues notes or bills of its own, intended to be a circulating medium of exchange or currency, instead of gold and silver. These bills are obtained from the bank by loans. When money is hired from a bank, it is the usual custom to deduct the interest at the time of receiving it. A man, therefore, who hires money from a bank, gives his note for a sum as much larger than he receives, as is the interest of the note for the given time. If a man therefore gives his note for $100, payable in 63 days, he *receives* only $98.95.

A promissory note is said to be *discounted*, when it is received at a bank as a security for money taken from it; and the interest deducted is the discount. The interest on every note discounted at a bank, is computed for 3 days more than the time specified in the note; that is, if the note is payable in 60 days, the interest is taken for 63 days; for the law allows 3 days to the debtor, after the time has expired for payment, which are called *days of grace.*

The rule for computing the discount is the same as in simple interest.

EXAMPLES.

1. What is the bank discount on $476, for 30 days and grace ?
　　　　　　　　　　　　　　　　　　　　Ans. $2.61.8.
2. What is the bank discount on $1000, for 60 days and grace ?　　　　　　　　　　　　　　　　Ans. $10.50.
3. What is the bank discount on $7800, for 90 days and grace ?　　　　　　　　　　　　　　　Ans. $120.90.
4. What is the bank discount on $8000, for 60 days and grace ?　　　　　　　　　　　　　　　Ans. $84.00.
5. How much money should be received on a note for $760, payable in 5 months, discounted at a bank when the interest is 6 per cent. ?　　　　　　　　　　　Ans. $740.62.
6. What sum is paid at a bank for a note of $1728, payable in 3 months ?　　　　　　　　　Ans. $1701.21.6.
7. A merchant sold a cargo of hemp for $7860, for which he received a note, payable in 6 months. How much money will he receive at a bank for this note ?　　　　Ans. $7620.27.

8. A merchant bought 450 quintals of fish at $3.50 cash, and sold them immediately for $4.00 on 6 months' credit, for which he received a note. If he should get this discounted at a bank, what will he gain on the fish ?   Ans. $170.10.

SECTION XXXVIII.

BARTER.

BARTER is the exchange of one kind of merchandise for another, without loss to either party.

Questions in this rule are solved, by finding what quantity of goods, at a given price of one kind, are equal in value to another kind of goods, whose price is also given.

EXAMPLES.

1. How much sugar at $12\frac{1}{2}$ cents per lb. must be given in barter, for 760 lbs. of raisins at 8 cents per lb. ?   Ans. $486\frac{2}{5}$ lbs.

2. What quantity of coffee at 17 cents per lb. must be given in barter for 760 lbs. of tea, at $62\frac{1}{2}$ cts. per lb. ?

Ans. $2794\frac{2}{17}$ lbs.

3. A merchant delivered 3 hogsheads of wine at $1.10 per gal. for 126 yds. of cloth; what was the cloth per yd. ?   Ans. $1.65.

4. A has 12 cwt. of sugar, worth 8 cents per lb. for which B gave him $1\frac{3}{4}$ cwt. of cinnamon ; what did B value his cinnamon?

Ans. 0.54\frac{6}{7}$ per lb.

5. A had 41 cwt. of hops at $6.70 per cwt. for which B gave him 17 cwt. 3 qrs. 4 lbs. of prunes, and $88 ; what were the prunes valued at per lb. ?   Ans. $0.09\frac{371}{996}$.

6. A has sugar, which he barters with B for 4 cents per lb. more than it cost him, against tea, which cost B 40 cents per lb., but which he puts in barter at 50 cents. What did A's sugar cost him per lb. ?   Ans. $0.16.

7. How many staves at $25 per thousand, must a merchant receive for 15 hogsheads of wine, at $1.25 per gallon ?

Ans. $47\frac{1}{4}$ M. staves.

8. Q has 670 bushels of oats, which cost him 35 cents per bushel ; these he barters with Z at 50 cents per bushel for flour, that cost Z $5.00 per barrel. What is the bartering price of the flour, and how much will Q receive ?

Ans. $46\frac{9}{10}$ barrels of flour at 7.14\frac{2}{7}$ per barrel.

SECTION XXXIX.

COMPOUND INTEREST.

THE law specifies, that the borrower of money shall pay a certain number of dollars, called *per cent.*, for the use of one hundred dollars for a year. Now, if this *borrower* does not pay to the *lender* this *per cent.* at the *end* of the year, it is no more than just, that he should pay interest for the use of it, so long as he shall keep it in his possession ; and this is called *Compound Interest.*

1. What is the compound interest of $300 for 3 years ?

Ans. $57.30.4.

```
   300 = Principal.
  1.06
  18.00 = Interest for 1 year.
   300
  318.00 = Amount for 1 year.
   1.06
  19.08 = Interest for second year.
   318
  337.08 = Amount for 2 years.
    1.06
  20.22.48 = Interest for third year.
  337.08
  357.30.48 = Amount for 3 years.
  300
$57,30.48 = Compound interest for 3 years.
```

SECOND METHOD.

```
5 = 1/20 )300
1 = 1/5 ) 15
          3
5 = 1/20 )318
1 = 1/5 ) 15.90
          3.18
5 = 1/20 )337.08
1 = 1/5 ) 16.85.4
          3.37.0
        357.30.4
        300.
       $57.30.4
```

NOTE. — 5 per cent is $\frac{1}{20}$ of the principal, and 1 per cent. is $\frac{1}{5}$ of 5 per cent.

Hence we see the propriety of the following

RULE.

Find the interest of the given sum for one year, and add it to the principal ; then find the interest of this amount for the next year; and so continue until the time of settlement. Subtract the principal from the last amount, and the remainder is the compound interest.

2. What is the amount of $500 for 3 years ? Ans. $595.50.8

3. What is the compound interest of $345 for 10 years ?
Ans. $272.84.2

4. What is the compound interest of $316 for 3 years, 4 months, and 18 days ? Ans. 69.01.7.

By the aid of the following Table, calculations are more easily effected, than by the preceding rule.

TABLE,

Showing the amount of 1 dollar, or pound, for any number of years under 50, at 5 and 6 per cent., compound interest.

Years.	5 per cent.	6 per cent.	Years.	5 per cent.	6 per cent
1	1.050000	1.060000	14	1.979931	2.260903
2	1.102500	1.123600	15	2.078928	2.396558
3	1.157625	1.191016	16	2.182874	2.540351
4	1.215506	1.262476	17	2.292018	2.692772
5	1.276281	1.338225	18	2.406619	2.854339
6	1.340095	1.418519	19	2.526950	3.025599
7	1.407100	1.503630	20	2.653297	3.207135
8	1.477455	1.593848	21	2.785962	3.399563
9	1.551328	1.689478	22	2.925260	3.603537
10	1.628894	1.790847	23	3.071523	3.819749
11	1.710339	1.898298	24	3.225099	4.048934
12	1.795856	2.012196	25	3.386354	4.291870
13	1.885649	2.132928	26	3.555672	4.549382

Years.	5 per cent.	6 per cent.	Years.	5 per cent.	6 per cent
27	3.733456	4.822345	39	6.704751	9.703507
28	3.920129	5.111686	40	7.039988	10.285717
29	4.116135	5.418387	41	7.391988	10.902860
30	4.321942	5.743491	42	7.761587	11.557032
31	4.538039	6.088100	43	8.149666	12.250454
32	4.764941	6.453386	44	8.557150	12.985431
33	5.003188	6.840589	45	8.985007	13.762610
34	5.253347	7.251025	46	9.434258	14.588367
35	5.516015	7.686086	47	9.905971	15.466669
36	5.791810	8.147252	48	10.401269	16.391489
37	6.081406	8.636087	49	10.921383	17.374978
38	6.385477	9.154252	50	11.467369	18.217477

To perform questions by the Table, multiply the amount of one dollar by the ... rate and time found in the Table, by the principal, and from the product ... the principal, and the remainder is the compound interest.

NOTE. — If there be months and days, find the amount for them on the number taken from the Table, *before* it is multiplied by the principal.

EXAMPLES.

5. What is the compound interest of $360, for 5 years, 6 months, and 24 days ?   Ans. $138.14.

          1.338225 = Amount of one dollar for 5 years.
              .034 = Ratio for 6 months and 24 days.

              5352900
              4014675

            .045499650
            1.338225

            1.383724650 = Amount of 1 dollar for 5 yrs. 6 mo. 24 days.
                   360

            83023479000 = Amount of principal for 5 yrs. 6 mo. 24 da.
            415117395

            498.14.0674000
            360

$138.14 = Compound interest of the principal for do.

6. What is the interest of $890, for 30 years ?
                                            Ans. $4221.70.6.
7. What is the amount of $480, for 40 years ?
                                            Ans. $4937.14.4.
8. What is the interest of $300 for 10 years, 7 months and 15 days ?   Ans. $257.40.1.
9. What is the amount of $586, for 12 years, 1 month, and 29 days, at 5 per cent. ?   Ans. $1060.99.5.
10. The probable number of blacks, at this time, (1835) in the United States, is 2.500.000. Now, supposing their increase to be 25 per cent. for every 10 years, what will be their number in the year 1935 ?   Ans. 23.283.057.
11. Supposing the annual increase of the people of color to be 2 per cent. ; how many must be sent out of the country, each year, that their numbers might not increase ?   Ans. 50.000.
12. What is the amount of $900 for 7 years at 5 per cent. ?
                                            Ans. $1266.39.
13. What is the interest of $350 for 3 years, 3 months, 24 days ?   Ans. $74.77.5+
14. What is the interest of $970 for 2 years, 9 months, and 24 days ?   Ans. $ 173.29.5.
14

To find the amount of a note by compound interest, when there have been partial payments.

Find the amount of the principal, and from it subtract the amount of the endorsements.

EXAMPLES.

13. A., by his note, dated January 1, 1830, promises to pay B. $500 on demand.

On this note are the following endorsements. July 16, 1830, received two hundred dollars. August 21, 1831, received two hundred dollars. December 1, 1832, received one hundred dollars.

What was the balance Sept. 1, 1834 ?   Ans. $52.73.0.

($100.)   Boston, Sept. 25, 1833.

For value received, I promise Peter Absalom, to pay him, or order, on demand, one hundred dollars, with interest after six months.   J. P. Jay.

On this note are the following endorsements. June 11, 1834, received fifty dollars. Sept. 25, 1834, received fifty dollars.

What was due August 25, 1835 ?   Ans. $2.24.7.

SECTION XL.

PRACTICE.

PRACTICE is an expeditious way of performing questions in Compound Multiplication and Proportion.

RULE.

Assume the price at some unit higher than the given price; that is, if the price be pence, or pence and farthings, assume the price at a shilling a yard, or pound, &c. If the price be in shillings, or shillings and pence, &c., assume the price at a pound, a yard, &c.; then take the aliquot parts of a pound.

TABLE,

Showing the aliquot parts of money and weights.

Parts of a £.

s.	d.	is	
10		is	1/2
6	8	"	1/3
5		"	1/4
4		"	1/5
3	4	"	1/6
2	6	"	1/8
2		"	1/10
1	8	"	1/12
1		"	1/20

Parts of a shilling.

d	is	
6	is	1/2
4	"	1/3
3	"	1/4
2	"	1/6
1½	"	1/8
1	"	1/12

Parts of a ton.

cwt.	qr.	is	
10		is	1/2
5		"	1/4
4		"	1/5
2	2	"	1/8
2		"	1/10
1		"	1/20

Parts of a cwt.

qrs.	lb.	is	
2		is	1/2
1		"	1/4
	16	"	1/7
	14	"	1/8
	8	"	1/14
	7	"	1/16
	4	"	1/28
	2	"	1/56

Parts of half a cwt.

lb.	is	
28	is	1/2
14	"	1/4
8	"	1/7
7	"	1/8
4	"	1/14
3½	"	1/16
2	"	1/28

Parts of a ¼ cwt.

lb.	is	
14	is	1/2
7	"	1/4
4	"	1/7
3½	"	1/8
2	"	1/14
1	"	1/28

EXAMPLES.

1. What will 368 yards of ribbon cost, at 6 pence a yard ?

Ans. 9£. 4s.

6d. = ½)368s.

20)184

9£. 4s.

We assume the price at a shilling a yard, and then say, if 368 shillings be the price, at a shilling a yard, at 6 pence it must be half as much, viz. 184 shillings. We then reduce the shillings to pounds.

2. What will 4785 yards of cotton cost, at 8 pence per yard ?

Ans. 159£. 10s. 0d.

6d. = ½)4785s. = Price at 1 shilling,
2d. = ⅓)2392 — 6 = Price at 6d.
　　　797 — 6 = Price at 2d.
20)3190 = 0
159£. 10s. 0d. = Price at 8d.

Having found the price at 6d. as before, we find it for the 2d., by saying, that 2d. is ⅓ of 6d.

3. What is the interest of $368, at 15 per cent. ?

Ans. $55.20.

10 per cent. = $\frac{1}{10}$)368

5. = $\frac{1}{2}$)36.80   10 per cent. is $\frac{1}{10}$ of the prin-
   18.40   cipal, and 5 per cent. is $\frac{1}{2}$ of
   _____   10 per cent.
   $55.20

4. What is the value of 17 acres, 3 roods, and 35 rods of land, at $80 per acre ?   Ans. $1437.50.

80
17
—
560
80
____

1360 = Price of 17 A.   By dividing the price
2 R. = $\frac{1}{2}$)   40 = do. of 2 R.   of 1 acre by 2, we obtain
1 R. = $\frac{1}{2}$)   20 = do. of 1 R.   the price of 2 R.; and by
20 r. = $\frac{1}{2}$)   10 = do. of 20 r.   halving this, we find the
10 r. = $\frac{1}{2}$)   5 = do. of 10 r.   price of 1 R.; and as 20
5 r. = $\frac{1}{2}$)   2.50 = do. of 5 r.   rods is half of a rood, its
value will be one half;
and in the same manner
10 rods will be half the
price of 20 rods, and 5
rods will be half the price
of 10 r.

$1437.50 = Price of 17 A. 3 R. 35rd.

5. What cost 14 tons, 15 cwt. 3 qr. 21 lbs. of iron, at $120 per ton ?   Ans. $1775.62.5.

120
14
____
1680.00 = price of 14 T.
10 cwt. = $\frac{1}{2}$)   60.00 = do. of 10 cwt.
_____
1740.00
5 cwt. = $\frac{1}{2}$)   30.00 = do. of 5 cwt.
2 qr. = $\frac{1}{10}$)   3.00 = do. of 2 qr.
1 qr. = $\frac{1}{2}$)   1.50 = do. of 1 qr.
14 lbs. = $\frac{1}{2}$)   75 = do. of 14 lbs.
7 lbs. = $\frac{1}{2}$)   375 = do. of 7 lbs.
_____

$1775.62.5 = do. 14 T. 15 cwt. 3 gr. 21 lb.

6. What cost 387 lbs. of sugar, at 9 pence a pound ?

Ans. 14£. 10s. 3d.

7. What cost 498 lbs. of green tea, at 2 shillings and 6 pence per pound ?   Ans. 62£. 5s. 0d.

8. What cost 384 yards of cloth, at 4 shillings and 9 pence a yard ?   Ans. 91£. 4s. 0d.

9. What cost 714 yards of broadcloth, at 15 shillings and 6 pence per yard ?   Ans. 553£. 7s. 0d.

10. What cost 16 cwt. 3 quarters, and 10 lbs. of copperas, at 12.50 per cwt. ?   Ans. $42.09.8.

11. What cost 27 cwt. 1 quarter, 21 lbs. of coffee, at $14 per cwt. ?   Ans. $384.12½.

12. What cost 7 tons, 13 cwt., 2 quarters, and 7 lbs. of hay, at $24.60 per ton ?   ` ,   Ans. $183.88.1⅞.

13. If 1 acre of land cost $80.50, what will 25 acres, 2 roods, and 35 rods cost ?   Ans. $2070.35.9⅜.

14. If 1 acre cost $32.32, what will 51 acres, 0 R. 15 rods cost ?   Ans. $1651.35.

15. If 1 yard of cloth cost $5.60, what will 7 yards, 3 qrs. and 2 nails cost ?   Ans. $44.10.

16. What is the premium on $6780, at 12½ per cent. ?
                                                            Ans. $847.50.

17. What is the interest of $1728, for 5 years, 7 months, and 20 days ?   Ans. $584.64.

18. What will 19 tons, 19 cwt. 3 qr. 27½ lbs. of copperas cost at 19£. 19s. 11¾d. per ton ?   Ans. 399£. 19s. 5¼¼¼¼d.

SECTION XLI.

EQUATION OF PAYMENTS.

When several sums of money, to be paid at different times, are reduced to a *mean* time, for the payment of the whole, without gain or loss to the debtor or creditor, it is called Equation of Payments.

EXAMPLES.

1. A owes B $19 ; $5 of which is to be paid in 6 months, $6 in 7 months, and $8 in 10 months. What is the equated time for the payment of the whole ?

OPERATION.

$5 × 6 = 30   By analysis. $5 for 6 months is
$6 × 7 = 42   the same as $1 for 30 months ; and
$8 × 10 = 80 $6 for 7 months, is the same as $1
—   —   for 42 months ; and $8 for 10 months,
19   19)152(8 months.   is the same as $1 for 80 months ;
        152   therefore $1, for 30 + 42 + 80 = 152

months, is the same as $ 5 for 6 months, $6 for 7 months, and $8 for 10 months ; but $5, $6, and $8 are $19 ; therefore $1 for 152 months, is the same as $19 for ¹⁄₁₉ of 152 months, which is 8 months, as before. Hence the propriety of the following
14*

RULE. *

Multiply each payment by the time, at which it is due; then divide the sum of the products by the sum of the payments, and the quotient will be the true time required.

2. A owes B $300, of which $50 is to be paid in 2 months, $100 in 5 months, and the remainder in 8 months. What is the equated time for the whole sum ?   - Ans. 6 months.

3. There is owing to a merchant $1000; $200 of it is to be paid in 3 months, $300 in 5 months, and the remainder in 10 months. What is the equated time for the payment of the whole sum ?   Ans. 7 months, 3 days.

4. A owes B $150, $50 to be paid in 4 months, and $100 in 8 months. B owes A $250 to be paid in 10 months. It is agreed between them, that A shall make present payment of his whole debt, and that B shall pay his so much sooner as to balance the favor. I demand the time at which B must pay the $250 ?
          Ans. 6 months.

5. A merchant has $144 due him, to be paid in 7 months, but the debtor agrees to pay ⅓ ready money, and ⅓ in 4 months. What time should be allowed him to pay the remainder ?
          Ans. 2 years, 10 months.

6. There is due to a merchant $800, ⅓ of which is to be paid in 2 months, ⅓ in 3 months, and the remainder in six months; but the debtor agrees to pay ½ *down.* How long may the debtor retain the other half so that neither party may sustain loss ?
          Ans. 8¾ months.

7. I have purchased goods of A. B. at sundry times and on various terms of credit, as by the statement annexed. When is the *medium* time of payment ?

Jan.   1, a bill amounting to $375.50 on 4 months' credit.
   "   20, do.   do.   168.75 on 5 months' credit.
Feb.   4, do.   do.   386.25 on 4 months' credit.
March 11, do.   do.   . 144.60 on 5 months' credit.
April   7, do.   do.   386.90 on 3 months' credit.

---

* This is the rule usually adopted by merchants, but it is not *perfectly* correct; for if I owe a man $200, and $100 of which I was to pay *down,* and the other $100 in two years, the equated time for the payment of both sums would be one year. It is evident that for deferring the payment of the first $100 for 1 year, I ought to pay the amount of $100 for that time, which is $106; but for the other $100, which I pay a year *before* it is due, I ought to pay the *present worth* of $100, which is $94.33⅓, whereas, by Equation of Payments, I only pay $200. *Strict justice* would therefore demand, that *interest* should be required on all sums from the time they become due, until the time of payment; and the *present worth* of all sums, paid *before* they are due. The better rule would be to find the present worth on each of the sums due, and then find 'n what time the sum of these present worths would amount to the payments.

Form of statement.

Due May 1, $875.50
    June 20, 168.75 × 50 = 843750
    June 4, 886.25 × 34 = 1813250
    Aug. 11, 144.60 × 102 = 1474920
    July 7, 386.90 × 67 = 2592230
           ——————
         1462.00   )6224150(42$\frac{1672}{1462}$ Answer.
                       584800

                       376150
                       292400

                        83750

The medium time of payment will therefore be 42$\frac{1672}{1462}$ days, that is, 43 days from May 1, which will be June 12.

8. I have sold to C. D. several parcels of goods, at sundry times, and on various terms of credit, as by the statement annexed.

January 1, a bill amounting to $600 on 4 months' credit.
Feb.   7, do.   do.   370 on 5 months' credit.
March 15, do.   do.   560 on 4 months' credit.
April 20, do.   do.   420 on 6 months' credit.

When is the equated time for the payment of all the bills?
                                    Ans. July 11.

SECTION XLII.

CUSTOM HOUSE BUSINESS.

In every port of the United States, where merchandise is either exported or imported, there is an establishment called a Custom House. Connected with this, are certain officers appointed by government, called custom house officers, whose business is to collect the duties on various kinds of merchandise, &c. imported into the United States.

The following article on allowances, &c. was very politely furnished the author by the officers of the Boston Custom House, and may therefore be relied on as perfectly correct.

Allowances.

Draft is an allowance made by the officers of the United States Government in the collection of duties on merchandise, liable to a specific duty, and ascertained by weight, and is also given by the usage of merchants in buying and selling. It is a deduction from the actual gross weight of the article paying duty by the pound or sold by weight.

For example, a box of sugar *actually* weighs 500 pounds. The draft upon this weight is 4 pounds.

500 gross.
4 draft.

496 difference. Upon this difference is made a further allowance of fifteen per cent. as *tare,* or as the actual weight of the box before the sugar was put into it. This tare is allowed by the government in the collection of the duty, and by the merchant in buying and selling. Take, then, the box of sugar, say

500 lbs. gross.
4 draft.

496 difference.
74 tare

422 net weight, upon which a duty is paid to the government, or price is paid to the merchant in his sale. This tare of 74 pounds or 15 per cent., is usually more than the *actual tare,* but is assumed as the *probable or actual tare,* by reason of the impossibility of starting every box to ascertain the *actual* weight of the sugar, and the *actual* weight of the box which contains it. This *tare* is sufficiently correct for the collector of the duty, and the merchant who deals in the article. It is intended to be a liberal allowance, and varies but little from the *actual* tare.

Drafts allowed at the Custom House in the collection of duties, and by the merchants in their purchases and sales are as follow—

Allowance for draft.

Draft is another name for *Tret,* which is an allowance in weight for waste.

lbs.				lbs.
On 112				1
Above 112 and not exceeding			224	2
" 224	"	"	336	3
" 336	"	"	1120	4
" 1120	"	"	2016	7
" 2016	"	"		9

EXPLANATION.—Many articles of merchandise are weighed separately ; for example,—boxes and casks of sugar, chests of tea and indigo. Upon each box or cask, or chest, an allowance should be made for draft, according to its weight, as by the above rule. Bags of sugar and coffee, or bars of iron and bundles of steel might be weighed together ; say, 10 bags of coffee at one draft might weigh 1121 pounds; from this gross weight must be deducted 7 pounds as draft; 35 bars of Russia iron might be weighed at one draft—weight 2250 pounds, upon which would be an allowance of 9 pounds draft, and by law and usage there can be no greater allowance than 9 pounds for draft. A greater or less number of bags of coffee or bars of iron, or any other article of merchandise is weighed, and the deduction is according to the weight of each draft. An old rule, and probably a better one among merchants, was the allowance of ½ per cent. on the gross weight of all merchandise weighed, as draft.

Allowance for leakage.

Two per cent. is allowed on the gauge of ale, beer, porter, brandy, gin, molasses, oil, wine and rum, and other liquors in casks; besides the *real* wants of the cask; for example,—a cask of molasses may gauge 140 gallons, gross gauge,—from this first deduct 5 gallons, the *actual* wants, or the quantity necessary to fill the cask—we have,

$$140 \text{ gross.}$$
$$5 \text{ out.}$$
$$\overline{135 \text{ difference.}}$$
$$3\text{—2 per cent. for leakage.}$$
$$\overline{132 \text{ gallons net.}}$$

Tare is an allowance made for the *actual or supposed weight* of the cask, box, case or bag, which contains the article of merchandise.

The usage of merchants is in conformity with the law and usage of the officers of the customs in their allowance for tare, directed by law, or found to be correct by their examination and experience.

The following table gives the legal tare; and when the law determines no tare, the *actual* tare is stated as usually allowed, and by experience found to be correct.

Tares according to law and actual weight.

A

Almonds, in bags, 3 per cent.
Alum, casks, 12 per cent.

B

Bristles, Cronstadt 12 per cent. *actual.*
   do. Archangel 14 do. do.
Beef, jerk'd hhds. 112 lbs. per bhd.
  do. do. drums 70 "

C

Cordage, mats	1¼ per cent.	*actual*	
Camphor, crude, in tubs	35	do.	do.
do. refined			
Candles, boxes	8	do.	*legal.*
do. chests, 160 lbs.	20	do.	*actual.*
Cinnamon, do. 16 " per chest.			
do. mats	8	do.	*legal.*
Cloves, casks	15	do.	*actual.*
Cocoa, in bags	1	do.	*legal, actual* 2 per ct.
do. casks	10	do.	do.
do. zeroons	10	do.	do.
Chocolate, boxes	10	do.	do.
Coffee, bags, W. India	2	do.	do.
do. East India, grass bags	2 lbs. per bag		*actual.*
do. bales	3 per cent.	*legal.*	
do. casks	12	do.	do.
Cotton, bales	2	do.	do.
do. zeroons	6	do.	do.
Currants, casks	12	do.	*actual.*
Cheese, hampers or baskets	10	do.	do.
do. boxes	20	do.	do.
Copperas, casks	12	do.	do.
Candy sugar, in baskets	5	do.	do.
do. do. in boxes	10	do.	*legal.*
Corks, bales	15	and 20 lbs.	*actual.*

F

Figs, boxes	60 lbs.	9 lbs.	*actual.*
do. half boxes	30 "	5¼ "	do.
do. quarter do.	15 "	3¼ "	do.
do. drums 10 per cent.			*legal.*
Frails	5	do.	*actual.*

G

Glue, bales, Russia   5 lbs. *actual.*

I

Indigo, bags or mats	3 per cent.	*legal.*	
do. zeroons	10	do.	do.
do. barrels	12	do.	do.
do. other casks	15	do.	do.
do. cases	20	do.	do.

L

Lines, bales	2	do.	*actual.*
Looking-glasses, French 30 per cent.			*actual.*

M

Mace, casks or kegs 33 per cent. *actual.*

N

Nutmegs 12 per cent. *actual.*
Nails, casks 8 per cent. *legal.*

O

Ochre, French, casks 12 per cent. *actual.*

P

Pepper, bags	2 per cent. *legal.*	.
do. bales	5 do.	do.
do. casks	12 do.	do.
Pimento, bags	3 do.	do.
do. bales	5 do.	*actual.*
do. casks	15 do.	*legal.*

Prunes, boxes, 7 lbs. per box.

R

Raisins, Malaga, boxes	6 lbs. *actual.*
do. do. jars	5 " do.
do. do. casks	12 " do.
do. Smyrna do.	12 per cent. *actual.*

S

Sugar, Java, in willow baskets 60 lbs. per basket *actual.*

Sugar, bags or mats	5 per cent. *legal.*	
do. casks	12 do.	do.
do. boxes	15 do.	do.
do. canisters	35 do.	*actual.*
Sugar candy, baskets	5 do.	do.
do. do. boxes	10 do.	*legal.*
Soap, boxes	10 do.	do.

Soap, Marseilles, in boxes 12 per cent. *actual.*

MEMORANDUM.—Actual tare of soap more than legal tare. Invoice tare usually claimed at Custom House.

Salts, glauber, in casks	8 do.	*legal tare*
Shot, casks	3 do.	do.
Steel, bundles 2 lbs. per bundle		*actual.*
do. cases 60 lbs. per case		*actual.*

T

| Tea chests | 22 lbs. each. |
| " half chests | 14 " do. |

Ten caddy boxes   6 lbs. each.
Twine, casks   12 per cent. *legal.*
 do.   bales   3   do.   do.
Tallow, zeroons   10   do.   *actual.*
  "   casks   12   do.   do.

W

Wool, bales, Smyrna 10 lbs. per bale   *actual.*
 do.   do.   Hamburg 3 per cent.   do.
 do.   do.   South America 15 lbs. per bale do.

EXAMPLES.

1. Find the net weight of a hogshead of sugar, weighing gross 1228 lbs.   Ans. 1075 lbs.

OPERATION.

                1228 lbs. gross weight.
                   7 lbs. draft.
                ────
                1221
12 per cent. of 1221 lbs.=146 lbs. tare.
                ────
                1075 net weight.

NOTE. — For draft and tare, let the pupil examine the preceding tables.

2. Required the net weight of 6 boxes of sugar, weighing gross, as follows—No. 1, 450 lbs.

            No. 2, 470 do.   OPERATION.
            No. 3. 510 do.   2914 lbs. gross.
            No. 4, 496 do.   6×4=   24 draft.
            No. 5, 468 do.   ────
            No. 6, 520 do.   2890
            ────   tare 15 per cent. 433
gross, 2914 lbs.   ────
                                    2457 net weight.

3. What is the net weight of 4 chests of tea, which weigh as follows.   Ans. 384 lbs.

                              OPERATION.
    No. 1, 120 lbs.   480 gross.
    No. 2, 116 do.   8 draft.
    No. 3, 126 do.   ────
    No. 4, 118 do.   472
    ────   22×4   88 tare.
    480 lbs.   ────
                            384 net.

4. What is the net weight of 5
bags of pepper, weighing as follows ; 108 lbs., 112 lbs., 100 lbs.,
120 lbs., and 98 lbs. ?

               Ans. 524.   2 per cent.

               538 gross.
                 3 draft.

               535
               11 tare.
               ---
               524 net.

5. What is the net weight of
4 kegs of mace, weighing as
follows, 112 lbs., 120 lbs., 118
lbs., and 110 lbs. ?

             Ans. 305 lbs.   33 per cent.

             460 lbs. gross.
               5 lbs. draft.

             455
             150 tare.
             ---
             305 net.

NOTE. — In making allowances, if there be a fraction of more than half a
pound, 1 pound is added to the tare.

AMERICAN DUTIES.

The duties on merchandise imported into the United States
are either specific or ad valorem duties.

Specific duty is a certain sum paid on a ton, hundred weight,
pound, square yard, gallon, &c. ; but when the duty is a certain
per cent. on the actual cost of the goods in the country, from
which they are imported, it is called an *ad valorem duty.*

6. What is the duty on 6 hogsheads of sugar, weighing gross,
as follows ; No. 1, 1276 lbs., No. 2, 1280 lbs., No. 3, 1178 lbs.,
No. 4, 1378 lbs., No. 5, 1570 lbs., No. 6, 1338 lbs., duty 2¼ cents.
per lb. ?   Ans. $175.52.5

7. What is the duty on an invoice of woolen goods, which
cost in London 986£ sterling, at 44 per cent. ad valorem ?
                                  Ans. $1928.17.+

8. Required the duty on 5 pipes of Port wine, gross gauge as
follows ; No. 1. 176 gallons, No. 2, 145 gallons, No. 3, 128 gallons, No. 4, 148 gallons, No, 5, 150 gallons ; *wants* of each pipe,
4 gallons ; duty 15 cents per gallon.   Ans. $106.80.

9. Required the duty on a cargo of iron, weighing 270 tons,
at $30 per ton ?   Ans. 8100.

10. Compute the duty on 7890 pounds of tarred cordage, at 4
cents per pound.   Ans. $310.08.

11. What duty should be paid on 10 casks of nails, weighing
each 450 lbs. gross at 4 cents per lb ?   Ans. $164.12.

SECTION XLIII.

RATIO.

RATIO is the mutual relation, which one magnitude or number, has to another of the same kind, without the intervention of a third. The ratio which the first of any two numbers has to the second, may be expressed by dividing the first number by the second. Thus the ratio of 10 to $5 = \frac{10}{5} = 2$; and the ratio of 7 to $21 = \frac{7}{21} = \frac{1}{3}$. Quantities of different kinds have no ratio to each other; as no one would inquire how often 5 dollars were contained in 15 miles, or what part of 8 bushels were 2 minutes.

A ratio shows how many times one number or term, is contained in another.

NOTE.—The French mathematicians express the ratio, which one number has to another by dividing the consequent by the antecedent. The ratio of 10 to 5 they would write thus, $\frac{5}{10} = \frac{1}{2}$; and the ratio of 7 to 21, thus, $\frac{21}{7} = 3$.

It is doubtful whether this innovation is productive of any good.

SECTION XLIV.

PROPORTION.

PROPORTION is the likeness or equalities of ratios. Thus, because 5 has the same relation or ratio to 10 that 8 has to 16, we say such numbers are in *proportion* to each other, and are therefore called *proportionals*.

If any four numbers whatever be taken, the first is said to have the same ratio or relation to the second, that the third has to the fourth, when the first number or term contains the second, as many times as the third contains the fourth; or when the second contains the first, as many times as the fourth does the third. Thus, 8 has the same ratio to 4 that 12 has to 6; because 8 contains 4, as many times as 12 does 6. And 3 has the same relation to 9, that 4 has to 12; because 9 contains 3, as many times as 12 does 4. Ratios are represented by colons; and the equalities of ratios by double colons.

3 : 9 :: 8 : 24 is read thus; 3 has the same ratio or relation to 9, as 8 to 24. The first and third numbers of a proportion are called *antecedents*, and the second and fourth are called *consequents*; also, the first and fourth are called *extremes*, and the second and third are called *means*.

Whatever four numbers are proportionals, if their antecedents or consequents be multiplied or divided by the same numbers, they are still proportionals; and if the terms of one proportion be multiplied or divided by the corresponding term of another proportion, their products and quotients are still proportionals.

This will appear evident, from the various changes, that the following example admits.

$$4 : 8 :: 3 : 6 \text{ Directly.}$$
$$8 : 4 :: 6 : 3 \text{ By inversion.}$$
$$4 : 3 :: 8 : 6 \text{ By permutation.}$$
$$4+8 : 8 :: 3+6 : 6 \text{ By composition.}$$
$$4 : 4+8 :: 3 : 3+6 \text{ By composition.}$$
$$4 : 8-4 :: 3 : 6-3 \text{ By division.}$$
$$8-4 : 8 :: 6-3 : 6 \text{ By division.}$$
$$4\times4 : 8\times8 :: 3\times3 : 6\times6 \text{ By compound ratios.}$$
$$\tfrac{4}{8} : \tfrac{8}{8} :: \tfrac{3}{6} : \tfrac{6}{6} \text{ By division.}$$

That the product of the extremes is equal to that of the means, is evident from the following consideration. Let the following proportionals be taken. $12 : 3 :: 8 : 2$. From the definition of proportion, the first term contains the second, as many times as the third does the fourth; therefore, $\tfrac{12}{3}=\tfrac{8}{2}$, but $\tfrac{12}{3}=\tfrac{24}{6}$, and $\tfrac{8}{2}=\tfrac{24}{6}$; and if 24, the numerator of the first fraction, which is a substitute for the first term, be multiplied by 6, the denominator of the second fraction, and a substitute for the fourth term, the product will be the same as if 6, the denominator of the first fraction, and a substitute for the second term, be multiplied by 24, the numerator of the second fraction, and a substitute for the third term. Thus $24 \times 6 = 6 \times 24$. Therefore the product of the extremes is, in all cases, equal to that of the means.

If, then, one of the extremes be wanting, divide the product of the means by one of the extremes; or, if one of the means be wanting, divide the product of the extremes by one of the means.

To apply this, we will take the following question. If 5 yards of cloth cost $15, what will 7 yards cost? It is evident, that twice the quantity of cloth would cost twice the sum, and that three times the quantity, three times the sum, &c.— That is, the price will be in *proportion* to the quantity purchased. We then have three terms of a proportion given; one of the extremes and the two means, to find the other extreme.

Thus, $5 : 7 :: 15$. Therefore to find the other extreme by the rule above stated, we multiply the two means, 7 and 15, and divide their product by the extreme given, and the quotient is the extreme required. $7 \times 15 = 105$. $105 \div 5 = 21$ dollars, the answer required.

To perform this question by analysis, we reason thus. If 5 yards cost 15 dollars, 1 yard will cost one-fifth as much, which is 3 dollars; and if one yard cost 3 dollars, 7 yards will cost 7 times as much, which is 21 dollars.

From the illustrations above given, we deduce the following

RULE.*

State the question by making that number, which is of the same name or quality of the answer required the third term; then, if the answer required is to be greater than the third term, make the second term greater than the first; but if the answer is to be less than the third term, make the second less than the first.

Reduce the first and second terms to the lowest denomination, mentioned in either, and the third term to the lowest denomination mentioned in it.

Multiply the second and third terms together, and divide their product by the first, and the quotient is the answer in the same denomination to which the third is reduced.

If any thing remains after division, reduce it to the next lowest denomination, and divide as before.

If either of the terms consists of fractions, state the question, as in whole numbers, and reduce the mixed numbers to improper fractions, compound fractions to simple ones, and invert the first term, and then multiply the three terms continually together, and the product is the answer to the question. Or the fractions may be reduced to a common denominator; and their numerators may be used as whole numbers. For when fractions are reduced to a common denominator, their value is as their numerators.

* This rule was formerly divided into the Rule of Three Direct, and Rule of Three Inverse. The Rule of Three Direct, included those questions, where *more required more and less required less*, thus; If 5 lbs. of coffee cost 60 cents, what would be the value of 10 lbs., would be a question in the Rule of Three Direct, because the more coffee there was, the more money it would take to purchase it.

But if the question were thus. If 4 men can mow a certain field in 12 days, how long would it take 8 men; it would be in inverse, because the more men, the less would be the time to perform the labor, that is, *more would require less*.

The method for stating questions was this; To make that number which is the demand of the question, the third term, that which is of the same name the first, and that which is of the same name as the answer required, the second term.

If the question was direct, the first and second terms must be multiplied together, and their product divided by the first; but if it was inverse, the first and second terms must be multiplied together, and their product divided by the third.

NOTE. — In the Rule of Three, the second term is the quantity whose price is wanted; the third term is the value of the first term; when, therefore, the second term is multiplied by the third, the answer is as much more than it should be, as the first term is greater than unity; therefore, by dividing by the first term, we have the value of the quantity required. Or by multiplying the third, by the number of times which the second contains the first, will produce the answer.

NOTE 2. — The pupil should perform every question by analysis, previous to his performing it by Proportion.

EXAMPLES.

1. If a man travels 243 miles in 9 days, how far will he travel in 24 days ?   Ans. 648 m.

D. ds.   miles.

9 : 24 : : 243
     24
   ——
  972
  586
  ——
9)5832
  ——
Ans. = 648 m.

As the answer to the question must be in miles, we make the third term miles (243); and from the nature of the question, we know that he will travel farther in 24 days than in 9 days, we therefore place 24 as the larger of the two remaining terms in the *second place*, and the remaining number, 9 days, in the *first place*.

To perform this question by *analysis*, we proceed thus. If he travel 243 miles in 9 days, he will in one day travel ⅑ of 243 miles, which is 27 miles ; then, if he travel 27 miles, in one day, in 24 days he will travel 24 times as far, which is 648 miles, the answer as before.

2. If 17 yards of broadcloth cost $102.00, what will 7 yards cost ?   Ans. $42.00.

Yds. yds.   .$

17 : 7 : : 102
     7
   ——
17)714(42
  68
  ——
  34
  34

As the answer is to be in dollars, we make the third term dollars (102); and as 7 yards will not cost so much as 17 yards, we make the second term, (7) less than the first, (17.)

To perform this question by *analysis*, we reason thus. If 17 yards cost $102, one yard will cost 1/17 of $102, which is $6, and if one yard cost $6.00, 7 yards will cost 7 times as much, which is $42.00, the answer as before.

15*

3. If 3 men drink a barrel of cider in 24 days, how long would it take 9 men ?   Ans. 8 days.

Mn. m. ds.
9 : 3 :: 24
            3
            —
      9)72
      8 days.

As the answer is to be in days, we make 24 days the third term, and because 9 men will drink the cider in less time than 3 men, the second term will be less than the first.

To perform this question *analytically*, we would say, that if 3 men drink a barrel of cider in 24 days, it would take one man 3 times 24 days, which is 72 days ; and if one man drink a barrel of cider in 72 days, 9 men would drink it in ⅑ of 72 days, which is 8 days, as before.

NOTE. — Questions of this kind were formerly performed by the Rule of Three Inverse, where more required less, and less required more.

4. If 12 yards of cloth cost $48, what will 15 yards cost ?
        Ans. $60.00.
5. If 17 lbs. of sugar cost $1.19, what will 365 lbs. cost ?
        Ans. $25.55.
6. If 16 acres of land cost $720, what will 197 acres cost ?
        Ans. $8865.00.
7. If $8865 buy 197 acres, how many acres may be bought for $720 ?   Ans. 16 acres.
8. What will 84 hhds. of molasses cost, if 15 hhds. can be purchased for $175.95 ?   Ans. $985.32.
9. If $100 gain $6 in 12 months, how much would it gain in 40 months ?   Ans. $20.00.
10. If a certain vessel has provisions sufficient to last a crew of 10 men 45 days, how long would the provisions last, if the vessel were to ship 5 new hands ?   Ans. 30 days.
11. If 7 and 9 were 12, what, on the same supposition, would 8 and 4 be ?   Ans. 9.
12. If 9 men can perform a certain piece of labor in 17 days, how long would it take 3 men ?   Ans. 51 days.
13. If 3 men can perform a piece of labor in 51 days, how many must be added to the number to perform the labor in 17 days ?   Ans. 6 m.
14. How much in length, that is 5½ rods in breadth, is sufficient for an acre ?   Ans. 29 1/11 rods.
15. If 2 barrels of flour cost $12.00, what will 24 barrels cost ?   Ans. $144.00.
16. If 5 quintals of fish cost $16.25, what is the value of 75 quintals ?   Ans. $243.75.
17. If 2 cords of wood cost $11.50, what will 17 cords cost ?
        Ans. $97.75.

18. If 7 cwt. of iron cost $56.85, what will 49 cwt. cost ? Ans. $397.95.

19. If 5 acres of land be valued at $375.75, what would be the cost of 35 acres ? Ans. $2630.25.

20. If 7 pairs of shoes cost $10.50, how many pair will $52.50 buy ? Ans. 35 pairs.

21. If $4.75 are paid for 19 lbs. of salmon, how many pounds will $25.50 buy ? Ans. 102 lbs.

22. If a man travels 48 miles in 6 hours, how far will he travel in 24 hours ? Ans. 192 m.

23. If 8 men drink a barrel of cider in 24 days, how long would it last 3 men ? Ans. 64 ds.

24. If 7 ounces of silver is sufficient to make 17 tea spoons, how many may be made from 42 ounces ? Ans. 102 spoons.

25. If $100 gain $6 in a year, how much will $850 gain ? Ans. $51.00.

26. If $100 gain $6 in a year, how much would be sufficient to gain $32 in a year ? Ans. 533.33\frac{1}{3}$.

27. If 20 gallons of water weigh 167 lbs., what will 180 gallons weigh ? Ans. 1503 lbs.

28. If a staff 3 feet long, cast a shadow 2 feet, how high is that steeple whose shadow is 75 feet ? Ans. 112$\frac{1}{2}$ feet.

29. If 5$\frac{3}{8}$ cwt. be carried 36 miles for $4.75, how far might it be carried for $160 ? Ans. 1212$\frac{12}{19}$ m.

30. If 100 workmen, can finish a piece of work in 12 days, how many men are sufficient to finish the work in 8 days ? Ans. 150 men.

31. If $\frac{7}{12}$ of a yard cost $\frac{7}{20}$ of a dollar, what will $\frac{4}{5}$ of a yard cost ? Ans. $0.48.

Yd. yd. $

$\frac{7}{12} : \frac{4}{5} :: \frac{7}{20}$ Ans. $\frac{12}{7} \times \frac{4}{5} \times \frac{7}{20} = \frac{336}{700} = \frac{48}{100} =$ $0.48.

To perform this question by analysis, we would proceed thus, and say, If $\frac{7}{12}$ of a yard cost $\frac{7}{20}$ of a dollar, $\frac{1}{12}$ would cost $\frac{1}{7}$ of \$$\frac{7}{20}$, which is $\frac{1}{20}$; and if $\frac{1}{12}$ cost $\frac{1}{20}$, $\frac{12}{12}$ will cost 12 times $\frac{1}{20}$, which is $\frac{12}{20} = \frac{3}{5}$ of a dollar. And if one yard cost $\frac{3}{5}$ of a dollar, $\frac{1}{5}$ of a yard will cost $\frac{3}{25}$ of a dollar, and $\frac{4}{5}$ will cost 4 times as much, that is, it will cost 4 times $\frac{3}{25}$, which is $\frac{12}{25} =$ $0.48, as before.

32. If $\frac{1}{2}$ of a yard cost $\frac{3}{5}$ of a £., what will $\frac{7}{8}$ of a yard cost ? Ans. 1£. 1s. 0d.

33. If 4$\frac{1}{2}$ yards cost $9.75, what will 13$\frac{1}{2}$ yards cost ? Ans. $29.25.

34. How much in length that is 2$\frac{1}{2}$ inches wide will make a square foot ? Ans. 57$\frac{3}{5}$ inches.

35. If $\frac{7}{16}$ of a ship cost 51£, what are $\frac{3}{32}$ of her worth ? Ans. 10£. 18s. 6$\frac{6}{7}$d.

36. If a merchant bought a number of bales of velvet, each containing 129$\frac{17}{47}$ yards, at the rate of $7 per 5 yards, and sold them out again at the rate of $11 per 7 yards, and gained $200 by the bargain; how many bales were there ?   Ans. 9 bales.

37. If the moon moves 13° 10′ 35″ in one day, in what time does she perform one revolution ?   Ans. 27 ds. 7 h. 43 m.+

38. If 7 lbs. of sugar cost $\frac{3}{4}$ of a dollar, what is 12 lbs. worth ?
                                            Ans. 1.28\frac{4}{7}$.

39. If $1.75 will buy 7 lbs. of loaf sugar, how much will $213.50 buy ?   Ans. 7 cwt. 2 qrs. 14 lbs.

40. If 7 ounces of gold is worth 30£., what is the value of 7lbs. 11oz. ?   Ans. 407£. 2s. 10$\frac{2}{7}$.

41. My friend borrowed of me $500 for 6 months, promising me like favor ; soon after I had occasion for $600 ; how long should I keep it to receive full compensation for the kindness ?
                                            Ans. 5 months.

42. If the penny loaf weighs 7 oz. when flour is $8 per barrel, how much should it weigh when flour is $7.50?
                                            Ans. 7$\frac{7}{15}$ ounces.

43. If a regiment of soldiers consisting of 1000 men, were to be clothed, each suit containing 3$\frac{1}{4}$ yards of cloth, that is 1$\frac{1}{2}$ yard wide, and to be lined with flannel 1$\frac{1}{4}$ yard wide, how many yards will it take to line the whole ?   Ans. 5625 yds.

44. If 9$\frac{3}{5}$ yards of broadcloth cost 11\frac{1}{2}$, what will 16$\frac{14}{103}$, ells English cost ?   Ans. $24.

45. A merchant failing in trade, owes in all $17280 ; his effects are sold for $15120. What does he pay on a dollar ; and what does A receive, to whom he owes $5670 ?
                          Ans. He pays 0.87\frac{1}{2}$ on the dollar.
                          A receives $4961.25.

46. Bought in London 57 yards of broadcloth for 49 guineas, 28 shillings each, what did it cost per ell English.
                                   Ans. 1£. 10s. 1$\frac{1}{15}$d.

47. Bought a cask of wine at $1.15 per gallon, for $100 ; how much did it contain ?   Ans. 86 gal. 3 qts. 1$\frac{16}{23}$ pts.

48. A merchant bought 9 packages of cloth for $34560, each package containing 8 parcels, each parcel 12 pieces, and each piece 20 yards ; what was the price per yard ?   Ans. $2.00.

49. If 75 gallons of water fall into a cistern containing 500 gallons, and by a pipe in the cistern 40 gallons run out in an hour ; in what time will it be filled ?   Ans. 14h. 17m. 8$\frac{4}{7}$ sec.

50. How many dozen pair of gloves, at $0.56 per pair, may be bought for $120.96 ?   Ans. 18 doz.

51. A certain cistern has three pipes, the first will empty it in 20 minutes, the second in 40 minutes, and the third in 75 minutes ; in what time would they all empty it ?
                                   Ans. 11 m. 19$\frac{13}{43}$ sec.

52. A can mow a certain field in 5 days, and B can mow it in 6 days ; in what time would they both mow it ?

Ans. $2\frac{8}{11}$ days.

53. A wall which was to be built 32 feet high, was raised 8 feet by 6 men in 12 days ; how many men must be employed to finish the wall in 6 days ?   Ans. 36 men.

54. A can build a boat in 20 days, but with the assistance of C, he can do it in 12 days ; in what time would C do it himself?

Ans. 30 days.

55. In a fort there are 700 men, provided with 184000 lbs. of provisions, of which they consume each man 5 lbs. a week ; how long can they subsist ?   Ans. 52 weeks, 4 days.

56. If 25 men have ¼ of a pound of beef, each three times in a week, how long will $150 lbs. last them ?   Ans. 56 weeks.

57. How many tiles 8 inches square, will lay a floor 20 feet long and 16 feet wide ?   Ans. 720.

58. How many stones 10 inches long, 9 inches broad, and 4 inches thick, would it take to build a wall 80 feet long, 20 feet high and 2¼ feet thick.   Ans. 17280 stones.

59. Bought 3 score pieces of Holland for three times as many dollars, and sold them again for four times as many dollars ; but if they had cost me as much as I sold them for, for what should I have sold them to gain after the same rate ?   Ans. $320.

60. A sets out on a journey and travels 27 miles a day ; 7 days after, B sets out and travels the same road 36 miles a day ; in how many days will B overtake A ?   Ans. 21 days.

61. If I sell coffee at 2s. 3d. per lb. and gain 35 per cent., what did I give per lb. ?   Ans. 20d.

62. Suppose 2000 soldiers had been supplied with bread, sufficient to last them 12 weeks, allowing each man 14 ounces a day, but, on examining, find 105 barrels, containing 200 lbs. each, wholly spoiled ; how much a day may each man eat, that the remainder may supply them 12 weeks ?

Ans. 12 ounces.

63. If 2000 soldiers were put on an allowance of 12 ounces of bread per day, for 12 weeks ; having a seventh part of their bread spoiled ; what was the whole weight of their bread, (good and bad,) and how much was spoiled ?

The whole, 147000 lbs.
Spoiled,   21000 lbs.

64. If 2000 soldiers, having lost 105 barrels of bread, weighing 200 lbs. each, were obliged to subsist on 12 ounces a day for 12 weeks. But had none been lost, they might have had 14 ounces a day, the same time. What was the whole weight, including what was lost ; and how much had they left to subsist on ?

The whole weight, 147000 lbs.
Left to subsist on, 126000 lbs.

65. If 2000 soldiers, after losing one seventh part of their bread, had each 12 ounces a day, for 12 weeks, what was the

whole weight of their bread, including that lost, and how much might they have had per day, each man, if none had been lost?

The weight was 147000 lbs.
The loss,   21000 lbs.
Had none been lost, they might have had 14 ounces per day.

66. If .85 of a gallon of wine cost $2.72, how much will .25 of a gallon cost?   Ans. $0.80.

67. If 61.3 pounds of tea cost $44.9942, what is the price per lb.?   Ans. $0.73.4.

68. What is the value of .15 of a hogshead of lime, at $2.39 per hhd.?   Ans. $0.35.85.

69. If .75 of a ton of hay cost $15, what is it per ton?   Ans. $20.00.

70. How many yards of carpeting that is half a yard wide, will cover a room that is 30 feet long and 18 feet wide?   Ans. 120 yards

71. If a man perform a journey in 15 days, when the day is 12 hours long, in how many will he do it when the day is but 10 hours?   Ans. 18 days.

72. If 450 men are in a garrison, and their provisions will last them but 5 months; how many must leave the garrison that the same provisions may be sufficient for them who remain 9 months?   Ans. 200 men.

73. The hour and minute hand of a watch are together at 12 o'clock, when are they next together?   Ans. 1 h. 5 m. 27$\frac{3}{11}$ sec.

74. A and B can perform a piece of work in 5$\frac{5}{11}$ days, B and C in 6$\frac{2}{3}$ days; and A and C in 6 days. In what time would each of them perform the work alone, and how long would it take them to do the work together?

A would do the work in 10 days,
B would do the work in 12 days,
C would do the work in 15 days
A, B and C, together in 4 days.

SECTION XLV.

COMPOUND PROPORTION,

OR

DOUBLE RULE OF THREE.

COMPOUND PROPORTION is the method of performing such operations in Proportion, as require two or more statements.

EXAMPLES.

1. If a man travel 117 miles in 30 days, employing only 9 hours a day, how far would he go in 20 days, travelling 12 hours a day?

The distance to be travelled depends on two circumstances —the number of *days* the man travels, and the number of *hours* he travels in each day.

We will first suppose the hours to be the same in each case; the question will then be,—If a man travel 117 miles in 30 days, how far will he travel in 20 days?

This will lead to the following proportion.

$$30 \text{ days} : 20 :: 117 \text{ miles} : \tfrac{117 \times 20}{30} = 78 \text{ miles.}$$

That is, if we multiply 117 by 20, and divide the product by 30, we obtain the number of miles he will travel in 20 days, which is 78.

Now, if we take into consideration the number of hours, we must say,—If a man, travelling 9 hours a day for a certain number of days, has travelled 78 miles, how far will he go in the same time, if he travel 12 hours a day? This will furnish the following proportion.

$$9 \text{ hours} : 12 \text{ hours} :: 78 \text{ miles} : \tfrac{12 \times 78}{9} = 104 \text{ miles, the}$$

answer to the question.

By this mode of resolving the question, we see that 117 miles have, to the answer 104 miles, the proportion that 30 days have to 20 days, and that 9 hours have to 12 hours. Stating this in compound proportion, we have

$$\left. \begin{array}{c} 30 : 20 \\ 9 : 12 \end{array} \right\} :: 117 : 104 \text{ miles, the answer.}$$

Thus it appears, that if 117 be multiplied by both 20 and 12, and the product be divided by 30 times 9, the quotient will be

104 miles ; or if we multiply 117 by 20, and divide the product by 30, and then multiply this quotient by 12 and divide by 9, it will produce the same answer as before.

This question may be performed by analysis, thus.   If he travel 117 miles in 30 days, in one day he will travel $\frac{1}{30}$ of 117 miles, which is $\frac{117}{30}$ miles ; and, travelling 9 hours a day, he will in one hour travel $\frac{1}{9}$ of $\frac{117}{30}$ miles ; which is $\frac{117}{270}$ miles ; and in a day of 12 hours, he will travel 12 times $\frac{117}{270}$ miles, which is $\frac{1404}{270}$ miles ; and in 20 days he will travel 20 times $\frac{117}{270}$ miles, which is 104 miles, the answer as before.

The answer to the above question might have been obtained by dividing the third term by the product of the two ratios, which the first two terms have to the second terms ; that is, by the ratio of 30 to 20, which is $\frac{30}{20} = \frac{3}{2}$; and of 9 to 12, which is $\frac{9}{12} = \frac{3}{4}$.

Thus, $117 \div \frac{3}{2} \times \frac{3}{4} = 117 \div \frac{9}{8} = \frac{117}{1} \times \frac{8}{9} = \frac{936}{9} = 104$ Ans.

2. If 6 men, in 16 days of 9 hours each, build a wall 20 feet long, 6 feet high, and 4 feet thick, in how many days of 8 hours each, will 24 men build a wall 200 feet long, 8 feet high, and 6 feet thick ?

In stating this question, there are several circumstances to be taken into consideration ; the number of men employed, the length of the days, length of the wall, and its height and breadth.

As the answer to the question is to be in days, we make the days the third term.

Were all the circumstances of the question alike, except the number of men and the number of days, the question would consist in finding in how many days 24 men would perform the same labor, that 6 men had done in 16 days ; that is, if 6 men had built a certain wall in 16 days, how many days would it take 24 men to perform the same labor. This would furnish the following proportion.

24 men : 6 men :: 16 days : $\frac{6 \times 16}{24} = 4$ days.

Or, if this were the question. If a certain number of men, by laboring 9 hours a day, perform a piece of work in 16 days, how many days would it take the same men to do the labor, by working 8 hours a day ?   The following would be the proportion.

8 hours : 9 hours :: 16 days : $\frac{9 \times 16}{8} = 18$ days.

Or, if this were the question. If a certain number of men build a wall 20 feet long in 16 days, how long would it take the same men to build a wall 200 feet long ; the following would be the statement.

20 feet : 200 feet :: 16 days : $\frac{16 \times 200}{20} = 160$ days.

Or, if only the days and height of the wall were considered, this would be the statement.

6 feet : 8 feet :: 16 days : $\frac{8 \times 16}{6}$ = 21⅓ days.

Lastly, were we to consider only the days and the thickness of the wall, it would furnish the following statement.

4 feet : 6 feet :: 16 days : $\frac{6 \times 16}{4}$ = 24 days.

We see, by this mode of resolving the question, that 16 days must have, to the *true answer*, the ratio compounded of the ratios,

That 24 men have to 6 men ;
That 8 hours have to 9 hours ;
That 20 feet have to 200 feet ;
That 6 feet have to 8 feet ; and
That 4 feet have to 6 feet.

Stating the above in Compound Proportion, we have

$$
\left.
\begin{array}{ll}
24\text{ men} & : 6 \ \ \text{men} \\
8\text{ hours} & : 9\text{ hours} \\
20\text{ feet} & : 200\text{ feet} \\
6\text{ feet} & : 8\text{ feet} \\
4\text{ feet} & : 6\text{ feet}
\end{array}
\right\}
:: 16\text{ days}: 90\text{ days} = \text{Answer.}
$$

The continued product of all the second terms by the third term, and this divided by the continued product of the first terms, will produce the answer.

Thus $\frac{6}{24} \times \frac{9}{8} \times \frac{200}{20} \times \frac{8}{6} \times \frac{6}{4} \times 16 = \frac{831840}{9240}$ = 90 days = Answer.

Hence the following

RULE.

Make that number which is of the same kind as the answer required, the third term; and of the remaining numbers, take any two, that are of the same kind, and consider whether an answer, depending upon these alone, would be greater or less than the third term, and place them as directed in Simple Proportion. Then take any other two, and consider whether an answer depending only upon them would be greater or less than the third term, and arrange them accordingly; and so on, until all are used. Multiply the continued product of the second terms by the third, and divide by the continued product of the first, and you produce the answer.

3. If $100 gain $6 in one year, how much would $500 gain in 4 months ? Ans. $10.

4. If $100 gain $6 in one year, what must be the sum to gain $10 in 4 months ? Ans. $500.

16

5. How long will it take $500 to gain $10, if $100 gain $6 in one year ? Ans. 4 months.

6. If $500 gain $10 in 4 months, what is the rate per cent. ? Ans. 6 per cent.

7. If 8 men spend $32 in 13 weeks, what will 24 men spend in 52 weeks ? Ans. $384.

8. If 12 men, in 15 days, can build a wall 30 feet long, 6 feet high, and 3 feet thick, when the days are 12 hours long ; in what time will 60 men build a wall 300 feet long, 8 feet high, and 6 feet thick, when they work only 8 hours a day ? Ans. 120 days.

9. If 16 horses consume 84 bushels of grain in 24 days, how many bushels will suffice 32 horses 48 days ? · Ans. 336 bushels.

10. If the carriage of 5 cwt. 3. qrs., 150 miles, cost $24.58, what must be paid for the carriage of 7 cwt. 2 qrs. 25 lbs., 64 miles, at the same rate ? Ans. $14.08.6.

11. If 7 oz. 5 dwt. of bread be bought at 4¼d., when corn is 4s. 2d. per bushel, what weight of it may be bought for 1s. 2d. when the price per bushel is 5s. 6d. ?

Ans. 1 lb. 4 oz. 3$\frac{113}{119}$ dwt.

SECTION XLVI.

CHAIN RULE.

THE CHAIN RULE consists in joining many proportions to gether ; and by the relation which the several antecedents have to their consequents, the proportion between the first antecedent and the last consequent is discovered.

This rule may often be abridged by cancelling equal quantities on both sides, and abbreviating commensurables.

NOTE. — The first numbers in each part of the question are called *antecedents*, and the following *consequents*.

1. If 20 lbs. at Boston, make 23 lbs. at Antwerp, and 155 lbs at Antwerp make 180 lbs. at Leghorn, how many pounds at Boston are equal to 144 lbs. at Leghorn ?

OPERATION.

20 lbs. of Boston = 23 of Antwerp,
155 lbs. of Antwerp = 180 of Leghorn,
144 lbs. of Leghorn.

$\frac{20 \times 155 \times 144}{23 \times 180} = \frac{446400}{4140} = 107\frac{19}{23}$ Answer.

It will be perceived in the operation, that the continued product of the antecedents is divided by the continued product of the consequents. Hence the following

RULE.

Write the numbers alternately, that is, the antecedents at the left hand, and the consequents at the right; and, if the last number stands at the left hand, multiply the numbers of the left hand column continually together for a dividend, and those at the right for a divisor; but, if the last number stands at the right hand, multiply the numbers of the right hand column continually together for a dividend, and those at the left for a divisor; and the quotient will be the answer.

NOTE. — The demonstration for this rule is the same as for Compound Proportion.

2. If 12 lbs. at Boston make 10 lbs. at Amsterdam, and 10 lbs. at Amsterdam make 12 lbs. at Paris, how many pounds at Boston are equal to 80 lbs. at Paris ?　　　　　　　Ans. 80 lbs.

3. If 25 lbs. at Boston are equal to 22 lbs. at Nuremburg, and 88 lbs. at Nuremburg are equal to 92 lbs. at Hamburg, and 46 lbs. at Hamburg are equal to 49 lbs. at Lyons, how many pounds at Boston are equal to 98 lbs. at Lyons ?　　　　Ans. 100 lbs.

4. If 24 shillings in Massachusetts are equal to 32 shillings in New York ; and if 48 shillings in New York are equal to 45 shillings in Pennsylvania ; and if 15 shillings in Pennsylvania are equal to 10 shillings in Canada ; how many shillings in Canada are equal to 100 shillings in Massachusetts ?

Ans. 83¼ shillings.

5. If 17 men can do as much work as 25 women, and 5 women do as much as 7 boys, how many men would it take to do the work of 75 boys ?　　　　　　　　　　Ans. 36³⁄₇ men.

6. If 10 barrels of cider will pay for 5 cords of wood, and 20 cords of wood for 4 tons of hay, how many barrels of cider will it take to purchase 50 tons of hay ?　　　　Ans. 500 bbls.

7. If 100 acres in Bradford be worth 120 in Haverhill, and 50 in Haverhill worth 65 in Methuen, how many acres in Bradford are equal to 150 in Methuen ?　　　　　Ans. 96²⁄₁₃ acres.

8. If 10 pounds of cheese are equal in value to 7 lbs. of butter, and 11 lbs. of butter to 2 bushels of corn, and 11 bushels of corn to 8 bushels of rye, and 4 bushels of rye to one cord of wood, how many pounds of cheese are equal in value to 10 cords of wood ?　　　　　　　　　Ans. 432½ lbs.

SECTION XLVII.

FELLOWSHIP, OR COMPANY BUSINESS.

FELLOWSHIP is a rule by which merchants, and others in partnership, estimate their gain or loss in trade. It is of two kinds, *single* and *double*.

Single Fellowship is, when merchants in partnership, employ their stock for equal times.

NOTE. — Partnership is the union of two or more persons in the same trade.

EXAMPLES.

1. Three men, A, B, and C, enter into partnership for two years, with a capital of $1080. A puts in $240; B $360; and C $480. They gain $54. What is each man's share of the gain?

As the whole stock in trade is $1080, of which $240 belongs to A, A's share of the stock therefore, will be $\frac{240}{1080}=\frac{2}{9}$, and as each man's gain is in proportion to his stock, $\frac{2}{9}$ of $54, which is $12, is A's gain. B's stock in trade is $360; therefore $\frac{360}{1080}=\frac{1}{3}$ of $54, which is $18, is B's gain. C's stock is $480; therefore his part of the whole stock is $\frac{480}{1080}=\frac{4}{9}$; consequently C's share of the gain is $\frac{4}{9}$ of $54, which is $24. Hence to find any man's gain or loss in trade we have the following

RULE.

Multiply the whole gain or loss by each man's fractional part of the stock.

NOTE. — The pupil, who may be desirous of performing the questions of this rule in the " *old way*," will adopt the following

RULE.

As the whole stock is to the whole gain or loss, so is each man's particular stock to his particular share of the gain or loss.

The following is the statement of the first question with the answers and proof.

As the stock $1080 : $54 :: $240 : $12 A's gain.
$1080 : $54 :: $360 : $18 B's gain.
$1080 : $54 :: $480 : $24 C's gain.
Proof, $12 + $18 + $24 = $ 54 whole gain.

2. A, B, and C, enter into partnership with a capital of $1100, of which A put in $250, B put in $300, and C $550; they lost by trading 5 per cent. on their capital. What was each man's share of the loss?
    A's loss, $12.50,
    B's loss, $15.00,
    C's loss, $27.50.

3. Two merchants, C and D, engaged in trade ; C put in $6780, D put in $12000; they gain $1000. What is each man's share?
    C's share, $361.02.2$\frac{114}{513}$.
    D's share, $638.97.7$\frac{1}{4}$$\frac{99}{13}$.

4. M, P, Q, trade in company with a capital of $10.000 ; M put in $3000, P $2000, and Q $5000 ; they gain $500. What is each man's share of the gain?
    M's gain, $150,
    P's gain, $100,
    Q's gain, $250.

5. A, B, and C, enter into partnership; A put in $500, B $350, and C put in 320 yards of broadcloth ; they gain $332.50, of which C's share is $120. What was A's and B's shares of the gain, and the value of C's cloth per yard ?
    A's gain, $125.00,
    B's gain, $ 87.50,
    C's cloth per yd. $ 1.50.

6. A, B, and C, trade in company ; A put in $5000, B put in $6500, and C put in $7500 ; they gain 40 per cent. on their capital, but receive the whole amount of their goods in bills, for which they are obliged to allow a discount of 10 per cent. How much was each man's net gain ?
    A's gain, $1800,
    B's gain, $2340,
    C's gain, $2700.

7. A merchant failing in trade, owes A $600, B $760, C $840, D $800. His effects are sold for $2275. What will each receive of the dividend ?
    A, $455.00
    B, 576.33\frac{1}{3}$
    C, $637.00
    D, 606.66\frac{2}{3}$

8. A bankrupt owes $5000. His effects, sold at auction, amount to $4000. What will his creditors receive on a dollar ?
    Ans. $0.80.

9. A merchant, having sustained many losses, is obliged to become a bankrupt. His effects are valued at $1728, with which he can pay only fifteen cents on the dollar. What did he owe ?
    Ans. $11520.

16*

SECTION XLVIII.

DOUBLE FELLOWSHIP.

WHEN merchants in partnership employ their stock for unequal times, it is called Double Fellowship.

EXAMPLES.

1. Two men, A, and B, trade in company. A puts in $420 for 5 months, and B puts in $350 for 8 months; they gain $84. What is each man's share of the gain?

Method of operation by analysis.

$420 for 5 months is the same as $5 \times \$420 = \$2100$ for 1 month; and $350 for 8 months, is the same as $8 \times \$350 = \$2800$ for 1 month. The question therefore, is the same, as if A had put in $2100 and B $2800 for 1 month each. The whole stock would then be $4900. A's share of the gains therefore will be $\frac{2100}{4900} = \frac{3}{7}$ of $84 = $36. And B's share will be $\frac{2800}{4900} = \frac{4}{7}$ of $84 = $48.

Hence we deduce the following

RULE.

Multiply each man's stock by the time it continued in trade, and consider each product a numerator to be written over their sums, as a common denominator; then multiply the whole gain or loss by each fraction, and the several products will be the gain or loss of each man.

NOTE.—It might be well for the student to be acquainted with the " *old way* " of performing these questions. The following is the

RULE.

Multiply each man's stock by the time it was continued in trade; then as the sum of all the products is to the whole gain or loss, so is each man's particular product to his share of the gain or loss.

The first question would be performed thus,

$420 \times 5 = 2100$ 4900 : 84 : : 2100 : $36 A's gain,
$350 \times 8 = \underline{2800}$ 4900 : 84 : : 2800 : $48 B's gain.
         4900

Proof, $36 + $48 = $84 = A and B's gain.

2. A commenced business January 1, with a capital of $3200. May 1, he took B into partnership with a capital of $4200; and at the end of the year they had gained $240. What was each man's share of the gain ?

$128, A's gain,
$112, B's gain.

3. A, B, and C traded in company. A put in $300 for 5 months, B put in $400 for 8 months, and C put in $500 for 3 months; they gain $100. What is the gain of each ?

24.19\frac{11}{31}$, A's gain,
51.61\frac{8}{31}$, B's gain,
24.19\frac{11}{31}$, C's gain.

4. Three men hire a pasture in common; for which they are to pay $26.40. A put in 24 oxen for 8 weeks; B put in 18 oxen for 12; and C put in 12 oxen for 10 weeks. What ought each to pay ?

A should pay $ 9.60,
B should pay $10.80,
C should pay $ 6.00.

5. Two men in Boston hire a carriage for $25, to go to Concord, N. H. the distance being 72 miles, with the privilege of taking in three more persons. Having gone 20 miles, they take in A; at Concord they take in B; and when within 30 miles of the city, they take in C. How much shall each man pay ?

First man pays $7.60.9$\frac{103}{108}$.
Second man pays $7.60.9$\frac{103}{108}$.
A pays $5.87.3$\frac{71}{108}$,
B pays $2.86.4$\frac{7}{13}$,
C pays $1.04.1$\frac{8}{13}$.

$25.00.0.

6. Three men engage in partnership for 20 months. A at first put into the firm $4000; and at the end of four months, he put in $500 more, but at the end of 16 months, he took out $1000. B at first put in $3000, but at the end of 10 months, he took out $1500, and at the end of 14 months he put in $3000; C at first put in $2000, and at the end of 6 months, he put in $2000 more, and at the end of 14 months he put in $2000 more, but at the end of 16 months, he took out $1500; they had gained by trade $4420. What is each man's share of the gain ?

A's gain, $1680,
B's gain, $1260,
C's gain, $1480.

7. John Jones, Samuel Eaton, and Joseph Brown, formed a partnership under the firm of Jones, Eaton, & Co., with a capital of $10.000; of which Jones put in $4000, Eaton put in $3500, and Brown put in $2500; but at the end of 6 months Jones withdrew $2000 of his stock; and at the end of 8 months Eaton withdrew $1500 from the firm; but at the end of 10

months Brown added $2000 to his stock. At the end of 2 years they found their gains to be $1041.80.   What was the share of each man ?

Jones's gain, 300.51\frac{11}{13}$.
Eaton's gain, 300.51\frac{11}{13}$.
Brown's gain, 440.76\frac{4}{13}$.

SECTION XLIX.

METHOD OF ASSESSING TOWN, STATE AND COUNTY TAXES.

RULE.

First. Say, as the whole amount of the State tax, which the town has to pay the State, is to the poll tax, which each individual pays the State; so is the whole amount of the town, State, and county taxes, which the town has to raise, to the poll tax, which is to be assessed on each individual; then multiply this poll tax by the number of polls in the town, subtract this product from the amount of all the taxes, and note the remainder; then say, as the valuation of the town is to this remainder, so is one dollar to the sum, that must be assessed on the dollar.

EXAMPLES.

1. Suppose the State Treasurer sends to the town of A, that their proportion of the State tax of $75.000, is $277.20, of which the polls are to pay 11 cents each, and the remainder to be assessed equally on the estates, real and personal. The county tax required is $554.40, and the town vote to raise $1940.40, to defray town charges. The valuation of the town is $154,000, and it contains 420 polls. What will be B's tax, who stands in the valuation as follows: real estate, $3678 ; personal estate, $526 ; and 3 polls ?

OPERATION.

```
        277.20              420 number of polls.
        554.40              1.10
       1940.40             ───────
       ───────
$277.20 : 11 cts. : : 2772.00       462.00
                       .11
              ─────────────────────
       277.20)3049200($1.10 Poll tax.   Carried forward.
```

277.20)3049200($1.10 Poll tax.　Brought forward
27720

　　　　　　　　　　　　　$2772.00
　　　　　27720　　　　　462.00
　　　　　27720　　　　　────────
　　　　　　　　　　　　　2310.00

$154000 : $231000 : : $1.
　　　　　1
　────── c. m.
154000)231000(.01.5 to be assessed on the dollar.
154000

770000
770000

TABLE.

	$10000	$1000	$100	$10	$1	.10 cts.	1 ct.
1	150.	15.	1.50	.15	.01.5	.00.15	.00.01
2	300.	30.	3.00	.30	.03.0	.00.30	.00.03
3	450.	45.	4.50	.45	.04.5	.00.45	.00.04
4	600.	60.	6.00	.60	.06.0	.00.60	.00.06
5	750.	75.	7.50	.75	.07.5	.00.75	.00.07
6	900.	90.	9.00	.90	.09.0	.00.90	.00.09
7	1050.	105.	10.50	1.05	.10.5	.01.05	.00.10
8	1200.	120.	12.00	1.20	.12.0	.01.20	.00.12
9	1350.	135.	13.50	1.35	.13.5	.01.35	.00.13

Now to know the amount of B's taxes, his real estate being $3678, I find by the table, that

　　$3000 pays　．．．　$45.00
　　$600 pays　．．．　$9.00
　　$70 pays　．．．　$1.05
　　$8 pays　．．．　　.12
　　　　　　　　　　　────────
　　　　　　　　　　　$55.17

His personal estate being $526, it is found by the table, that

　　$500 pays　．．．　$7.50
　　$20 pays　．．．　　.30
　　$6 pays　，．．　　　9
　$526　　　　　　　　　────────
　　　　　　　　　　　$7.89
Three polls at $1.10 each,　　3.30
His real estate must then pay　55.17
His personal estate,　　　　　7.89
Three polls at $1.10 each,　　3.30
Total B's tax for town, state,　────────
　　and county,　　　　　　$66.36

NOTE.— The tax on personal estate is often specific.

SECTION L.

LOSS AND GAIN.

LOSS AND GAIN, is a rule, by which merchants estimate their *profit and loss*, in buying and selling goods.

The questions are performed by Proportion and the other preceding rules.

The pupil should give an analysis of each question.

EXAMPLES.

1. Sold broadcloth at $6.12½ per yard, and by so doing lost 12½ per cent. What was the original cost per yard?

Ans. $7.00.

BY ANALYSIS.

If 12½ per cent. be lost, 87½ per cent. will remain. It is now required to find of what number $6.12½ is $\frac{87\frac{1}{2}}{100}$. This is done by multiplying $6.12½ by 100 and dividing by 87½, and it produces the answer, $7.00.

2. Bought cloth at $7.00 per yard, and sold it at $6.12½. What did I lose per cent.? Ans. 12½ per cent.

3. Bought cloth at $7.00 per yard, and sold it for 12½ per cent. less than what it cost. What did I receive? Ans. $6.12½.

4. Bought cloth at $3.60 per yard. For how much must I sell it to gain 12½ per cent.? Ans. $4.05.

5. Sold cloth at $4.05 per yard, and by so doing I gained 12½ per cent. What did it cost? Ans. $3.60.

6. Bargained for cheese at $8.50 per cwt. For how much must I sell it to gain 10 per cent.? Ans. $9.35 per cwt.

7. Sold cheese at $9.35 per cwt. and gained 10 per cent. What did I give for it? Ans. $8.50 per cwt.

8. Sold wine at $1.25 per gallon, and by so doing lost 15 per cent. For what should I have sold it to gain 12 per cent.? Ans. $1.64.7$\frac{1}{17}$ per gallon.

9. Sold wine at $1.25 per gallon, and lost 15 per cent. What per cent. should I have gained, had I sold it for $1.64.7$\frac{1}{17}$ per gallon? Ans. 12 per cent.

10. Sold wine for $1.64.7$\frac{1}{17}$ per gallon, and gained 12 per cent. For what should I have sold it to lose 15 per cent.? Ans. $1.25 per gallon.

11. Sold wine for $1.64.7$\frac{1}{17}$ per gallon, and gained 12 per cent. What should I have lost, had I sold it for $1.25 per gallon? Ans. 15 per cent.

12. A buys corn for $0.90 per bushel, and sells it for $1.20. B buys for $1.12½, and sells for $1.50 ; who gains the most per cent. ? 　　　Ans. both gain alike.

13. If I buy cotton cloth at 13 cents per yard, on 8 months' credit, and sell it again at 12 cents cash, do I gain or lose, and how much per cent. ? 　　　Ans. lose 4 per cent.

14. If 24 yards of cloth are sold at $2.50 per yard, and there is 7½ per cent. loss in the sale, what is the prime cost of the whole ? 　　　Ans. $64.86.4$\frac{32}{37}$.

15. Bought 24 yards of cloth for $64.86.4$\frac{32}{37}$. For what must I sell it per yard to lose 7½ per cent. ? 　　　Ans. $2.50.

16. Bought a certain quantity of cloth for $64.86.4$\frac{32}{37}$, and by selling it at $2.50 per yard, I lost 7½ per cent. How many yards were bought ? 　　　Ans. 24 yards.

17. Bought 24 yards of cloth for $64.86.4$\frac{32}{37}$, and sold it at $2.50 per yard ; what is lost per cent.? 　　　Ans. 7½ per cent. ?

18. If 27½ cwt. of sugar be sold at $12.50 per cwt. and there is gained 17 per cent. what was the first cost ?

Ans. $10.68.3$\frac{89}{117}$.

19. Sold a horse for $75, and by so doing, I lost 25 per cent. ; whereas, I ought to have gained 30 per cent. How much was it sold under its real value ? 　　　Ans. $55.00.

20. Bought a horse which was worth 30 per cent. more than I gave for him ; but having been injured, I sold him for 25 per cent. less than what he cost, and thereby lost $55 on his real value. What was received for the horse ? 　　　Ans. $75.00.

21. Bought molasses at 42 cents per gallon, but not proving so good as I expected, I am willing to lose 5 per cent. For what must I sell it per gallon ? 　　　Ans. $0.39.9.

22. Bought a hogshead of wine for $112, but 15 gallons having leaked out, I am willing to lose 5 per cent. on the cost. For how much per gallon must I sell it ? 　　　Ans. $2.21.6$\frac{2}{3}$.

23. Bought a hogshead of wine for $112, but a number of gallons having leaked out, I sell the remainder for $2.21.6$\frac{2}{3}$ per gallon, and by so doing I lose 5 per cent. on the cost. How many gallons leaked out? 　　　Ans. 15 gallons.

24. Bought a hogshead of wine for a certain sum, but 15 gallons having leaked out, I sell the remainder for $2.21.6$\frac{2}{3}$, per gallon, and thereby lose 5 per cent. on the cost. What was the cost ? 　　　Ans. $112.00.

25. Bought a hogshead of wine for $112.00, but 15 gallons having leaked out, I sell the remainder at $2.21.6$\frac{2}{3}$ per gallon. What is my loss per cent. ? 　　　Ans. 5 per cent.

26. If I sell cloth at $5.60 per yard, and thereby lose 7 per cent. should I gain or lose, and how much per cent. by selling it at $6.25 per yard ? 　　　Ans. 3$\frac{89}{112}$ per cent. gain.

27. Sold a watch which cost me $30 for $35, on a credit of 8 months. What did I gain by the bargain ? 　　　Ans. $3.65.3$\frac{11}{11}$.

28. When tea is sold at $1.25 per lb. there is lost 25 per cent. ; what would be the gain or loss per cent. if it should be sold at $1.40 per lb. ?   Ans. 16 per cent. loss.

29. A exchanges with B 50 lbs. of indigo at $1.00 per lb. cash, and in barter $1.35 ; but he is willing to lose 12 per cent. to have ½ ready money. What is the cash price of 1 yard of cloth delivered by B at $5.00 per yard to equal A's bartering price, and how many yards were delivered ?

Ans. $4.20.⅒ cash price of 1 yard.

7⅞ yards delivered by B.

SECTION LI.

DUODECIMALS AND CROSS MULTIPLICATION.

DUODECIMALS are so called, because they decrease by twelves from the place of feet towards the right.

Inches are called *primes* and are marked thus ' ; the next division after is called *seconds*, marked thus " ; the next is *thirds*, marked thus ''' ; and so on.

Duodecimals are commonly used by workmen and artificers in finding the contents of their work.

EXAMPLES.

1. Multiply 6 feet 8 inches by 4 feet 5 inches.

OPERATION.

6 8'	As feet are the integers or units, it is evident,
4 5	that feet multiplied by feet will produce feet ; and
———	as inches are twelfths of a foot, the product of
26 8'	inches by feet will be twelfths of a foot. For the
2 9 4"	same reason, inches multiplied by inches will pro-
———	duce twelfths of an inch, or one hundred and forty-
29 5' 4"	fourths of a foot. Hence we deduce the following

RULE.

Under the multiplicand write the same names or denominations of the multiplier; that is, feet under feet, inches under inches, &c. Multiply each term in the multiplicand, beginning at the lowest by the feet of the multiplier, and write each result under its respective term, observing to carry a unit for every 12 from each denomination to its next superior. In the same manner the multiplicand by the inches of the multiplier, and write result of each term one place further towards the right of se in the multiplicand. Proceed in the same manner with the seconds and all the rest of the denominations, and the sum of all the lines will be the product required.

The denomination of the particular products will be as follows.

Feet multiplied by feet, give feet.
Feet multiplied by primes, give primes.
Feet multiplied by seconds, give seconds.
Primes multiplied by primes, give seconds.
Primes multiplied by seconds, give thirds.
Primes multiplied by thirds, give fourths.
Seconds multiplied by seconds, give fourths.
Seconds multiplied by thirds, give fifths.
Seconds multiplied by fourths, give sixths.
Thirds multiplied by thirds give sixths.
Thirds multiplied by fourths, give sevenths.
Thirds multiplied by fifths, give eighths, &c.

2. Multiply 4 feet 7′ by 6 feet 4′.   Ans. 29ft. 0′ 4′ .
3. Multiply 14 feet 9′ by 4 feet 6′.   Ans. 66ft. 4′ 6″.
4. Multiply 4 feet 7′ 8″ by 9 feet 6′.   Ans. 44ft. 0′ 10″.
5. Multiply 10 feet 4′ 5″ by 7 feet 8′ 6″.
                                Ans. 79ft. 11′ 0″ 6‴ 6⁗.
6. Multiply 39 feet 10′ 7″ by 18 feet 8′ 4″.
                                Ans. 745 ft. 6′ 10″ 2‴ 4⁗.
7. How many square feet in a floor 48 feet 6′ long 24 feet 3′ broad ?   Ans. 1176ft. 1′ 6″.
8. What are the contents of a marble slab, whose length is 5 feet 7′, and breadth 1 foot 10′ ?   Ans. 10ft. 2′ 10″.
9. The length of a room being 20 feet, its breadth 14 feet 6 inches, and height 10 feet 4 inches ; how many yards of painting are in it, deducting a surplus of 4 feet by 4 feet 4 inches, and 2 windows, each 6 feet by 3 feet 2 inches.   Ans. 73⅔₇ yards.
10. Required the solid contents of a wall 53 feet 6 inches long, 10 feet 3 inches high, and 2 feet thick ?   Ans. 1096ft. 9′.
11. There is a house with four tiers of windows, and 4 windows in a tier ; the height of the first is 6 feet 8 inches ; of the second 5 feet 9 inches ; of the third 4 feet 6 inches ; of the fourth 3 feet 10 inches ; and the breadth is 3 feet 5 inches ; how many square feet do they contain in the whole ?
                                Ans. 283ft. 7in.
12. How many feet of boards would it require to make 15 boxes, each of which is 7 feet 9 inches long, 3 feet 4 inches wide, and 2 feet 10 inches high ; and how many cubic yards would they contain ?   Ans. 1717ft. 1in. 40³¹¹⁄₃₂₄ cubic yds.
13. A mason has plastered 3 rooms, the ceiling of each is 20 feet by 16 feet 6 inches, and the walls of each 9 feet 6 inches high ; there is to be 90 yards deducted for doors, windows, &c. How many yards must he be paid for ?
17   Ans. 251 yards. 1ft. 6in.

14. How many feet in a board which is 17 feet 6 inches long and 1 foot 7 inches wide ?   Ans. 27 ft. 8′ 6″.

15. How many feet in a board 27 feet 9 inches long, 29 inches wide ?   Ans. 67 ft. 0′ 9″.

16. How many feet of boards will it take to cover the side of a building 47 feet long, 17 feet 9 inches high ?   Ans. 834 ft. 3′.

NOTE. — A board to be merchantable should be 1 inch thick; therefore to reduce a plank to board measure, the superficial contents of the plank should be multiplied by its thickness.

17. How many feet, board measure, are in a plank 18 feet 9 inches long, 1 foot 6 inches wide, and 3 inches thick ?   Ans. 84 ft. 4′ 6″.

18. How many feet, board measure, are in a plank 20 feet long, 1 foot 6 inches wide, and 2½ inches thick ?   Ans. 75 ft.

19. How many feet in a plank 40 feet 6 inches long, 30 inches wide, and 2¾ inches thick ?   Ans. 278 ft. 5′ 3″.

NOTE. — A pile of wood that is 8 feet long, 4 feet high, and 4 feet wide, contains 128 cubic feet, or a cord; and every cord contains 8 *cord-feet ;* and as 8 is 1-16 of 128, every *cord-foot* contains 16 cubic feet; therefore, dividing the cubic feet in a pile of wood by 16, the quotient is the cord-feet; and if cord-feet be divided by 8, the quotient is cords.

20. How many cords of wood in a pile 18 feet long, 6 feet nigh, and 4 feet wide ?   Ans. 3⅜ cords.

21. How many cords in a pile 10 feet long, 5 feet high, 7 feet wide ?   Ans. 2 cords, 94 cubic feet.

22. How many cords in a pile 35 feet long, 4 feet wide, 4 feet high ?   Ans. 4⅜ cords.

23. How many cords in a pile that measures 8 feet on each side ?   Ans. 4 cords.

24. How many cords in a pile that is 10 feet on each side ?   Ans. 7¹¹⁄₁₆ cords.

NOTE. — When wood is 'corded' in a pile 4 feet wide, by multiplying its length by its height, and dividing the product by 4, the quotient is the cord-feet; and if a load of wood be 8 feet long, and its height be multiplied by its width, and the product divided by 2, the quotient is the cord-feet.

25. How many cords of wood in a pile 4 feet wide, 70 feet 6 inches long, and 5 feet 3 inches high ?   Ans. 11³⁄₁₆ cords.

NOTE. — Small fractions are rejected.

26. How many cords in a pile of wood, 97 feet 9 inches long, 4 feet wide, and 3 feet 6 inches high ?   Ans. 10¹⁷⁷⁄₂₅₆ cords.

27. Required the number of cords of wood in a pile 100 feet long, 4 feet wide, and 6 feet 11 inches high ?   Ans. 21⅞.

28. Agreed with a man for 10 cords of wood, at $5.00 a cord; *it was to* be cut 4 feet long, but by *mistake* it was cut only 46

inches long. How much in *justice* should be deducted from the *stipulated* price ? Ans $2.08¼.

29. If a load of wood be 8 feet long, 3 feet 8 inches wide, and 5 feet high ; how much does it contain ? Ans. 9⅜ cord feet.

30. If a load of wood be 8 feet long, 3 feet 10 inches wide, and 6 feet 6 inches high ; how much does it contain ?

Ans. 12¹¹⁄₂₄ cord-feet.

31. If a load of wood be 8 feet long, 3 feet 6 inches wide ; how high should it be to contain 1 cord ? Ans. 4 ft. 6' 10''⅔.

32. If a load of wood be 12 feet long and 3 feet 9 inches wide ; how high should it be to contain 2 cords. Ans. 5 ft. 8' 3'' ⅕.

CROSS MULTIPLICATION.

CROSS MULTIPLICATION differs from duodecimals, in not having the inferior denominations confined to *twelves;* for any number, whether its inferior denominations decrease from the integer in the same ratio or not, may be multiplied crosswise ; and for the better understanding it, the learner must observe the following problem.

33. Multiply 3£. 6s. 8d. by 2£. 5s. 7d.

OPERATION.

	£.	s.	d.	
	3	6	8	
	2	5	7	
3£.×2£.= 6£.=6		0	0	
6s. ×2£.=12s. =0		12	0	
8d. ×2£.=16d.=0		1	4	
3£.×5s.=15s. =0		15	0	
6s. ×5s. =$\frac{30}{20}$s. =0		1	6	
8d. ×5s. =$\frac{40}{20}$d.=0		0	2	
3£.×7d. =21d =0		1	9	
6s. ×7d. = $\frac{42}{20}$q.=0		0	2¹⁄₁₀	
8d. ×7d. =$\frac{56}{240}$d.=0		0	0$\frac{7}{30}$	

7£. 11s. 11¼d.

The pupil will perceive by the operation, that pounds multiplied by pounds produce pounds ; pounds multiplied by shillings produce shillings ; pounds multiplied by pence produce pence ; and, as shillings are twentieths of a pound, if they be multiplied together, their product will be four hundredths of a pound, or twentieths of a shilling.

Thus $\frac{1}{20}$£.×$\frac{1}{20}$£.=$\frac{1}{400}$£.=$\frac{1}{20}$s. And, if pence be multiplied by pence, the product will be two hundred and fortieths of a penny, &c. Therefore we see the propriety of the following

RULE.

That if we multiply any denomination by an integer, the value of an unit in the product will be equal to an unit in the multiplicand; but, if we multiply by any number of an inferi-

or denomination, the value of an unit in the product will be so much inferior to the value of an unit in the multiplicand, as an unit of the multiplier is less than an integer.

34. Multiply 1£. 19s. 11¾d. by 1£. 19s. 11¾d.

Ans. 3£. 19s. 11d. 0 ¹⁄₅₅₆ qr.

35. Multiply 3 miles, 4 furlongs and 12 rods, by 2 miles, 6 furlongs, 8 rods, miles being the integer or unit.

Ans. 9 m. 6 fur. 21 rd. 4 ft. 11 in. 1½br.

36. What is the difference in time between London and Boston, the latter place being 71° 4′ west of the former, the sun passing 1° in 4′ of a solar day ?　　　Ans. 4h. 44′ 16″.

37. If a mill be multiplied by a mill, what will be the product, a dollar being the unit ?　　　Ans. ₁.₀₀₀.₀₀₀ ¹ of a dollar.

38. A, B and C bought a drove of sheep in company ; A paid 14£. 5s., B 13£. 10s., and C 11£. 5s. They agreed to dispose of them at the market ; that each man should take 18s. as pay for his time, &c. ; and that the remainder should be divided in proportion to their several stocks. At the close of the sale, they found themselves possessed of 46£. 5s. What was each man's gain, exclusive of the pay for his time, &c.?

Ans. { 1£. 13s. 3d. A's gain.
　　{ 1£. 11s. 6d. B's gain.
　　{ 1£.　6s. 3d. C's gain.

SECTION LII.

INVOLUTION.

INVOLUTION is the raising of powers from any given number, as a root.

A power is a quantity produced by multiplying any given number, called a *root*, a certain number of times continually by itself, thus,—

$2 =$ 2 is the root, or 1st power of　　$2 = 2^1$.
$2 \times 2 =$ 4 is the 2d power, or square of　　$2 = 2^2$.
$2 \times 2 \times 2 =$ 8 is the 3d power, or cube of　　$2 = 2^3$.
$2 \times 2 \times 2 \times 2 = 16$ is the 4th power, or biquadrate of $2 = 2^4$.

The number denoting the power, is called the *index* or *exponent* of the power. Thus, the fourth power of 3 is 81, or 3^4 ; the second power of 7 is 49, or 7^2.

To raise a number to any power required.—

RULE

Multiply the given number continually by itself, till the number of multiplications be one less than the index of the power to be found, and the last product will be the power required.

EXAMPLES.

1. What is the 5th power of 4 ? $4 \times 4 \times 4 \times 4 \times 4 = 1024$ Ans.
2. What is the 3d power of 8 ? Ans. 512.
3. What is the 10th power of 7 ? Ans. 282475249
4. What is the 6th power of 5 ? Ans. 15625.
5. What is the 3d power of $\frac{3}{4}$? Ans. $\frac{27}{64}$.
6. What is the 5th power of $\frac{1}{3}$? Ans. $\frac{1}{243}$.
7. What is the 4th power of $2\frac{2}{3}$? Ans. $50\frac{49}{81}$.
8. What is the 6th power of $1\frac{2}{3}$? Ans. $16\frac{1211}{15625}$.
9. What is the 4th power of .045 ? Ans. .000004100625.

EVOLUTION,

OR THE EXTRACTION OF ROOTS.

EVOLUTION, or the reverse of Involution, is the extraction or finding the roots of any given powers.

The *root* is a number, whose continual multiplication into itself produces the power and is denominated the square, cube, biquadrate, or 2d, 3d, 4th, &c. power, equal to that power. Thus, 4 is the square root of 16, because $4 \times 4 = 16$; and 3 is the cube root of 27, because $3 \times 3 \times 3 = 27$; and so on.

Although there is no number of which we cannot find any power exactly, yet there are many numbers of which precise roots can never be determined ; but by the help of decimals, we can approximate towards the root to any assigned degree of exactness.

The roots, which approximate, are called *surd roots;* and those, which are perfectly accurate, are called *rational roots.*

Roots are sometimes denoted by writing the character $\sqrt{}$ before the power, with the index of the root over it ; thus, the 3d root of 36 is expressed, $\sqrt[3]{36}$, and the second root of 36 is $\sqrt{36}$, the index 2 being omitted when the square root is designed.

If the power be expressed by several numbers with the sign $+$ or $-$ between them, a line is drawn from the top of the sign over all the parts of it ; thus, the 3d root of $\overline{42 + 22}$ is $\sqrt[3]{42 + 22}$, and the second root of $\overline{59 - 17}$ is $\sqrt{59 - 17}$, &c.

Sometimes roots are designed like powers with fractional indices. Thus the square root of 15 is $15^{\frac{1}{2}}$, the cube root of 21 is $21^{\frac{1}{3}}$. and the 4th root of $\overline{37 - 20}$ is $\overline{37 - 20}^{\frac{1}{4}}$, &c.

17*

A TABLE OF POWERS.

	1	2	3	4	5	6	7	8	9
1st Power	1	2	3	4	5	6	7	8	9
2d Power	1	4	9	16	25	36	49	64	81
3d Power	1	8	27	64	125	216	343	512	729
4th Power	1	16	81	256	625	1296	2401	4096	6561
5th Power	1	32	243	1024	3125	7776	16807	32768	59049
6th Power	1	64	729	4096	15625	46656	117649	262144	531441
7th Power	1	128	2187	16384	78125	279936	823543	2097152	4782969
8th Power	1	256	6561	65536	390625	1679616	5764801	16777216	43046721
9th Power	1	512	19683	262144	1953125	10077696	40353607	134217728	387420489
10th Power	1	1024	59049	1048576	9765625	60466176	282475249	1073741824	3486784401

SECTION LIII.

THE EXTRACTION OF THE SQUARE ROOT.

1. What is the square root of 576 ?

OPERATION.

```
576(24 Ans.
400
———
44)176
   176
```

To illustrate this question, I will suppose, that I have 576 tile, each of which is one foot square ; I wish to know the side of a square room, whose floor they will pave or cover.

If we find a number multiplied by itself, that will produce 576, that number will be the side of a square room, which will require 576 tiles to cover its floor. We perceive, that our number (576) consists of three figures ; therefore there will be two figures in its root ; for the product of any two numbers can have at most, but just so many figures as there are in both factors ; and at least, but one less. We will therefore, for convenience, divide our number (576) into two parts, called periods, writing a point over the right hand figure of each period ; thus 576. We now find by the table of powers, that the greatest square number in the left hand period 5 (hundred) is 4 (hundred ;) and that its root is 2, which we write in the quotient. (See operation.) As this 2 is in the place of tens, its value must be 20, and its square 400.

Let this be represented by a square, whose sides measure 20 feet each, and whose contents will therefore be 400 square feet. See figure D. We now subtract 400 from 576 and there remain 176 square feet to be arranged on two sides of the figure D, in order that its form may remain square. We therefore double the root 20, one of the sides, and it gives the length of the two sides to be enlarged : viz. 40. We then inquire how many times 40 as a divisor

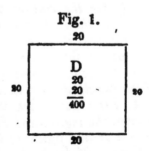

Fig. 1.

is contained in the dividend (except the right hand figure) and find it to be 4 times : this we write in the root and also in the divisor.

This 4 is the breadth of the addition to our square. (See figure 2.) And this breadth, multiplied by the length of the two additions, (40,) gives the contents of the two figures, E and F, 160 square feet, which is 80 feet for each.

There now remains the figure G, to complete the square, each side of which is 4 feet ; it being equal to the breadth of the additions E and F. Therefore, if we square 4 we have the contents of the last addition G=16. It is on account of this last addition, that the last figure of the root is placed in the divisor. If we now multiply the divisor, 44, by the last figure in the root, (4,) the product will be 176, which is equal to the remaining feet after we had formed our first square, and equal to the additions E, F, and G, in figure 2. We therefore perceive, that figure 2 may represent a floor 24 feet square, containing 576 square feet. From the above we infer the following

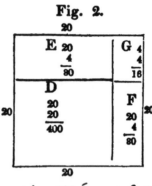

Fig. 2.

D contains 400 square feet.
E   "   80   do.   do.
F   "   80   do.   do.
G   "   16   do.   do.

Proof 576
or
24×24=576

. **RULE.**

1. *Distinguish the given number into periods of two figures each, by putting a point over the place of units, another over the place of hundreds, and so on, which points show the number of figures the root will consist of.*

2. *Find the greatest square number in the first or left hand period, place the root of it at the right hand of the given number, (after the manner of a quotient in division,) for the first figure of the root, and the square number under the period, and subtract it therefrom, and to the remainder bring down the next period for a dividend.*

3. *Place the double of the root already found, on the left hand of the dividend for a divisor.*

4. *Seek how often the divisor is contained in the dividend (except the right hand figure) and place the answer in the root for the second figure of it, and likewise on the right hand of the divisor. Multiply the divisor with the figure last annexed by the figure last placed in the root, and subtract the product from the dividend. To the remainder join the next period for a new dividend.*

5. *Double the figures already found in the root for a new divisor, (or bring down your last divisor for a new one, doubling the right hand figure of it,) and from these find the next*

figure in the root as last directed, and continue the operation in the same manner, till you have brought down all the periods.

NOTE 1st. — If, when the given power is pointed off as the power requires, the left hand period should be deficient, it must nevertheless stand as the first period.

NOTE 2d. — If there be decimals in the given number, it must be pointed both ways from the place of units. If, when there are integers, the first period in the decimals be deficient, it may be completed by annexing so many ciphers as the power requires. And the root must be made to consist of so many whole numbers and decimals as there are periods belonging to each; and when the periods belonging to the given numbers are exhausted, the operation may be continued at pleasure by annexing ciphers.

2. What is the square root of 278784 ?

$$278784(528 \text{ Answer.}$$
$$25$$
$$\overline{}$$
$$102)287$$
$$204$$
$$\overline{}$$
$$1048)8384$$
$$8384$$

3. What is the square root of 776161 ?   Ans. 881.
4. What is the square root of 806404 ?   Ans. 898.
5. What is the square root of 998001 ?   Ans. 999.
6. What is the square root of 10342656 ?   Ans. 3216.
7. What is the square root of 9645192360241 ?

Ans. 3105671.

8. Extract the square root of 234.09.   Ans. 15.3.
9. Extract the square root of .000729.   Ans. .027.
10. Required the square root of 17.3056.   Ans. 4.16.
11. Required the square root of 373.   Ans. 19.3132079+.
12. Required the square root of 8.93.   Ans. 2.98831055+.
13. What is the square root of 1.96 ?   Ans. 1.4.
14. Extract the square root of 3.15.   Ans. 1.77482393+.
15. What is the square root of 572199960721 ?

Ans. 756439.

If it be required to extract the square root of a vulgar fraction, reduce the fraction to its lowest terms, then extract the square root of the numerator for a new numerator, and of the denominator for a new denominator; or reduce the vulgar fraction to a decimal, and extract its root.

16. What is the square root of $\frac{144}{1521}$?   Ans. $\frac{4}{1}$.
17. What is the square root of $\frac{273}{550}$?   Ans. $\frac{7}{1}$.
18. What is the square root of $42\frac{1}{4}$?   Ans. $6\frac{1}{2}$.

19. What is the square root of $52\frac{1}{16}$?   Ans. $7\frac{1}{4}$.

20. What is the square root of $95\frac{1}{16}$?   Ans. $9\frac{3}{4}$.

21. What is the square root of $363\frac{1}{361}$?   Ans. $19\frac{1}{19}$.

22. Extract the square root of $6\frac{2}{5}$.   Ans. 2.5298+.

23. Extract the square root of $8\frac{2}{3}$.   Ans. 2.9519+.

24. Required the square root of 2.   Ans. 1.41421+.

APPLICATION OF THE SQUARE ROOT.

The following propositions are demonstrably true.

25. Circles are to each other, as the squares of their diameter.

26. In a right-angled triangle, the square of the longest side is equal to the sum of the squares of the other two sides.

27. In all similar triangles, that is, in all triangles whose corresponding angles are equal, the sides about the equal angles are in direct proportion to each other; that is, as the longest side of one triangle is to the longest side of the other, so is the corresponding side of the former triangle to the corresponding side of the latter.

28. If the diameter of a circle be multiplied by 3.14159, the product is the circumference.

29. If the square of the diameter of a circle be multiplied by .785398, the product is the area.

· 30. If the square root of half the square of the diameter of a circle be extracted, it is the side of an inscribed square.

31. If the area of a circle be divided by .785398, the quotient is the square of the diameter.

32. A certain general has an army of 141376 men. How many must he place in rank and file to form them into a square ?

Ans. 376.

33. If the area of a circle be 1760 yards, how many feet must the side of a square measure to contain that quantity ?

Ans. 125.857+feet.

34. If the diameter of a round stick of timber be 24 inches, how large a square stick may be hewn from it ?

Ans. 16.97+inches.

35. I wish to set out an orchard of 2400 mulberry trees, so that the length shall be to the breadth as 3 to 2; and the distance of each tree, one from the other, 7 yards. How many trees must be in length, and how many in breadth; and how many square yards of ground do they stand on ?

Trees in length,   60)

Trees in breadth,   40 } Answer.

Square yards,   112749)

36. If a lead pipe $\frac{3}{4}$ of an inch in diameter, will fill a cistern in 3 hours, what should be its diameter to fill it in 2 hours ?

Ans. .918+inches.

37. If a pipe 1½ inches in diameter will fill a cistern in 50 minutes, how long would it require a pipe, that is 2 inches in diameter, to fill the same cistern ?   Ans. 28m. 7½sec.

38. If a pipe 6 inches in diameter, will draw off a certain quantity of water in 4 hours ; in what time would it take 3 pipes of four inches in diameter, to draw off twice the quantity ?
.   Ans. 6 hours.

39. If a line, 144 feet long, will reach from the top of a fort to the opposite side of a river, that is 64 feet wide ; what is the height of the fort ?   Ans. 128.99+.

40. A certain room is 20 feet long, 16 feet wide, and 12 feet high ; how long must be a line to extend from one of the lower corners to an opposite upper corner ?   Ans. 28.28 feet.

41. Two ships sail from the same port ; one goes due north, 128 miles, the other due east, 72 miles ; how far are the ships from each other ?   Ans. 146.86+.

42. There are two columns in the ruins of Persepolis, left standing upright ; one is 70 feet above the plane, and the other 50 ; in a straight line between these, stands an ancient small statue, 5 feet in height ; the head of which is 100 feet from the summit of the higher, and 80 feet from the top of the lower column. Required the distance between the tops of the two columns.   Ans. 143.537+feet.

43. The height of a tree, growing in the centre of a circular island, 100 feet in diameter, is 160 feet ; and a line extending from the top of it to the further shore is 400 feet. What is the breadth of the stream, provided the land on each side of the water be level ?   Ans. 316.6+feet.

44. A ladder 70 feet long is so planted as to reach a window 40 feet from the ground, on one side of the street, and without moving it at the foot, will reach a window 30 feet high on the other side ; what is the breadth of the street ?
          Ans. 120.69+feet.

45. If an iron wire, $\frac{1}{10}$ inch in diameter, will sustain a weight of 450 pounds ; what weight might be sustained by a wire an inch in diameter ?   Ans. 45000 lbs.

46. Suppose a tree to stand on a horizontal plane 80 feet in height ; at what height from the ground must it be cut off, so that the top of it may fall on a point 40 feet from the bottom of the tree, the end where it was cut off resting on the stump ?
          Ans. 30 feet.

47. Four men, A, B, C, D, bought a grindstone, the diameter of which was 4 feet ; they agreed that A should grind off his share first, and that each man should have it alternately until he had worn off his share ; how much will each man grind off ?
          Ans. A 3.22+, B 3.81+, C 4.97+, D 12 inches.

48. What is the length of a rope that must be tied to a horse's neck, that he may feed on an acre ?   Ans. 7.136+rods.

SECTION LIV.

EXTRACTION OF THE CUBE ROOT.

A CUBE is a regular body, with six equal sides.

A number is said to be cubed, when it is multiplied into its square.

To extract the cube root, is to find a number, which, multiplied into its square, will produce the given number.

RULE.

1. *Separate the given number into periods of three figures each, by putting a point over the unit figure, and every third figure beyond the place of units.*

2. *Find by the table the greatest cube in the left hand period, and put its root in the quotient.*

3. *Subtract the cube thus found, from this period, and to the remainder bring down the next period; call this the dividend.*

4. *Multiply the square of the quotient by 300, calling it the triple square; multiply also the quotient by 30, calling it the triple quotient; the sum of these call the divisor.*

5. *Find how many times the divisor is contained in the dividend, and place the result in the quotient.*

6. *Multiply the triple square by the last quotient figure, and write the product under the dividend; multiply the square of the last quotient figure by the triple quotient, and place this product under the last; under all, set the cube of the last quotient figure, and call their sum the subtrahend.*

7. *Subtract the subtrahend, from the dividend and to the remainder bring down the next period for a new dividend, with which proceed as before, and so on, till the whole is completed.*

NOTE.—The same rule must be observed for continuing the operation, and pointing for decimals, as in the square root.

To illustrate this rule, let the pupil obtain the following apparatus.

1. A cubical block, whose side measures two inches; which let him call No. 1.

2. Three blocks, two inches square and half an inch thick. These let him call No. 2.

3. Three blocks, two inches long and half an inch square. lese he should call No. 3.

. One cubic block each side of which is half an inch in gth. Call this No. 4.

We now consider the following question.

I have 46.656 cubic blocks of granite, which measure 1 foot on each side. With these I wish to erect a cubical monument. It is required to ascertain how many blocks or feet will be the length of one side of the monument.

It is evident, that the number of blocks will be equal to the cube root of 46.656. As the given number consists of five figures, its cube root will contain two places. For the cube of any number can never contain more than three times that number, and at least but two less. We therefore *separate the given number into periods of three figures each, by putting a point over the unit figure, and every third figure beyond the place of units*; thus 46.656. We find by the table of powers the greatest power in the left hand period, 46 (thousand) is 27 (thousand,) the root of which is 3. This root we write in the quotient, and as it will occupy the place of *tens*, its real value is 30. If this be considered the side of a cube, it will contain 27.000 cubic feet, 30×30×30= 27.000.

Let this cube be represented by our apparatus No. 1. From 46.656 we take 27.000 and there remain 19.656 cubic feet. That is, *we should subtract the cube of the quotient from the first period, and to the remainder, we should bring down the next period.* (See operation.) The cubic feet or blocks, that remain must be applied to three sides of the monument; for unless a cube be equally increased on three sides, it ceases to be a cube. To effect this, we must find the superficial contents of three sides of the cube, and with this we must divide the remaining number of cubic feet or blocks, and the quotient will show the *thickness* of the additions. As the length of a side is 30, the superficial contents will be 30×30=900; and this multiplied by 3, will give the whole surface of the three sides; thus, 900×3=2700. This is equivalent to *multiplying the square of the quotient* by 900; thus, 3×3×900=8100. *With this as a divisor, we inquire how many times it is contained in 19.656, and find it to be 6 times. (One or two units are generally to be allowed on account of the other deficiencies in enlarging the cube.) This 6 is the thickness of each of the three additions to the cube; and by multiplying their superficial contents by it, we have the amount of the additions to be made; thus, 2700×6=16200; that is, we *multiply the triple square by the last quotient figure.* (See operation.)

If we now apply our blocks of the apparatus No. 2. to the three sides of the block, No. 1, we shall find there are three other deficiencies to complete the cube, the length of which is equal to that

OPERATION.

$$46.656(36 \text{ Ans.}$$
$$27$$

2700)19656

16200
3240
216

19656

* It is not necessary, that the *triple quotient* should be added to form a divisor,

18

of the addition, 30 ; and the height and breadth of each are equal
to the thickness of the additions, 6. To find the contents of these,
we multiply the continued product of the length, breadth, and
thickness of each by their number ; thus, 30×6×6×3=3240.
Or, which is the same thing, we *multiply the triple quotient by the
square of the last quotient figure* ; thus, 90×6×6=3240. Let the
contents of these be represented by the apparatus No. 3, and let
these, with No. 2, be applied to No. 1, and we shall find one
other small deficiency to form a complete cube ; the length,
breadth, and thickness of which are equal to the thickness of the
former additions, viz. 6. The contents of this may be found by
multiplying its length, breadth, and thickness together ; that is,
by *cubing the last quotient figure*, 6×6×6=216. Let this last
deficiency of the cube be supplied by adding the apparatus No.
4, to the former additions, and we shall find the cube is completed, and our cubical monument is finished ; and that the contents
of the first cube, together with the several additions to it, is equal
to the number of cubical blocks, 46.656.

Proof.

27000=contents of No. 1.
16200= do.   first additions.
3240= do.   second additions.
 216= do.   third addition.

46656=contents of the whole monument.

1 Required the cube root of 77308776.

OPERATION.

77308776(426 root.	4×4×300=4800
64	4× 30= 120
———	———
4920)13308=1st dividend.	1st divisor= 4920
———	———
9600	4800×2=9600
480	120×2×2= 480
8	2×2×2= 8
———	———
10088=1st subtrahend.	1st subtrahend=10088
———	
530460)3220776=2d dividend.	42×42×300=529200
	42×30= 1260
———	———
3175200	
45360	2d divisor= 530460
216	
———	———
3220776=2d subtrahend.	529200×6=3175200
	1260×6×6= 45360
	6×6×6= 216
	———
	2d subtrahend=3220776

2. What is the cube root of 34965783 ?　　　Ans. 327.
3. What is the cube root of 436036824287.　　　Ans. 7583.
4. What is the cube root of 84.604519 ?　　　Ans. 4.39.
5. Required the cube root of 54439939.　　　Ans. 379.
6. Extract the cube root of 60236288.　　　Ans. 392.
7. Extract the cube root of 109215352,-　　　Ans. 478.
8. What is the cube root of 116.930169 ?　　　Ans. 4.89.
9. What is the cube root of .726572699 ?　　　Ans. .899.
10. Required the cube root of 2.　　　Ans. 1.2599.+
11. Find the cube root of 11.　　　Ans. 2.2239.+
12. What is the cube root of 122615327232 ?　　　Ans. 4968.
13. What is the cube root of ₁₂₇/₃₄₃ ?　　　Ans. ⅘.
14. What is the cube root of ₁₃₃₁/₁₇₂₈ ?　　　Ans. ¹¹⁄₁₂.
15. What is the cube root of ₂₇/₆₄ ?　　　Ans. ⅜.
16. What is the cube root of ₂₃₉₁₈/₃₃₇₅₁ ?　　　Ans. ³⁷⁄₃₃.

APPLICATION OF THE CUBE ROOT.

Principles assumed.
Spheres are to each other, as the cubes of their diameter.
Cubes and all *similar* solid bodies, are to each other, as the cubes of their diameter, or *homologus* sides.

17. If a ball, 3 inches in diameter, weigh 4 pounds, what will be the weight of a ball that is 6 inches in diameter ?

Ans. 32 lbs.

18. If a cube of gold 1 inch in diameter, is worth $120, what is the value of a globe 3¼ inches in diameter ?　　　Ans. $5145.

19. If the weight of a well-proportioned man, 5 feet 10 inches in height, be 180 pounds, what must have been the weight of Goliath of Gath, who was 10 feet 4½ inches in height ?

Ans. 1015.1.+lbs.

20. If a bell, 4 inches in height, 3 inches in width, and ¼ of an inch in thickness, weigh 2 pounds, what should be the dimensions of a bell that would weigh 2000 pounds ?

Ans. 3 ft. 4 in. high, 2 ft. 6 in. wide, and 2¼ in. thick.

21. Having a small stack of hay 5 feet in height weighing 1 cwt., I wish to know the weight of a similar stack, that is 20 feet in height.　　　Ans. 64 cwt.

22. If a man dig a small square cellar, which will measure 6 feet each way, in one day, how long would it take him to dig a similar one that measured 10 feet each way ?

Ans. 4.629+days.

23. If an ox, whose girt is 6 feet, weighs 600lbs., what is the weight of an ox whose girt is 8 feet ?　　　Ans. 1422.2+lbs.

A GENERAL RULE FOR EXTRACTING THE
ROOTS OF ALL POWERS.

RULE.

1. *Prepare the given number for extraction, by pointing off from the unit's place, as the required root directs.*

2. *Find the first figure of the root by trial, or by inspection, in the table of powers, and subtract its power from the left hand period.*

3. *To the remainder, bring down the first figure in the next period, and call it the dividend.*

4. *Involve the root to the next inferior power to that which is given, and multiply it by the number denoting the given power for a divisor.*

5. *Find how many times the divisor is contained in the dividend, and the quotient will be another figure of the root.*

6. *Involve the whole root to the given power, and subtract it from the given number, as before.*

7. *Bring down the first figure of the next period to the remainder for a new dividend, to which find a new divisor, as before; and in like manner proceed till the whole is finished.*

1. What is the cube root of 20346417 ?

OPERATION.

$$20346417(273$$

$2^3=$	8	=1st subtrahend.
$2^2 \times 3=12$)	123	=1st dividend.
$27^3=$	19683	=2d subtrahend.
$27^2 \times 3=2187$)	6634	=2d dividend.
$273^3=$	20346417	=3d subtrahend.

2. What is the fourth root of 34828517376 ?   Ans. 432.
3. What is the fifth root of 281950621875 ?   Ans. 195.
4. Required the sixth root of 1178420166015625.   Ans. 325.
5. Required the seventh root of 1283918464543864.
                 Ans. 144.
6. Required the eighth root of 218340105584896.   Ans. 62.

SECTION LV.

ARITHMETICAL PROGRESSION.[*]

WHEN a series of quantities, or numbers, increase or decrease by a constant difference, it is called arithmetical progression or progression by difference. The constant difference is called the *common difference*, or *ratio* of the progression.

Thus, let there be the two following series,—

$$1, \quad 5, \quad 9, \quad 13, \quad 17, \quad 21, \quad 25, \quad 29, \quad 33,$$
$$25, \quad 22, \quad 19, \quad 16, \quad 13, \quad 10, \quad 7, \quad 4, \quad 1.$$

The first is called an ascending series, or progression. The second is called a descending series, or progression.

The numbers which form the series, are called the *terms* of progression.

The first and last terms of progression, are called the *extremes*, and the other terms, the *means*.

Any three of the five following things being given, the other two may be found.

1. The first term.
2. The last term.
3. The number of terms.
4. The common difference.
5. The sum of the terms.

PROBLEM I.

The first term, last term, and the number of terms being given, to find the common difference.

RULE.

Divide the difference of the extremes by the number of terms less one, and the quotient is the common difference.

1. The extremes are 3 and 45, and the number of terms is 22. What is the common difference?

OPERATION.

$$\frac{45-3}{22-1} = 2 \text{ Answer.}$$

[*] As a demonstration of this rule is so much better understood by Algebra than Arithmetic, the author has thought best to omit it.

18*

2. A man is to travel from Boston to a certain place, in 11 days, and to go but 5 miles the first day, increasing every day by an equal increase, so that the last day's journey may be 45 miles. Required the daily increase.   Ans. 4 miles.

3. A man had 10 sons, whose several ages differed alike ; the youngest was 3 years, and the oldest 48.. What was the common difference of their ages ?   Ans. 5 years.

4. A certain school consists of 19 scholars ; the youngest is 3 years old, and the oldest 39. What is the common difference of their ages ?   Ans. 2 years.

PROBLEM II.

The first term, the last term, and the number of terms given to find the sum of all the terms.

RULE.

Multiply the sum of the extremes by the number of terms, and half the product will be the sum.

5. The extremes of an arithmetical series are 3 and 45, and the number of terms 22. Required the sum of the series.

OPERATION.

$$\frac{45+3\times22}{2}=528 \text{ Answer.}$$

6. A man going a journey, travelled the first day 7 miles, the last day 51 miles, and he continued his journey 12 days. How far did he travel ?   Ans. 348 miles.

7. In a certain school there are 19 scholars ; the youngest is 3 years old, and the oldest 39. What is the sum of their ages ?
Ans. 399 years.

8. Suppose a number of stones were laid a rod distant from each other, for thirty miles, it being the distance from Boston to Haverhill ; and the first stone a rod from a basket. What length of ground will that man travel over who gathers them up singly, returning with them one by one to the basket ?
Ans. 288090 miles, 2 rods.

PROBLEM III.

Given the extremes and the common difference, to find the number of terms.

RULE.

Divide the difference of the extremes by the common difference, and the quotient increased by one, will be the number of terms required.

9. If the extremes are 3 and 45, and the common difference 2, what is the number of terms ?

$$\frac{45-3}{2}+1=22 \text{ Answer.}$$

10. In a certain school where the ages of the scholars all differ alike; the oldest is 39 years old, and the youngest is 3 years, and the difference between the ages of each is 2 years. Required the number of scholars ?   Ans. 19 scholars.

11. A man going a journey, travelled the first day 7 miles and the last day 51 miles, and each day increased his journey by 4 miles. How many days did he travel ?

Ans. 12 days.

PROBLEM IV.

The extremes and common difference given, to find the sum of all the series.

RULE.

Multiply the sum of the extremes by their difference, increased by the common difference, and the product divided by twice the common difference will give the sum.

12. If the extremes are 3 and 45, and the common difference 2, what is the sum of the series ?   Ans. 528.

13. A owes B a certain sum, to be discharged in a year, by paying 6 cents the first week, and 18 cents the second week, and thus to increase every week by 12 cents till the last payment should be $6.18. What is the debt ?   Ans. $162.24.

PROBLEM V.

The extremes and sum of the series given, to find the common difference.

RULE.

Divide the product of the sum and the difference of the extremes, by the difference of twice the sum of the series, and the sum of the extremes, and the quotient will be the common difference.

14. The extremes are 3 and 45, and the sum of the series 528. What is the common difference ?   Ans. 2.

PROBLEM VI.

Given the first term, number of terms, and the sum of the series, to find the last term.

RULE.

Divide twice the sum by the number of terms, from the quotient take the first term, and the remainder will be the last.

15. A merchant being indebted to 22 creditors $528, ordered his clerk to pay the first $3, and the rest increasing in arithmetical progression. What is the difference of the payments, and the last payment?   Ans. 2 difference. $45 last payment.

SECTION LVI.

GEOMETRICAL SERIES.*

If there be three or more numbers, and if there be the same quotient, when the second is divided by the first, and the third divided by the second, and the fourth divided by the third, &c. those numbers are in *geometrical progression.* If the series *increase,* the quotient is more than unity, if it *decrease,* it is less than unity.

The following series are examples of this kind.

2,   6,   18,   54,   162,   436
64,   32,   16,   8,   4,   2

The former is called an ascending series, and the latter a descending series.

In the first, the quotient is 3, and is called the *ratio;* in the second, it is ½.

The first and last terms of a series are called *extremes,* and the other terms *means.*

PROBLEM I.

Given one of the extremes, the ratio, and number of terms, to find the other extreme.

RULE.

If the series be ascending, multiply the given extreme by such power of the ratio, whose exponent is equal to the number of

* This rule cannot well be demonstrated except by Algebra.

terms less 1; *and the product will be the other extreme. Or, if the series be descending, divide the given extreme by such power of the ratio, whose exponent is equal to the number of terms less* 1; *and the quotient will be the other extreme.*

But, as the last term, or any term near the last, is very tedious to be found by continual multiplication, it will often be necessary, in order to ascertain it, to have a series of numbers in Arithmetical Proportion, called *indices* or *exponents*, beginning either with a cipher, or an unit, whose common difference is *one*. When the *first* term of the series and *ratio* are *equal*, the *indices* must begin with an unit, and, in this case, the product of any two terms, is equal to that term, signified by the *sum* of their *indices*.

Thus, $\begin{cases} 1 & 2 & 3 & 4 & 5 & 6 & \text{&c. indices or arithmetical series.} \\ 2 & 4 & 8 & 16 & 32 & 64 & \text{&c. geometrical series.} \end{cases}$

Now 6+6=12=the index of the twelfth term, and 64×64= 4096=the twelfth term.

But, when the *first* term of the series and the *ratio* are *different* the *indices* must begin with a cipher, and the sum of the *indices* made choice of, must be *one less* than the *number of terms* given in the question ; because 1 in the *indices* stands over the *second term*, and 2 in the *indices* over the *third* term, &c. And, in this case, the *product* of any *two* terms, divided by the *first*, is equal to that term *beyond* the first, signified by the sum of their *indices*.

Thus $\begin{cases} 0 & 1 & 2 & 3 & 4 & 5 & 6 & \text{&c. indices.} \\ 1 & 3 & 9 & 27 & 81 & 243 & 729 & \text{&c. geometrical series.} \end{cases}$

Here 6+5=11, the index of the 12th term.

729×243=177147, the 12th term, because the first term of the series and ratio are different, by which means a cipher stands over the first term.

Thus by the help of these *indices*, and a few of the first terms in any geometrical series, any term, whose distance from the first term is assigned, though it were ever so remote, may be obtained without producing all the terms.

1. If the first term be 4, the ratio 4, and the number of terms 9, what is the last term ?

OPERATION.

1. 2. 3. 4. + 4 = 8
4. 16. 64. 256 ×256=65536=power of the ratio, whose exponent is less by 1, than the number of terms. 65536×4 the first term=262144=last term.

Or, 4×4^8=262144=last term as before.

2. If the last term be 262144, the ratio 4, and the number of terms 9, what is the first term ?

$$4=65536)262144(4=\text{the first term.}$$

3. If the last term is 72, the ratio 3, and the number of terms 6, what is the first term ?   Ans. ⅓.

4. If I were to buy 30 oxen, and to give 2 cents for the first ox, and 4 cents for the second, 8 cents for the third, &c., what would be the price of the last ox ?   Ans. $10737418.24.

5. If the first term be 5, the ratio 3, what is the seventh term ?   Ans. 3645.

6. If the first term be 50, the ratio 1.06, and the number of terms 5, what is the last term ?   Ans. 63.123848.

7. What is the amount of $160.00 at compound interest for 6 years ?   Ans. $226.96.3.05796096.

8. What is the amount of $300.00 at compound interest at 5 per cent. for 8 years ?   Ans. $443.23.6+.

9. What is the amount of $100.00 at 6 per cent. for 30 years ? Ans. $574.34.91172913250116264106332310802645846357252196069357387776.

PROBLEM II.

Given the first term, the ratio, and the number of terms, to find the sum of the series.

RULE.

Find the last term, as before, multiply it by the ratio, and from the product subtract the first term. Divide the remainder by the ratio less 1, and the quotient will be the answer.

Or, raise the ratio to a power, whose index shall be equal to the number of terms; from which subtract 1; divide the remainder by the ratio less 1, and the quotient, multiplied by the first term, will give the sum of the series.

10. If the first term be 10, the ratio 3, and the number of terms 7 ; what is the sum of the series ?

OPERATION.

$$\frac{3^7}{\square}\times 10 = 10930 \text{ Answer.}$$

11. How large a debt may be discharged in a year, by paying $1 the first month, $10 the second, and so on, in a ten-fold proportion, each month ?   Ans. 111111111111.

12. A gentleman offered a house for sale, on these moderate terms ; that, for the first door, he should charge 10 cents, for the second 20 cents, for the third 40 cents, and so on in a geometrical ratio ; there being 40 doors. What was the price of the house? Ans. $109951162777.50.

13. If the first term be 50, the ratio 1.06, and the number of terms 4, what is the sum of the series ?   Ans. 218.7808.

14. A gentleman deposited annually $10 in a bank from the time his son was born, until he was 20 years of age. Required the amount of the deposites at 6 per cent., compound interest, when his son was 21 years old.   Ans. $423.92.2+.

15. If the first term be 7, the ratio ⅓, and the number of terms 5, what is the sum of the series ?   Ans. 9₈₁.

16. If one mill had been put at interest at the commencement of the Christian era, what would it amount to at compound interest, supposing the principal to have doubled itself every 12 years, January 1, 1837 ?

Ans. $1141798154164767904846628775559596109106197.99.2.

If this sum was all in dollars, it would take the present inhabitants of the globe more than 1,000,000 years to count it. If it was reduced to its value in pure gold, and was formed into a globe, it would be many million times larger than all the bodies that compose the solar system.

SECTION LVII.

INFINITE SERIES.

An INFINITE SERIES is such, as being continued, would run on ad infinitum; but the nature of its progression is such, that by having a few of its terms given, the others to any extent may be known. Such are the following series.

1,   2,   4,   8,   16,   32,   64,   128, &c. ad infinitum.
125, 25, 5,   1,   ⅕,   ₂₅,   ₁₂₅,   ₆₂₅ &c. ad infinitum.

To find the sum of a decreasing series.

RULE.

Multiply the first term by the ratio, and divide the product by the ratio less 1, and the quotient is the sum of an infinite decreasing series.

1. What is the sum of the series, 4, 1, ¼, ₁₆, ₆₄, &c. continued to an infinite number of terms ?

OPERATION.
$$\frac{4 \times 4}{3} = 5\tfrac{1}{3}$$ Answer.

2. What is the sum of the series 5, 1, $\frac{1}{5}$, $\frac{1}{25}$, &c. continued to infinity ?   Ans. 6$\frac{1}{4}$.

3. If the following series, 8, $\frac{4}{9}$, $\frac{2}{15}$, $\frac{1}{33}$, &c. were carried to infinity, what would be their sum ?   Ans. 9$\frac{1}{4}$.

4. What is the sum of the following series, carried to infinity ?   1, $\frac{1}{3}$, $\frac{1}{9}$, $\frac{1}{27}$, $\frac{1}{81}$, &c.   Ans. 1$\frac{1}{2}$.

5. What is the sum of the following series, carried to infinity ?   11, $\frac{11}{2}$, $\frac{11}{4}$, &c.   Ans. 12$\frac{1}{4}$.

6. If the series $\frac{2}{3}$, $\frac{1}{3}$, $\frac{1}{6}$, $\frac{1}{12}$, $\frac{1}{24}$, &c. were carried to infinity, what would be their sum ?   Ans. 1$\frac{1}{4}$.

SECTION LVIII.

DISCOUNT BY COMPOUND INTEREST.

What is the present worth of \$600, due 3 years hence at 6 per cent. compound interest ?

OPERATION.   BY ANALYSIS.

$1.06)^3$=1.191016)600.00(\$503.77.+ Ans.   We find the amount of \$1, at compound interest for 3 years to be \$1.191016; therefore \$1 is the present worth of \$1.191016 due 3 years hence.   And if \$1 is the present worth of \$1.191016, the present worth of

$$\$600. = \frac{600}{1.191016} = \$503.77.1+.$$

Hence we see the propriety of the following

RULE.

Divide the debt by the amount of one dollar for the given time, and the quotient is the present worth, which, if subtracted from the debt, will leave the discount.

2. What is the present worth of \$500.00, due 4 years hence, at 6 per cent. compound interest ?   Ans. \$396.04.6+.

3. What is the present worth of \$1000.00, due 10 years hence, at 5 per cent. compound interest ?   Ans. \$613.91.3+.

4. What is the discount on \$800.00, due 2 years hence, at 6 per cent. compound interest ?   Ans. \$88.00.3+.

SECTION LXIX

ANNUITIES AT COMPOUND INTEREST.

To find the amount of an annuity.

An *annuity* is a certain sum of money to be paid at regular periods, either for a limited time or for ever.

The *present worth* or *value* of an annuity is that sum, which being improved at compound interest will be sufficient to pay the annuity.

The *amount* of an annuity is the compound interest of each payment added to their sum.

Therefore to find the amount of an annuity at compound interest, we adopt the following

RULE.

Make $1.00 the first term of a geometrical series, and the amount of $1.00 at the given rate per cent., the ratio. Carry the series to so many terms as the number of years, and find its sum.

Multiply the sum thus found by the given annuity and the product will be the amount

EXAMPLES.

1. What will an annuity of $60 per annum, payable yearly, amount to in 4 years, at 6 per cent. ?

$$1 + 1.06 + \overline{1.06}^2 + \overline{1.06}^3 = 4.374616$$
$$4.374616 \times 60 = \$262.47.6 + \text{Answer.}$$

Or, $$\frac{\overline{1.06}^4 - 1}{1.06 - 1} \times 60 = \$262.47.6 +. \text{ Answer.}$$

2. What will an annuity of $500.00 amount to in 5 years, at 6 per cent. ?   Ans. $2818.54.6+.

3. What will an annuity of $1000.00, payable yearly, amount to in 10 years ?   Ans. $13180.79.4+.

4. What will an annuity of $30.00, payable yearly, amount to in 3 years ?   Ans. $95.50.8+.

To find the present worth of an annuity.

As the first payment is made at the end of the year, its present worth or value is a sum, that will amount in one year to that payment; and as the second payment is made at the end of the

19

second year, its value is a sum, that will, at compound interest, amount in two years to that payment; and the same principal is adopted for the third, fourth, &c. years. This may be illustrated in the following question.

5. What is the present worth of an annuity of $1, to continue 5 years at compound interest?

The present worth of $1.00 for 1 year =$0.943396
Do. do. do. 2 years= 0.889996
Do. do. do. 3 years= 0.839619
Do. do. do. 4 years= 0.792094
Do do. do. 5 years= 0.747258

$4.212363

By the above illustration, we perceive, that the present worth of an annuity of $1 to continue 5 years, is $4.21.2+. Hence, having found the present worth of an annuity of $1 for any given time by Section LVIII, the present worth of any other sum may be found by multiplying it by the present worth of $1 for that time; therefore we induce the following

RULE.

Multiply the present worth of the annuity of one dollar for the given time by the given annuity, and the product is the present worth required. Or,

Find the amount of the annuity by the last rule, and then find its present worth.

6. What is the present worth of an annuity of $60 to be continued 4 years at compound interest?

First Method.

Present worth of $1.00 for 1 year $0.943396
Do. do. 2 do. 0.889996
Do. do. 3 do. 0.839619
Do. do. 4 do. 0.792093

$3.465104.
$3.465104×60=$207.90.6+Answer

Second Method.

$$\frac{1.06^4-1}{1.06-1}=\$4.374616\times.792093=\$3.465102\times60=\$207.90.6+\text{Ans.}$$

7. A gentleman wishes to purchase an annuity, which shall afford him at 6 per cent. compound interest $500 a year for ten years. What sum must he deposit in the annuity office to produce it?   Ans. $3680.04+

By the assistance of the following tables, questions in annuities may be easily performed.

TABLE I.

Showing the amount of $1. *annuity* from 1 year to 40.

Years.	5 per cent.	6 per cent.	Years.	5 per cent.	6 per cent.
1	1.000000	1.000000	21	35.719252	39.992727
2	2.050000	2.060000	22	38.505214	43.392290
3	3.152500	3.183600	23	41.430475	46.995828
4	4.310125	4.374616	24	44.501999	50.815577
5	5.525631	5.637093	25	47.727099	54.864512
6	6.801913	6.975319	26	51.113454	59.156383
7	8.142008	8.393838	27	54.669126	63.705766
8	9.549109	9.897468	28	58.402583	68.528112
9	11.026564	11.491316	29	62.322712	73.639798
10	12.577893	13.180795	30	66.438347	79.058186
11	14.206787	14.971643	31	70.760790	84.801677
12	15.917127	16.369941	32	75.298829	90.889778
13	17.712983	18.882138	33	80.063771	97.343165
14	19.598632	21.015066	34	85.066959	104.183755
15	21.578564	23.275970	35	90.220307	111.434780
16	23.657492	25.672528	36	95.836323	119.120867
17	25.840366	28.212880	37	101.628139	127.268119
18	28.132385	30.905653	38	107.709546	135.904206
19	30.539004	33.759992	39	114.095023	145.058458
20	33.065954	36.785591	40	120.799774	154.761966

TABLE II.

Showing the present value of an *annuity* of $1 from 1 year to 40

Years.	5 per cent.	6 per cent.	Years.	5 per cent.	6 per cent.
1	0.952381	0.943396	21	12.821153	11.764077
2	1.859410	1.833393	22	13.163003	12.041582
3	2.723248	2.673012	23	13.488574	12.303379
4	3.545950	3.465106	24	13.798642	12.550358
5	4.329477	4.212364	25	14.093945	12.783356
6	5.075692	4.917324	26	14.375185	13.003166
7	5.786373	5.582381	27	14.643034	13.210534
8	6.463213	6.209794	28	14.898127	13.406164
9	7.107822	6.801692	29	15.141074	13.590721
10	7.721735	7.360087	30	15.372451	13.764831
11	8.306414	7.886875	31	15.592810	13.929096
12	8.863252	8.388844	32	15.802677	14.084043
13	9.393573	8.852683	33	16.002549	14.230230
14	9.898641	9.294984	34	16.192904	14.368141
15	10.379658	9.712249	35	16.374194	14.498246
16	10.837770	10.105895	36	16.546852	14.620987
17	11.274066	10.477260	37	16.711287	14.736780
18	11.689587	10.827603	38	16.867893	14.846019
19	12.085321	11.158116	39	17.017041	14.949075
20	12.462216	11.469921	40	17.159086	15.046297

8. What is the present worth of an annuity of $200 at 5 per cent. compound interest for 7 years ?   Ans. $1157.27+.

9. What is the present worth of an annuity of $300 to continue 8 years at 6 per cent. compound interest ?

Ans. $1862.93.8+.

10. What is the present value of an annuity of $100 at 6 per cent. for 9 years ?   Ans. $680.16.9+.

Questions to be performed by preceding tables.

11. What will an annuity of $30 amount to in 11 years at 6 per cent. ?

By Table I., the amount of $1 for 11 years is $14.971643; therefore, $14.971643×30=$449.14.9+ Answer.

12. What is the present worth of an annuity of $80 for 30 years at 5 per cent.

By Table II., the present worth of $1 for 30 years is $15.372451, therefore, $15.372451×80=$1229.79.6+ Answer.

13. What will an annuity of $800 amount to in 25 years at 5 per cent. ?   Ans. $38181.67.9+.

14. What will an annuity of $40 amount to in 30 years at 6 per cent. ?   Ans. $3162.32.7+.

15. Required the present worth of an annuity of $500 to continue 40 years at 6 per cent.   Ans. 7523.14.8+.

16. A certain parish in the town of B., having neglected for 6 years to pay their minister's salary of $700, what in justice, provided he has preached the truth, should he receive ?

Ans. $4882.72.3.

SECTION LX.

ALLIGATION.

ALLIGATION teaches how to compound or mix together several simples of different qualities, so that the composition may be of some intermediate quality or rate.   It is of two kinds. Alligation Medial, and Alligation Alternate.

ALLIGATION MEDIAL.

Alligation Medial teaches how to find the mean price of several articles mixed, the quantity and value of each being given.

RULE.

As the sum of the quantities to be mixed is to their value, so is any part of the composition to its mean price.

EXAMPLES.

1. A grocer mixed 2 cwt. of sugar, at $9.00 per cwt., and 1 cwt., at $7.00 per cwt., and 2 cwt. at $10.00 per cwt. ; what is the value of 1 cwt. of this mixture ?

```
2 cwt., at 9.00=18.00
1          7.00= 7.00
2         10.00=20.00
___                _____
5       :       45.00 :: 1 cwt. : $9.00 Answer.
```

2. If 19 bushels of wheat, at $1.00 per bushel, be mixed with 40 bushels of rye, at $0.66 per bushel, and 11 bushels of barley, at $0.50 per bushel ; what is a bushel of this mixture worth ?

Ans. $0.72.7½.

3. If 3 pounds of gold, of 22 carats fine, be mixed with 3 pounds, of 20 carats fine ; what is the fineness of the mixture ?

Ans. 21 carats.

4. If I mix 20 pounds of tea, at 70 cents per pound, with 15 pounds, at 60 cents per pound, and 80 pounds, at 40 cents per pound ; what is the value of 1 pound of this mixture ?

Ans. 0.47\frac{13}{23}$.

NOTE. — If an ounce, or any other quantity of pure gold, be divided into 24 equal parts; these parts are called *carats*. But gold is often mixed with some baser metal, which is called the *alloy;* and the mixture is said to be so many carats fine, according to the proportion of pure gold contained in it; thus, if 22 carats of pure gold, and 2 of alloy be mixed together, it is said to be 22 carats fine.

ALLIGATION ALTERNATE.

This rule teaches us how, from the prices of several articles given, to find how much of each must be mixed, to bear a certain price.

CASE I.

RULE.—*Place the prices under each other, in the order of their value; connect the price of each ingredient, which is less in value, than the intended compound, with one, which is of greater value than the compound. Place the difference between the price and that of each simple, opposite to the price with which they are connected.*

19*

EXAMPLES.

5. A merchant has spices, some at 18 cents a lb., some at 24 cents, some at 48 cents, and some at 60 cents. How much of each sort must he mix that he may sell the mixture at 40 cents a lb. ?

$$
\text{Mean rate } 40 \begin{cases} 18 \\ 24 \\ 48 \\ 60 \end{cases}
\begin{array}{l}
\text{lbs.} \quad \text{cts.} \\
20 \text{ at } 18 \\
8 \text{ `` } 24 \\
16 \text{ `` } 48 \\
22 \text{ `` } 60
\end{array} \Bigg\} \text{ Answers. ,}
$$

$$
\text{Mean rate } 40 \begin{cases} 18 \\ 24 \\ 48 \\ 60 \end{cases}
\begin{array}{l}
\text{lbs.} \quad \text{cts.} \\
8 \text{ at } 18 \\
20 \text{ `` } 24 \\
22 \text{ `` } 48 \\
16 \text{ `` } 60
\end{array} \Bigg\} \text{ Answers}
$$

$$
\text{Mean rate } 40 \begin{cases} 18 \\ 24 \\ 48 \\ 60 \end{cases}
\begin{array}{l}
\text{lbs.} \quad \text{lbs.} \quad \text{lbs.} \quad \text{cts.} \\
20 + 8 \mid 28 \text{ at } 18 \\
20 + 8 \mid 28 \text{ `` } 24 \\
22 + 16 \mid 38 \text{ `` } 48 \\
22 + 16 \mid 38 \text{ `` } 60
\end{array} \Bigg\} \text{ Answers.}
$$

The various answers arise from the different ways of connecting the rates of the ingredients. Almost every question will admit of seven different ways of connecting, and will produce as many answers ; and by increasing the answers of any one question *proportionally*, an indefinite number of answers might be produced.

NOTE. — By connecting the less rate with the greater, and placing the difference between them and the mean rate, alternately, the quantities are such, that there is precisely as much gained by one quantity, as is lost by the other; and, therefore, the *gain* and *loss* upon the whole are equal, and are precisely the *proposed rate.*

6. How much barley, at 50 cents a bushel, and rye, at 75 cents, and wheat, at $1.00, must be mixed, that the composition may be worth 80 cents a bushel ?
Ans. 20 bushels of rye, 20 of barley, and 35 of wheat.

7. A goldsmith would mix gold, of 19 carats fine, with some of 15, 23 and 24 carats fine, that the compound may be 20 carats fine. What quantity of each must he take ?
Ans. 4oz. of 15 carats, 3oz. of 19, 1oz. of 23, and 5oz. of 24.

8. It is required to mix several sorts of wine, at 60 cents, 80 cents, and $1.20, with water, that the mixture may be worth 75 cents per gallon ; how much of each sort must be taken ?
Ans. 45 gals. of water, 5 gals. of 60 cents, 15 gals. of 80 cents, and 75 gals. of $1.20.

CASE II.

When one of the ingredients is limited to a certain quantity.

RULE.

Take the difference between each price, and the mean rate as before; then say, as the difference of that simple, whose quantity is given, is to the rest of the differences severally; so is the quantity given, to the several quantities required.

EXAMPLES.

9. How much wine, at 5s., at 5s. 6d. and 6s. a gallon, must be mixed with 3 gallons, at 4s. per gallon, so that the mixture may be worth 5s. 4d. per gallon ?

$$64 \begin{cases} 48 \\ 60 \\ 66 \\ 72 \end{cases}$$
$$\begin{aligned} 8+2&=10 \\ 8+2&=10 \\ 16+4&=20 \\ 16+4&=20 \end{aligned}$$
Then $10:10::3:3$
$10:20::3:6$
$10:20::3:6$

Ans. 3 gals. at 5s., 6 at 5s. 6d., and 6 at 6s.

10. A grocer would mix teas, at 12s., 10s., and 6s. per pound, with 20 pounds, at 4s. per pound ; how much of each sort must he take to make the composition worth 8s. per pound ?
Ans. 20lbs. at 4s.; 10lbs. at 6s., 10lbs. at 10s., and 20lbs. at 12s.

11. How much Port wine, at $1.75, and Temperance wine, at $1.25, must be mixed with 20 gallons of water, that the whole may be sold at $1.00 per gallon ?
Ans. 20gals. Port wine, and 20gals. Temperance wine.

CASE III. *

When the sum and quality of the ingredients are given.

* To this case belongs that curious fact of king Hiero's crown.

Hiero, king of Syracuse, gave orders for a crown to be made of pure gold, but suspecting that the workmen had debased it, by mixing it with silver or copper, he recommended the discovery of the fraud to the famous Archimides, and desired to know the exact quantity of alloy in the crown.

Archimides, in order to detect the imposition, procured two other masses, the one of pure gold, the other of copper, and each of the same weight of the former; and by putting each separately, into a vessel full of water, the quantity of water expelled by them, determined their specific gravities; from which, and their given weights, the exact quantities of gold and alloy in the crown may be determined.

Suppose the weight of each crown to be 10 pounds, and that the water expelled by the copper, was 92 pounds, by the gold 52 pounds, and by the compound crown, 64 pounds: what will be the quantities of gold and alloy in the crown ? The rates of the simples are 92 and 52, and of the compound 64; therefore,

$$64 \begin{vmatrix} 92 \\ 62 \end{vmatrix}$$
12 of copper, $40:10::12:$ 3lbs. of copper $\}$
28 of gold, $40:10::28:$ 7lbs. of gold $\}$ Answer.

RULE.

Find an answer as before, by linking; then say, as the sum of the quantities or differences, thus determined, is to the given quantity, so is each ingredient, found by linking to the required quantity of each.

EXAMPLES.

12. How many gallons of water must be mixed with wine, at $1.50, so as to fill a vessel containing 100 gallons, and that it might be sold at $1.20 per gallon?

$$120 \begin{cases} 0 \text{———} & 30 \\ 150 \text{——} & 120 \end{cases} \quad \begin{matrix} \text{gals.} & \text{gals. gals.} \\ 150 : 100 :: 30 : 20 \text{ water} \\ 150 : 100 :: 120 : 80 \text{ wine} \end{matrix} \Big\} \text{Answer.}$$

$$\overline{\quad 150 \quad}$$

13. A merchant has sugar, at 8 cents, 10 cents, 12 cents, and 20 cents; with these he would fill a hogshead, that would contain 200 pounds. How much of each kind must he take, that he may sell the mixture at 15 cents per pound?
Ans. $33\frac{1}{3}$lbs., of 8 cts., 10 cts., and 12 cts., and 100 lbs. of 20 cts.

14. How much gold, of 15, 17, and 22 carats fine, must be mixed with 5 ounces of 18 carats fine, so that the composition may be 20 carats fine?

Ans. 5oz. of 15, 5oz. of 17, and 25oz. of 22.

The last question should be performed by Rule under Case II., p. 228.

SECTION LXI.

PERMUTATIONS AND COMBINATIONS.

THE permutation of quantities, is the showing how many different ways the order or position of any given number of things may be changed.

The combination of quantities, is the showing how often a less number of things can be taken out of a greater, and combined together, without considering their places or the order they stand in.

CASE I.

To find the number of permutations or changes, that can be made of any given number of things, all different from each other.

RULE.

Multiply all the terms of the natural series of numbers, from 1 up to the given number, continually together, and the last product will be the answer required.

EXAMPLES.

1. How many changes may be rung on 6 bells;
   $1 \times 2 \times 3 \times 4 \times 5 \times 6 = 720$ changes, Answer.

2. For how many days can 10 persons be placed in a different position at dinner ?   Ans. 3628800.

3. How many changes may be rung on 12 bells, and how long would they be in ringing, supposing 10 changes to be rung in one minute, and the year to consist of 365 days, 5 hours, and 49 minutes ?   Ans. 479001600, and 91y. 26d. 22h. 41m.

4. How many changes or variations will the alphabet admit of ? .   Ans. 403291461126605635584000000.

CASE II.

Any number of different things being given, to find how many changes can be made out of them, by taking any given number of quantities at a time.

RULE.

Take a series of numbers, beginning at the number of things given, and decreasing by 1, to the number of quantities to be taken at a time; the product of all the terms will be the answer required. .

EXAMPLES.

5. How many changes can be rung with 4 bells out of 8 ?
   $8 \times 7 \times 6 \times 5 = 1680$ changes, Ans.

6. A butcher wishing to buy some sheep, asked the owner how much he must give him for 20 ; on hearing his price, he said it was too much ; the owner replied, that he should have 10, provided he would give him a cent for each different choice of 10 in 20, to which he agreed.   How much did he pay for the 10 sheep, according to the bargain ?   Ans. $1847.56

7. How many words can be made with 6 letters out of 26 of the alphabet, admitting a word might be made from consonants ?
                                                Ans. 165765600.

CASE III.

To find the compositions of any, in an equal number of sets, the things being all different.

RULE.

Multiply the number of things in every set continually together, and the product will be the answer required.

EXAMPLES.

8. Suppose there are 4 companies, in each of which there are 9 men ; it is required to find how many ways 4 men may be chosen, one out of each company. $9\times9\times9\times9=6561$ Ans.

9. There are 4 companies, in one of which there are 6 men, in another 8, and in each of the other two 9 men. What are the choices, by a composition of 4 men, one out of each company ? Ans. 3888.

10. How many changes are there in throwing 5 dice ?
Ans. 7776.

11. If the gambler make use of 3 dice, what advantage has he over his opponent ?

We will suppose the bet to be on the ace. Then by the rule, there will be $6\times6\times6=216$ different ways which the dice may present themselves, that is, with and without an ace. Then, if we exclude the ace side of the die, there will be 5 sides left, and $5\times5\times5=125$ ways without the ace ; there will, therefore, remain only $216-125=91$ ways, in which there could be an ace. The chance for the gambler, then, is as 125 to 91 ; that is, out of 216 throws, the probability is, that he will win 125 times, and lose 91 times.

Ans. 125 chances for him, 91 against him.

Corollary. Dishonesty is always connected with gambling.

SECTION LXII.

POSITION.

Position is a method of performing such questions as cannot be resolved by the common direct rules, and is of two kinds, called *single* and *double*.

SINGLE POSITION.

Single Position teaches us to resolve those questions, whose results are proportional to their suppositions.

RULE.

Take any number, and perform the same operations with it,

*as are desirable to be performed in the question. Then say,
as the result of the operation is to the position, so is the result
in the question to the number required.*

EXAMPLES.

1. A schoolmaster being asked how many scholars he had,
replied, that if he had as many more as he now had, and half as
many more, he should have 200; of how many did his school
consist ?

Suppose he had   60   As 150 : 200 : : 60
Then, as many more 60   60
½ half as many more 30   ——
          —   150)12000(80 scholars, Ans.
        150   12000

By analysis. By having as many more, and half as many
more, he must have 2½ times the original number; therefore,
by dividing 200 by 2½, we obtain the answer, 80, as before.

NOTE. — Having performed all the following questions, by *position*, the stu
dent should then perform them by *analysis*.

2. A person after spending ⅓ and ¼ of his money, had $60
left, what had he at first ?   Ans. $144.
3. What number is that, which being increased by ½, ⅓, and
¼ of itself, the sum shall be 125 ?   ●   Ans. 60.
4. A's age is double that of B, and B's is triple that of C, and
the sum of all their ages is 140. What is each person's age ?
                Ans. A's 84, B's 42, C's 14 years.
5. A person lent a sum of money, at 6 per cent., and at the
end of 10 years received the amount, $560. What was the sum
lent ?   Ans. $350.
6. Seven-eighths of a certain number exceed ⅓ by 81, what
is the number ?   Ans. 120.
7. What number is that whose ⅔ exceed ½ by 2¹¹⁄₁₂?
                      Ans. 87.

DOUBLE POSITION. *

DOUBLE POSITION teaches to resolve questions, by making
two suppositions of false numbers.

Those questions in which the results are not proportional to
their positions, belong to this rule.

* This rule is founded on the supposition, that the first error is to the second,
as the difference between the true and first supposed number is to the difference
between the true and second supposed number. When this is not the case, the
exact answer to the questions cannot be found by this rule.

RULE.

Take any two convenient numbers, and proceed with each according to the conditions of the question. Find how much the results are different from the result in the question. Multiply each of the errors by the contra supposition, and find the sum and difference of the products. If the errors are alike, divide the difference of the products by the difference of the errors, and the quotient will be the answer. If the errors are unlike, divide the sum of the products by the sum of the errors, and the quotient will be the answer.

NOTE.—The errors are said to be alike, when they are both too great, or both too small; and unlike, when one is too great and the other too little.

EXAMPLES.

1. A lady purchased a piece of silk for a gown, at 80 cents per yard, and lining for it, at 30 cents per yard; the gown and lining contained 15 yards, and the price of the whole was $7.00. How many yards were there of each?

Suppose 6 yards of silk, value	$4.80
She must then have 9 yards of lining, value	2.70
	———
Sum of their values,	$7.50
Which should have been	7.00
	———
So the first error is 50 too much,	+.50
Again; suppose she had 4 yards of silk, value	$3.20
Then she must have 11 yards of lining, value	3.30
	———
Sum of their values,	$6.50
Which should have been	7.00
	———
So that the second error is 50 too little,	—.50
First supposition multiplied by last error,	$6 \times 50 = 3.00$
Last supposition multiplied by first error,	$4 \times 50 = 2.00$
	———
Add the products, because *unlike*,	5.00

$500 \div \overline{50 + 50} = 5$ yards of silk, $\Big\}$ Ans. $5 \times 80 = \$4.00$

$15 - 5 = 10$ yards of lining, $10 \times 30 = 3.00$

———

Proof $7.00

By analysis. As the silk and lining contain 15 yards, and cost $7.00, the average price per yard is 46⅔; and this taken from

80, leaves 33⅓ ; and 30 taken from 46⅔, leaves 16⅔ ; and as the quantity of lining will be to that of the silk, as 33⅓ to 16⅔, it is therefore evident, that the quantity of lining is twice the quantity of silk. Wherefore, if 15, the number of yards, be divided into three parts, two of those parts (10) will be the number of yards for the lining, and the other part (5) will be the yards for the silk, as before.

NOTE. — The student should perform each question by analysis.

2. A and B laid out equal sums in trade ; A gained a sum equal to ¼ of his stock, and B lost $225 ; then A's money was double that of B's. What did each lay out ? Ans. $600.

3. A person being asked the age of each of his sons, replied, that his eldest son was 4 years older than the second, his second 4 years older than the third, his third son 4 years older than the fourth, or youngest, and his youngest was half the age of the oldest. What was the age of each of his sons ?

Ans. 12, 16, 20, 24 years.

4. A gentleman has two horses and a saddle worth $50. Now, if the saddle be put on the first horse, it will make his value double that of the second horse ; but, if it be put on the second it will make his value triple that of the first. What was the value of each horse ? Ans. The first $30, second $40.

5. A gentleman was asked the time of day, and replied, that ⅔ of the time past from noon, was equal to $\frac{8}{33}$ of the time to midnight. Required the time. Ans. 12 minutes past 3.

6. A and B have the same income. A saves $\frac{1}{12}$ of his ; but B by spending $100 per annum more than A, at the end of 10 years, finds himself $600 in debt. What was their income ?

Ans. $480.00.

7. A gentleman hired a laborer for 90 days on these conditions ; that, for every day he wrought he should receive 60 cts.; and for every day he was absent he should forfeit 80 cents. At the expiration of the term he received $33. How many days did he work, and how many days was he idle ?

Ans. He labored 75 days, and was idle 15 days.

The following question, with some variations, in the language, is taken from Fenn's Algebra, page 62. It is believed, however, that Sir Isaac Newton was the author of it.

8. If 12 oxen eat 3⅓ acres of grass in 4 weeks ; and 21 oxen eat up 10 acres in 9 weeks ; how many oxen would it require to eat up 24 acres in 18 weeks, the grass to be growing uniformly ?

Ans. 36 oxen.

OPERATION BY ANALYSIS.

Each ox eats a certain quantity in each week, which we may suppose to be 100 pounds ; and of the whole quantity eat up in each case, a part must have already grown during the time of eating.

Then by the first conditions of the question,

$12\times4\times100=4800$ lbs.=whole quantity on $3\frac{1}{3}$ acres for 4 weeks.

$4800\div3\frac{1}{3}=1440$ lbs. = whole quantity on 1 acre for 4 weeks.

By the second conditions of the question,

$21\times9\times100=18900$ lbs.=whole quantity on 10 acres for 9 weeks.

$18900\div10=1890$ lbs. = whole quantity on 1 acre for 9 weeks.

$1890-1440=450$ lbs. = the quantity grown on an acre for 9 $-4=5$ weeks.

$450\div\overline{9-4}=90$ lbs.=the quantity which grows on each acre for 1 week.

$90\times3\frac{1}{3}\times4=1200$ lbs.=quantity grown on $3\frac{1}{3}$ acres for 4 weeks.

$4800-1200=3600$ lbs.=original quantity of grass on $3\frac{1}{3}$ acres.

$3600\div3\frac{1}{3}=1080$ lbs.=original quantity on 1 acre.

Then, by the last condition of the question,

$24\times1080=25920$ lbs. = original quantity on 24 acres.

$24\times90\times18=38880$ lbs.=quantity which grows on 24 acres in 18 weeks.

$25920+38880=64800$ lbs. = whole quantity on 24 acres for 18 weeks.

$64800\div18=3600$ lbs.=quantity to be eat from 24 acres each week.

$3600\div100=36=$ number of oxen required to eat the whole, and the answer to the question.

9. There is a fish whose head weighs 15 pounds, his tail weighs as much as his head, and $\frac{1}{4}$ as much as his body, and his body weighs as much as his head and tail. Required the weight of the fish.　　　　　　　　　　Ans. 72lbs.

10. A certain grocer has only 5 weights; with these he can weigh any quantity, from 1 pound to 121 pounds. Required the weights.

SECTION LXIII.

EXCHANGE.

Exchange is the act of paying or receiving the money of one country for its equivalent in the money of another country, by means of bills of exchange. This operation, therefore, comprehends both the reduction of moneys, and the negotiation of bills. It determines the comparative value of the currencies of all nations; and shows how foreign debts are discharged, loans and subsidies paid and other remittances made from one country to another, without the risk, trouble, or expense of transporting specie or bullion.

BILLS OF EXCHANGE.

A Bill of Exchange is a written order for the payment of a certain sum of money, at an appointed time. It is a mercantile contract, in which four persons are mostly concerned ; viz.

1. The *drawer*, who receives the value, and is also called the *maker* and *seller* of the bill.

2. His debtor, in a distant place, upon whom the bill is drawn, and who is called the *drawee*. He is also called the acceptor, when he accepts the bill, which is an engagement to pay it when due.

3. The person, who gives value for the bill, who is called the *buyer, taker,* and *remitter.*

4. The person to whom it is ordered to be paid, who is called the *payee*, and who may, by indorsement, pass it to any other person.

Most mercantile payments are made in Bills of Exchange, which generally pass from hand to hand, until due, like any other circulating medium ; and the person, who at any time has a bill in his possession, is called the *holder.*

When the holder of a bill disposes of it, he writes his name on the back, which is called *indorsing ;* and the payee should be the first indorser. If the bill be indorsed in favor of any particular person, it is called a *special indorsement ;* and the person to whom it is thus made payable, is called the *indorsee*, who must also indorse the bill, if he negotiates it. Any person may indorse a bill, and every indorser (as well as the acceptor, or payee) is a security for the bill, and may therefore be sued for payment.

The *term* of a bill varies according to the agreement between the parties, or the custom of countries. Some bills are drawn at sight ; others, at a certain number of days, or months, after sight or after date ; and some, at *usance,* which is the customary or usual term between different places.

Days of grace, are a certain number of days granted to the acceptor, after the term of a bill is expired. Three days are usually allowed.

In reckoning when a bill, payable after date, becomes due, the day on which it is dated is not included ; and if it be a bill payable after *sight,* the day of presentment is not included. When the term is expressed in months, calendar months are understood ; and when a month is longer than the preceding, it is a rule not to go in the computation, into a third month.

Thus, if a bill be dated the 28th, 29th, 30th, or 31st of January, and payable one month after date, the term equally expires on the last day of February, to which the days of grace must, of course, be added ; and, therefore, the bill becomes due on the third of March.

Form of a Bill of Exchange.

Boston, September 25, 1835.

Exchange for 5.000£ Sterling.

At ninety days sight of this, my first Bill of Exchange, (second and third of the same tenor and date unpaid,) pay to James Ayer, or order, five thousand pounds sterling, with or without further advice from me.

John L. French.

Messrs. Dana & Hyde,
  Merchants, Liverpool.

ACCEPTING BILLS.

When a bill is presented for acceptance, it is generally left till the next day ; and the common way of accepting, is, for the drawee to write his name at the bottom, or across the body of the bill with the word *accepted.*

When two or more persons are in partnership, the accept-ance of one binds all the others, if the bill concerns their joint trade ; but if it should be made known to the person, who re-ceives the bill, that it concerns the acceptor only, in a distinct interest, he alone, as acceptor, can be sued.

A clerk or servant may accept a bill for his master, when he has authority for that purpose ; or, if he usually transacts busi-ness of this nature for him ; and his acceptance binds the mas-ter. But if the bill be drawn nominally on the servant, direct-ing him to place it to the account of his master, and if the ser-vant should accept it generally, without specifying that he does it for his master's account, the acceptance binds the servant only, and not his employer.

When a bill is drawn for the account of a third person, and is accepted as such, and he fails without making provision for its payment, the acceptor must discharge the bill, and can have no recourse against the drawer.

A bill may be accepted to be paid at a longer period than is mentioned in the bill, or to pay a part of the sum only ; such an acceptance is binding on him who made it ; but the holder is at liberty to take it as it is offered, or to act as if acceptance had been entirely refused.

INDORSING BILLS.

Bills, payable to bearer, are transferred by simple delivery, and without any indorsement ; but, in order to transfer a bill, payable to order, the holder must express his order of paying to another person, which is always done by an indorsement.

An indorsement may be blank or special. A *blank indorsement* consists only of the indorser's name, and the bill becomes then transferable by simple delivery ; a *special indorsement* orders the money to be paid to some particular person, or to his order ; a blank indorsement may always be filled up with any person's name, so as to make it special.

An indorsement may take place at any time after the bill is issued, even after the day of payment is elapsed.

A person, who receives a bill with a blank indorsement, may take it as indorsee, negotiate it again, or demand payment on his own account, or he may receive the money as agent, or for the account of the indorser ; and the latter, notwithstanding his indorsement, may still appear as holder, in an action against the drawer or acceptor.

A special indorsement need not contain the words, *to order*, and the bill is negotiable ; it may, also, be restrictive, giving authority to the indorsee to receive the money for the indorser, but not to transfer the bill again to another.

An indorsement for part of the money only, is not valid, except with regard to him, who makes it ; the drawer and acceptor are not bound by it.

When the holder of a bill dies, his executors may indorse it ; but by so doing, they become answerable to their indorsee, personally, and not as executors.

PROTESTING BILLS.

When acceptance or payment has been refused, the holder of the bill should give regular and immediate notice to all the parties, to whom he intends to resort for payment ; and if, on account of unnecessary delay, a loss should be incurred by the failure of any of the parties, the holder must bear the loss.

With respect to the manner in which notices of non-acceptance, or non-payment are to be given, a difference exists between inland and foreign bills.

For foreign bills a protest is indispensably necessary ; thus, a public notary is to appear with the bill, and to demand either acceptance or payment ; and, on being refused, he is to draw up an instrument, called a *protest*, expressing that acceptance or payment has been demanded, and refused, and that the holder of the bill intends to recover any damages, which he may sustain in consequence. This instrument is admitted in foreign countries, as legal proof of the fact.

It it customary, as a precaution against accidents or miscarriage, to draw three copies of a foreign bill, and to send them by different posts. They are denominated the *first, second*, and

20*

third of exchange; and, when any one of them is paid, the rest become void and of no value. When the acceptor of a bill becomes insolvent, or absconds before the term of payment is expired, the holder may cause a notary to demand better security, and, on that being refused, to protest the bill for want of it. In such cases, however, the most general practice is to wait the regular time, till the bill becomes due.

The damages incurred by non-acceptance and non-payment, besides interest, consist, usually, of the exchange, or reëxchange, commission, and postage together, with the expenses of protest, and interest. The exchange is reckoned according to the course at sight, from the place where the protest is made, to the place where the bill is to be paid by the drawer; and, if it be not paid there, the exchange is then reckoned from the same place, to that, where the bill is paid, and also, double commission. The interest commences from the day, when the demand was made.

RECOVERING BILLS.

The drawer, accepter, and even indorser of a bill, are equally liable to the payment of it; and though the holder can have but one satisfaction, yet, until such satisfaction is actually had, he may sue any of them, or all of them, either at the same time, or in succession, and obtain judgment against them all, till satisfaction be made. Proceedings cannot be staid in any action, but on payment of the debt and costs, not only in that action, but in all the others in which judgment has not been obtained; and though the principal sum should be paid by one of the parties, still costs may be recovered in the several actions against the others.

When acceptance is refused, and the bill is returned by protest, an action may be commenced immediately against the drawer, though the regular time of payment be not arrived. His debt, in such a case, is considered as contracted the moment the bill is drawn. Thus, if before the bill is returned, the drawer should become a bankrupt, the debt was contracted before the commission of bankruptcy took place.

Nothing will discharge an indorser from his engagement, but the absolute payment of the money; not even a judgment recovered against the drawer, or any previous indorser, or any execution against any of them, unless the money be paid in consequence.

INLAND EXCHANGE, OR DRAFTS.

By Inland Exchange, is understood, the act of remitting bills to places in the same country; by which means, debts are discharged more conveniently than by cash remittances.

Suppose, for example, A, of Boston, is creditor to B, of Baltimore, $100, and C, of Boston, debtor to D, of Baltimore, $100,
both these debts may be discharged by means of one bill. Thus,
A draws for this sum on B, and sells his bill to C, who remits it
to D, and the latter receives the amount, when due from B.
Here, by a transfer of claims, the Boston debtor pays the Boston creditor; and the Baltimore debtor the Baltimore creditor;
and no money is sent from one place to the other. The same
would take place if D, of Baltimore, drew on C, of Boston, and
sold his bill to B, of Baltimore, who should send it to A, of Boston; the effect, in either case, being merely a transfer of debtors
and creditors.

NOTE. — In this operation, A is the *drawer* and *seller*, B the *drawee* and accepter, C the *buyer* and *remitter*, and D the *payee*, if his name be mentioned in
the bill; and he is the *holder* when he receives the bill from A. When D, or
any other holder, presents the bill for acceptance or payment, he is called the
presenter.

By the foregoing example, it appears, that reciprocal and
equal debts, due between two places, may be discharged without remitting specie; and it may be supposed, that such an operation is of equal convenience to all parties concerned; but
when the debts are unequal, the advantage must be different, as
the obligation of remittance is no longer mutual, because the
debtor place must pay its balance, either by sending cash, or
bills; and as the latter mode is generally preferred, an increased demand for bills must be the consequence, which enhances
their price, as it would that of any other article of sale, or purchase.

PAR OF EXCHANGE.

The Par of Exchange, may be considered under two general
heads; viz. the *intrinsic par*, and the *commercial par*, each of
which admits of suborbinate divisions and distinctions.
The *intrinsic par* is the value of the money of one country,
compared with that of another, with respect both to weight and
fineness.
The *commercial par* is the comparative value of the moneys of
different countries, according to the weight, fineness, and market prices of the metals.
Thus, two sums of different countries are *intrinsically* at par,
when they *contain* an equal quantity of the same kind of pure
metal; and two sums of different countries are commercially at
par, when they can purchase an equal quantity of the same kind
of pure metal.

COURSE OF EXCHANGE.

The Course of Exchange, is the variable price of the money of one country, which is given for a fixed sum of money of another country; the latter is called the *certain*, and the former, the *uncertain* price, as before stated.

When the market price of foreign bills is above par, the exchange is said to be favorable to the place, that gives the certain for the uncertain.

It should, however, be recollected, that when the exchange is favorable to a place, it is only so to the buyers and remitters of bills, but it is unfavorable to the drawers and sellers.

Thus, the interest of the remitter is identified with that of the place where he purchases the bill, and the interest of the drawer with that of the place where his funds are established, and on which he draws.

It is natural to inquire, why such prices are considered favorable, or unfavorable, if the drawers and remitters, whose interests are opposite, are natives of the same country. The usual answer is, that when the exchange is against a place, it becomes the interest of remitters to pay their foreign debts in specie instead of bills, and the exportation of the precious metals is often considered a national disadvantage.

The fluctuations of exchange are occasioned by various circumstances, both political and commercial. The principal cause is generally stated to be the balance of trade; that is, the difference between the commercial exports and imports of any one country with respect to another. Experience, however, shows, that the exchange may be unfavorable to a country, when the balance of trade is greatly in its favor; for the demand for bills must chiefly depend on the balance of such debts as come into immediate liquidation; that is to say, on the *balance of payments.*

Besides, it does not follow, that large exports are always successful, or quick in their returns; and even should it be the case, the balance of payments may still be unfavorable from political causes.

When any alteration takes place in the coin or currency of a country, the exchange will, of course, vary so as to keep pace or correspond to such alteration. This, however, cannot be considered a change in the price of bills, but in the money in which they are bought or sold.

In times of peace, the course of exchange seldom remains long unfavorable to any country, at least, beyond the expense that might be incurred by the transportation of the precious metals; for bullion is considered the universal currency of merchants, and exchange gives it circulation, and thus tends to maintain the level of money throughout the commercial world.

An unfavorable course of Exchange may, therefore, be corrected, either by the exportation of bullion, or the shipment of goods,—and another method sometimes offers, by negotiating bills through several places ; but the latter remedy must fail if the exchange be universally unfavorable.

From what has been said of the causes, both commercial and political, which produce the fluctuations of exchange, and which sometimes counteract or balance each other, the following simple conclusion may be drawn ;—that bills rise or fall in their prices, like any other saleable articles, according to the proportion that exists between the demand and supply.

AMSTERDAM.

Accounts are kept in florins, stivers, and pennings, or in pounds, shillings and pence flemish.

        16 pennings = 1 stiver.
        20 stivers   = 1 florin or guilder.
    Also, 12 grotes or pence flemish, or
         6 stivers = 1 shilling flemish.
        20 shillings flemish, or
        6 florins = 1 pound flemish.
        2½ florins, or 50 stivers = 1 rix dollar.

1. United States on Amsterdam. Reduce 896 florins, 10 stivers, to federal money, exchange at 38 cents per florin.
                                        Ans. $340.67.

2. Amsterdam on the United States. Reduce $340.67 to the money of Amsterdam, exchange at 38 cents per florin.
                                Ans. 896 florins, 10 stivers

CONSTANTINOPLE.

Accounts are kept in piastres, paras, and aspers ; or in piasters and aspers ; sometimes in piasters and half paras, or in piastres and minas.

        3 aspers = 1 para.
        40 paras = 1 piastre or Turkish dollar.
    Also, 80 half paras, or 100 minas = 1 piastre.

3. United States on Constantinople. Reduce 78 piastres, 20 paras to federal money, exchange at 40 cents per piastre.
                                        Ans. $31.40.

4. Constantinople on the United States. Reduce $31.40 to the money of Constantinople.   Ans. 78 piastres, 20 paras.

COPENHAGEN.

Accounts are kept here, in rix dollars, marks, and skillings Danish, but sometimes in rix dollars, marks, and skillings lubs. Pfenings are also occasionally reckoned.

12 pfenings = 1 skilling.
16 skillings = 1 mark.
 6 marks Danish, or 3 marks lubs = 1 ryksdaler, or rix dollar.

5. United States on Copenhagen. Reduce 896 rix dollars, 3 marks, to federal money, exchange at 50 cents per rix dollar.
Ans. $448.25.

DANTZIC.

Accounts are computed here, in florins, groschen, and pfenings.

3 pfenings = 1 groschen.
30 groschen = 1 florin.
3 florins　 = 1 rix dollar.

6. United States on Dantzic. Reduce 196 rix dollars, 2 florins, to federal money, exchange at 17 cents per florin.
Ans. $100.30.
7. Dantzic on the United States. Reduce $100.30 to the currency of Dantzic. Ans. 196 rix dollars, 2 florins.

FRANCE.

Computations are made in francs and centimes ; or in livres, sous, and denires.

. 10 centimes = 1 decime.
10 décimes = 1 franc.
12 denires　 = 1 sou or sol.
20 sous　　 = 1 livre tournois.
80 francs　 = 81 livres.
 3 livres　 = 1 ecu of exchange.

8. United States on France. Reduce 8763 francs, 30 centimes, to federal money, exchange at $1 per 5 francs, 35 centimes. Ans. $1638.
9. France on the United States. Reduce $1638 to French money, exchange at 5 francs, 35 centimes to the dollar.
Ans. 8763 francs, 30 centimes.

HAMBURGH.

Computations are made in marks, schillings, and pfenings, banco or current,—banco bears an agio on currency, from 20 to 25 per cent.

12 pfenings   = 1 schilling or sol lubs.
16 schillings lubs = 1 mark.
3 marks   = 1 rix dollar.

10. United States on Hamburgh. Reduce 675 rix dollars, 2 marks, to federal money, exchange at 30 cents per mark.
                                              Ans. $608.10.
11. Hamburgh on the United States. Reduce $608.10 to the currency of Hamburgh, exchange at 30 cents per mark.
                                   Ans. 675 rix dollars, 2 marks.

LEGHORN.

Accounts are kept in pezze, soldi, and denari di pezza.

12 denari di pezza = 1 soldo di pezza.
20 soldi di pezza = 1 pezza.
12 denari di lira = 1 soldi di lira.
20 soldi di lira   = 1 lira.

12. United States on Leghorn. Reduce 286 pezza, 10 soldi, to federal money, exchange at 90 cents per pezza.
                                              Ans. $257.85.
13. Leghorn on the United States. Change $257.85 to the currency of Leghorn.   Ans. 286 pezza, 10 soldi.

MILAN.

Accounts are kept in lire soldi, and denari correnti or imperiali.

12 denari   = 1 soldo.
20 soldi   = 1 lira.
106 soldi or lire imperiali = 150 soldi or lire correnti = 1 filippo.
117 soldi imperiali   = 1 scudo or crown.

14. United States on Milan. Change 176 lira, 10 soldi, to federal money, the lira being valued at 20 cents. Ans. $35.30.
15. Milan on the United States. Reduce $35.30 to the currency of Milan.   Ans. 176 lira, 10 soldi.

NAPLES.

Computations are made in ducats of 100 grains.

$$10 \text{ grani } = 1 \text{ carlino.}$$
$$10 \text{ carlini } = 1 \text{ ducato di regno.}$$

16. United States on Naples. Change 769 ducati de regno, 5 carlini to federal money, the value of the ducato being 80 cents.
Ans. $615.60.

17. Naples on the United States. Reduce $615.60 to the currency of Naples.   '   Ans. 769 ducati de regno, 5 carlini.

SICILY.

Accounts are kept here, in oncie, tari and grani ; and also in scudi, tari, and grani.

$$20 \text{ grani } = 1 \text{ tari.}$$
$$30 \text{ tari } = 1 \text{ oncia.}$$
$$\text{Also, } 12 \text{ tari } = 1 \text{ scudo or Sicilian crown.}$$
$$5 \text{ scudi } = 2 \text{ oncia.}$$

18. United States on Sicily. Change 876 oncia, 3 scudi, to federal money, the oncia being valued at $2.38.3.
Ans. $2090.36.7⅔.

19. Sicily on the United States. Change $2090.36.7⅔ to the currency of Sicily.   Ans. 876 oncia, 3 scudi

RUSSIA.

Computations are made here, in rubles and copecks.

$$10 \text{ copecks} = 1 \text{ grieve or grievener.}$$
$$10 \text{ grieves } = 1 \text{ ruble.}$$

20. United States on Russia. What is the value of 7684 rubles, 8 grieves, in federal money, the value of the ruble in the United States being 75 cents ?   Ans. $5763.60.

21. Russia on the United States. What is the value of $5763.60 in Russian currency ?   Ans. 7684 rubles, 8 grieves.

ROME.

Accounts are kept in ecudi moneta, and bajocchi ; or in scudi di stampa soldi, and denari d'oro quattrini, and mezzi quattrini are sometimes reckoned.

2 mezzi quattrini = 1 quattrino.
5 quattrini　　　= 1 bajoccho.
10 bajocchi,　　= 1 paolo.
10 paoli　　　　= 1 scudo moneta, or Roman crown.

22. United States on Rome. Change 7689 scudi moneta to federal money, the value of the scudi being $1.00.0$^{67}_{100}$.

Ans. 7694.15^{163}_{1000}$.

23. Rome on the United States. Change 7694.15^{163}_{1000}$ to Roman currency.　　　Ans. 7689 scudo moneta.

SPAIN.

The general mode of keeping accounts in Spain, is in maravedis and reals.

34 maravedis = 1 real.
8 reals = 1 dollar of plate.
375 maravedis = 1 ducat of exchange.
4 dollars of plate = 1 pistole of exchange.

Exchanges are generally computed in denominations of plate, which is always understood to be *old plate*, if *new plate* be not mentioned.

There are three principal denominations of these imaginary moneys in which exchanges are generally transacted ; viz., dollars, doubloons and ducats, and they are divided into reals and maravedis of plate, and sometimes converted into vellon and other denominations.

The dollar of exchange, also called *peso* or *piastre de cambio*, or *de plata*, is divided into 8 reals of 34 maravedis of plate each, and sometimes into 16 quartos.

The *doubloon de plata*, or pistole of exchange, is four times the dollar, and therefore contains 32 reals, or 1088 maravedis of plate.

The ducat of plate, also called *ducado de cambio*, contains 11 reals 1 maravedi, or 375 maravedis of plate.

At Alicant, Valencia, and Barcelona, exchanges are transacted in libras of 20 sueldos, or 240 dineros.

The libra of Alicant and Valencia is the dollar of plate. This is sometimes divided into 10 reals of new plate, which are, of course, equal to 8 reals of old plate.

The libra of Barcelona, commonly called *libra catalan*, is worth 5⅝ reals of plate ; hence 7 of those libras equal 5 dollars of plate, and therefore, 28 sueldos catalan equal 1 dollar.

The hard dollar of 20 reals vellon, is occasionally used in exchanges, and is also divided into 12 reals, each of 16 quartos. The current dollar, which is an imaginary money, valued at two-thirds of the hard dollar, is divided into 8 reals, and the real into 16 quartos.   The two latter are the principal moneys of exchange used at Gibraltar.

24. United States on Spain.   What is the value in federal money of 7600 dollars of plate, exchange at 75 cents for a plate dollar ?   Ans. $5700.

25. Spain on the United States.   What is the value of $5700 in Spanish money, exchange of 75 cents per dollar of plate.
                                      Ans. 7600 dollars of plate.

SWEDEN.

Accounts are kept in rix dollars, skillings and fenings.

12 fenings = 1 skilling.
48 skillings = 1 rix dollar.

26. United States on Sweden.   Reduce 476 rix dollars, 24 skillings to federal money, the rix dollar being valued at 107 cents.   Ans. $509.85.5.

27. Sweden on the United States.   Change $509.85.5 to the currency of Sweden.   Ans. 476 rix dollars, 24 skillings.

TURIN.

Accounts are kept in lire, soldi and denari.

12 denari = 1 soldi.
20 soldi   = 1 lira.

28. United States on Turin.   Change 462 lira, 10 soldi, to federal money, exchange at 20 cents per lira.   Ans. $92.50.

29. Turin on the United States.   Change $92.50 to the currency of Turin, exchange at 20 cents per lira.
                                      Ans. 462 lira, 10 soldi.

VIENNA.

Computations are made in florins and creutzers, or in rix dollars and creutzers.

4 pfenings = 1 creutzer.
60 creutzers = 1 florin.
1½ florins   = 1 rix dollar of account.
2 florins   = 1   "   "   specie.

30. United States on Vienna.   Reduce 876 rix dollars, specie, florin to federal money, the specie rix dollar being equal to 97 cents.   Ans. $850.20.5.

31. Vienna on the United States. Change $850.20.5 to the currency of Vienna.　　　Ans. 876 rix dollars, specie, 1 florin.

EAST INDIES, BENGAL, CALCUTTA, ECT.

12 pice　　　=1 anna.
16 annas　　=1 rupee.
1 sica rupee =2s. 6d. sterling.

32. United States on Calcutta. Reduce 432 rupees, 12 annas, to federal money, exchange at 50 cents per rupee.
　　　　　　　　　　　　　Ans. $216.37½.

33. Calcutta on the United States. Reduce $216.37½ to the currency of Calcutta, exchange at 50 cents per rupee.
　　　　　　　　　　　Ans. 432 rupees, 12 annas.

BOMBAY.

100 rees　　=1 quarter.
4 quarters =1 rupee.
1 rupee　　=2s. 4d. sterling.

34. United States on Bombay. Change 678 rupees, 2 quarters, to federal money, the rupee being 50 cents. Ans. $339.25.

35. Bombay on the United States. Change $339.25 to Bombay money, reckoning the rupee at 50 cents.
　　　　　　　　　　Ans. 678 rupees, 2 quarters.

MADRAS.

Accounts are kept here, in pagodas, fanams, and cash.

80 cash　　　　=1 fanam
45 fanams　　　=1 star pagoda.
1 star pagoda = 8s. sterling.

36. United States on Madras. Change 375 stars pagoda, to federal money, the star pagoda valued at $1.77⅞.
　　　　　　　　　　　Ans. $666.66+.

37. Madras on the United States. Reduce $896 to the currency of Madras.　　　　Ans. 504 pagodas.

TRIESTE.

Accounts are here kept, in pfenings, florins, and creutzers.

4 pfenings =1 creutzer.
60 creutzers =1 florin.
1½ florins　=1 rix dollar.

38. United States on Trieste.   What is the value 'in federal money of 769 rix dollars, 40 creutzers, the value of the rix dollar being 92 cents.   '   Ans. $707.88⅔.

39. Trieste on the United States.   Reduce $707.88⅔ to the currency of Trieste.   Ans. 769 rix dollars, 40 creutzers.

GREAT BRITAIN.

Accounts are here kept, in pounds, shillings, pence, and farthings, sterling. The principal coins are guineas and crowns, and their fractional parts.

The guinea sterling is 21 shillings.
Crown,   5   "

40. United States on England.   A merchant in Boston has a bill on London for 1728£. 15s. 6d. ; what is the value of this, the exchange being 5 per cent. above par ?   Ans. $8067.61⅔.

41. England on the United States.   What should a merchant in London give for a bill on Boston, for $8967.75, the exchange being 2 per cent. below par ?   Ans. 1977£. 7s. 9$\frac{33}{100}$d.

SECTION LXIV.

VALUE OF GOLD COINS,

ACCORDING TO THE LAWS OF MAY AND JUNE, 1834.

NAMES OF COINS.	Weight.		Former Standard.			Standard of July 31st, 1834.		
	dw.	gr.	$.	c.	m.	$.	c.	m
UNITED STATES.								
Eagle coined before July 31, 1834,	11	6	10			10	66	5
Shares in proportion . . .								
Foreign Gold.								
AUSTRIAN DOMINIONS.								
Souverein, ...	3	14	3	17	6	3	37	7
Double ducat, ...	4	12	4	29	9	4	58	9
Hungarian ducat, . . .	2	5¾	2	15	4	2	29	6
BAVARIA.								
Carolin, ...	6	5¼	4	64	6	4	95	7
Max d'or, or maximilian, . .	4	4	3	11	1	3	31	8
Ducat, ...	2	5¾	2	13	3	2	27	5
BERNE.								
Ducat, double in proportion, .	1	23	1	85	4	1	98	6
Pistole, ...	4	21	4	26	2	5	54	2

BRAZIL.			
Johannes, ½ in proportion, .	18	16	17 06 4
Dobraon, ...	34 12	30 66 6	32 70 6
Dobra, ...	18 6	16 22 2	17 30 1
Moidore, ½ in proportion, . .	6 22	6 14 9	6 55 7
Crusade, ...	16½	59 8	63 5
BRUNSWICK.			
Pistole, double in proportion, .	4 21½	4 27 1	4 54 8
Ducat, ...	2 5¾	2 09 2	2 23
COLOGNE.			
Ducat, ...	2 5¾	2 12 5	2 26 7
COLUMBIA.			
Doubloons, ...	17 9	14 56	15 53 5
DENMARK.			
Ducat, current, ...	2	1 70 5	1 81 2
Ducat, specie, ...	2 5¾	2 12 5	2 26 7
Christian d'or, ...	4 7	3 77	4 02 1
EAST INDIES.			
Rupee, Bombay, 1818, . .	7 11	6 65 4	7 09 6
Rupee, Madras, 1818, . .	7 12	6 66 7	7 11
Pagoda, star, ...	2 4¾	1 68 9	1 79 8
ENGLAND.			
Guinea, ½ in proportion, . .	5 8½	4 79 9	5 07 5
Sovereign, ...	5 2½	4 57	4 84 6
Seven shilling piece, ...	1 19	1 60	1 69 8
FRANCE.			
Double Louis, coined before 1786,	10 11	9 08 7	9 69 7
Louis, " "	5 5½	4 54 1	4 84 6
Double Louis, coined since 1786,	9 20	8 59	9 15 3
Louis, " "	4 22	4 29 5	4 57 6
Double Napoleon, or 40 francs,	8 7	7 23 2	7 70 2
Napoleon, or 20 "	4 3½	3 61 6	3 85 1
Same as the new Louis guinea,	5	4 36 2	4 65 5
FRANKFORT ON THE MAINE.			
Ducat, ...	2 5¾	2 13 7	2 27 9
GENEVA.			
Pistole old, ...	4 7½	3 73 7	3 98 5
Pistole new, ...	3 15¼	3 23 2	3 44 4
HAMBURGH.			
Ducat, double in proportion, .	2 5¾	2 13 7	2 27 9
GENOA.			
Sequin, ...	2 5¾	2 15 8	2 30 2

HANOVER.								
Double Geo. d'or, single in pro.	8	13	7	48	2	7	87	9
Ducat, ...	2	5¾	2	15	4	2	29	6
Gold florin, double in proportion, .	2	2	1	57	6	1	67	
HOLLAND.								
Double ryder, ...	12	21	11	44	2	12	20	5
Ryder, ...	6	9	5	66	5	6	04	3
Ducat, ...	2	5¾	2	13	3	2	27	5
Ten guilder piece, 5 duc'. in pro.	4	8	3	78		4	03	4
MALTA.								
Double Louis, ...	10	16	8	69	9	9	27	8
Louis, ...	5	8	4	36	4	4	85	2
Demi Louis, ...	2	16	2	20	2	2	33	6
MEXICO.								
Doubloons, shares in proportion,	17	9	14	56		15	53	5
MILAN.								
Sequin, ...	2	5¾	2	15	6	2	29	
Doppia, or pistole, ...	4	1½	3	57	2	3	80	7
Forty livre piece, 1808, . .	8	8	7	26	1	7	74	2
NAPLES.								
Six Ducat piece, 1783, . .	5	16	4	92	5	5	24	9
Two " " or sequin, 1762, .	1	20¼	1	51	1	1	59	1
Three " " or oncetta, 1818,	2	10¼	2	34	7	2	49	
NETHERLANDS.								
Gold lion, or 14 Florence piece,	5	7¾	4	73	1	5	04	6
Ten Florence piece, 1820, . .	4	7¾	3	76	6	4	01	9
PARMA.								
Quadruple pistole, double in pro.	18	9	15	59	6	16	62	8
Pistole or dappia, 1787, . . .	4	14	3	93	5	4	19	4
" " 1796, . .	4	14	3	87	5	4	13	5
Maria Theresa, 1818, . .	4	3½	3	62	4	3	86	1
PIEDMONT.								
Pistole, coined since 1785, . .								
Half pistole in proportion, .	5	20	5	07	5	5	41	1
Sequin, ½ in proportion, . .	2	5	2	13	7	2	28	
Carlino, coined since 1785, ½ pro.	29	6	25	63	2	27	34	
Piece of 20 francs, called marengo,	4	3¼	3	34	1	3	56	4
POLAND.								
Ducat, ...	2	5¾	2	13	7	2	27	5
PORTUGAL.								
Dabraon, ...	34	12	30	66	6	32	70	6
Dabra, ...	18	6	16	22	2	17	30	1
Johannes, ...	18		16			17	06	4
Moidore, ½ in proportion, . .	6	22	6	14	9	6	55	7

Piece of 16 testoons or 1600 rees,	2	6	·1	99	2	2	12	1
Old crusado of 400 rees,		15		84	9		58	5
New " 480 "		16½		59	8		63	5
Milrea, coined in 1775,		19¾		73	2		78	
Russia.								
Ducat, 1748,	2	5¾	2	13	7	2	27	9
" 1787,	2	5¾	2	12	5	2	26	7
Frederick, double, 1769,	8	14	7	47	5	7	95	5
" " 1800,	8	14	7	45	4	7	95	1
" single 1778,	4	7	3	74	9	3	99	7
" " 1800,	4	7	3	72	5	3	97	5
Rome.								
Sequin, coined since 1760,	2	4¼	2	10	9	2	25	1
Scudo of republic,	17	0¼	14	82	8	15	81	1
Russia.								
Ducat, 1796,	2	6	2	15		2	29	7
" 1763,	2	5¾	2	12	5	2	26	7
Gold ruble, 1756,	1	0½		90	9		96	7
Gold ruble, 1799,		18¾		69	1		73	7
" palten, 1777,		9		33	1		35	5
Imperial, 1801,	7	17¼	7	34	9	7	82	9
Half " "	4	3½	3	68	9	3	93	3
Sardinia.								
Carlino, ½ in proportion,	10	7½	8	83	1	9	47	2
Saxony.								
Ducat, 1784,	2	5¾	2	12	5	2	26	7
" 1797,	2	5¾	2	13	7	2	27	9
Augustus, 1754,	4	6½	3	68	5	3	92	5
" 1784,	4	6½	3	72	5	3	97	4
Sicily.								
Ounce, 1751,	2	20½	2	35	1	2	50	4
Double ounce, 1758,	5	17	4	72	7	5	04	4
Spain.								
Doubloons, 1772, parts in pro.	17	8½	15	03		16	02	8
Doubloon,	17	9	14	56		15	53	5
Pistole,	4	8¼	3	64		3	83	4
Coronilla, gold dol or vinteren 1801	1	3		92	1		98	3
Sweden.								
Ducat,	2	5	2	09	7	2	23	5
Switzerland.								
Pistole of the Helvetic Republic, 1800,	4	21½	4	27	9	4	56	
Treves.								
Ducat,	2	5¾	2	02	5	2	26	7

TURKEY.

Sequin, fonducli of Constantinople,								
" 1773, ...	2	5¾	1	74	9	1	86	8
" 1789,	2	5½	1	73	3	1	84	8
Half misseir, 1818, . . .		18½		49	1		52	1
Sequin fonducli, . . .	2	5	1	71	7	1	83	
Yeermeeblekblek, ...	3	1¾	2	84		3	02	8

TUSCANY.

Zechino or sequin, . . .	2	5¾	2	16	6	2	31	8
Ruspone of the kingdom of Etruria,	6	17¼	6	50	5	6	93	8

VENICE.

Zechino or sequin, shares in pro. .	2	6	2	16		2	31	

WIRTEMBURG.

Carolin, ...	6	3½	4	59	4	4	89	8
Ducat, ...	2	5	2	09	7	2	23	5

ZURICH.

Ducat, double, and ½ in proportion,	2	5¾	2	12	5	2	26	7

SECTION LXV.

GEOMETRY.

DEFINITIONS.

1. A point is that which has position, but no magnitude, nor dimensions ; neither length, breadth, nor thickness.

2. A line is length without breadth, or thickness.

3. A surface, or superficies, is an extension, or a figure of two dimensions, length and breadth ; but without thickness.

4. A body or solid, is a figure of three dimensions ; viz. length, breadth and thickness.

5. Lines are either right or curved, or mixed of these two.

6. A right or straight line, lies all in the same direction between its extremities, and is the shortest distance between two points. When a line is mentioned simply, it means a right line.

7. A curve continually changes its direction between its extreme points.

. 8. Lines are either parallel, oblique, perpendicular, or tangential.

9. Parallel lines are always at the same perpendicular distance, and never meet, though ever so far produced.

10. Oblique lines change their distance, and would meet if produced on the side of the least distance.

11. One line is perpendicular to another, when it inclines not more on the one side than the other, or when the angles on both sides of it are equal.

12. An angle is the inclination, or opening of two lines having different directions, and meeting in a point.

13. Angles are right or oblique, acute or obtuse.

14. A right angle is that which is made by one line perpendicular to another ; or, when the angles on each side are equal to one another, they are right angles.

15. An oblique angle is one, which is made by two oblique lines, and is either less or greater than a right angle.

16. An acute angle is less than a right angle.

17. An obtuse angle is greater than a right angle.

18. Superficies are either plane or curved.

19. A plane superficies, or plane, is that with which a right line may every way coincide ; or, if the line touch the plane in two points, it will touch it in every point ; but if not, it is curved.

20. Plane figures are bounded either by right lines or curves.

21. Plain figures, that are bounded by right lines, have names according to the number of their sides, or of their angles ; for they have as many sides as angles, the least number being three.

22. A figure of three sides and angles, is called a triangle ; and it receives particular denominations from the relations of its sides and angles.

23. An equilateral triangle is that, whose three sides are equal.

24. An isoscles triangle is that, which has two sides equal.

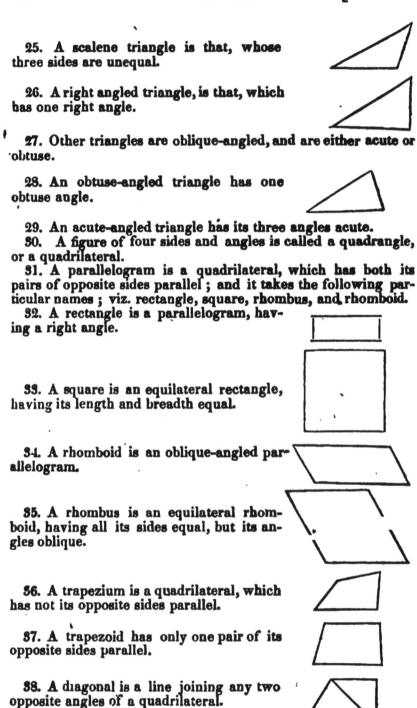

25. A scalene triangle is that, whose three sides are unequal.

26. A right angled triangle, is that, which has one right angle.

27. Other triangles are oblique-angled, and are either acute or obtuse.

28. An obtuse-angled triangle has one obtuse angle.

29. An acute-angled triangle has its three angles acute.

30. A figure of four sides and angles is called a quadrangle, or a quadrilateral.

31. A parallelogram is a quadrilateral, which has both its pairs of opposite sides parallel ; and it takes the following particular names ; viz. rectangle, square, rhombus, and rhomboid.

32. A rectangle is a parallelogram, having a right angle.

33. A square is an equilateral rectangle, having its length and breadth equal.

34. A rhomboid is an oblique-angled parallelogram.

35. A rhombus is an equilateral rhomboid, having all its sides equal, but its angles oblique.

36. A trapezium is a quadrilateral, which has not its opposite sides parallel.

37. A trapezoid has only one pair of its opposite sides parallel.

38. A diagonal is a line joining any two opposite angles of a quadrilateral.

39. Plane figures, that have more than four sides, are in general called polygons; and they receive other particular names, according to the number of their sides or angles. Thus,

40. A pentagon is a polygon of five sides; a hexagon, of six sides; a heptagon, of seven; an octagon, of eight; a nonagon, of nine; a decagon, of ten; an undecagon, of eleven; and a dodecagon, of twelve sides.

41. A regular polygon has all its sides and all its angles equal. If they are not both equal, the polygon is irregular.

42. An equilateral-triangle is also a regular figure of three sides, and the square is one of four; the former being also called a trigon, and the latter a tetragon.

43. Any figure is equilateral, when all its sides are equal; and it is equilangular, when all its angles are equal. When both these are equal, it is a regular figure.

44. A circle is a plane figure, bounded by a curve line, called the circumference, which is every where equidistant from a certain point, called its centre. The circumference itself is often called a circle, and also the periphery.

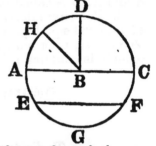

45. The radius of a circle is a line drawn from the centre to the circumference, as A, B.

46. The diameter of a circle is a line drawn through the centre and terminating at the circumference on both sides, as A, C.

47. An arc of a circle is any part of the circumference, as A, D.

48. A chord is a right line joining the extremities of an arch, as E, F.

49. A segment is any part of a circle bounded by an arc, and its chord, as E, F, G.

50. A semicircle is half the circle, or segment cut off by a diameter. The half circumference is sometimes called the semicircle, as A, G, C.

51. A sector is any part of the circle bounded by an arc and two radii drawn to its extremities, as A, B, H.

52. A quadrant, or quarter of a circle is a sector, having a quarter of its circumference for its arc, and its two radii are perpendicular to each other. A quarter of the circumference is sometimes called a quadrant, as A, B, D.

53. The height or altitude of a figure is a perpendicular, let fall from an angle, or its vertex, to the opposite side, called the base, as C, D.

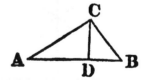

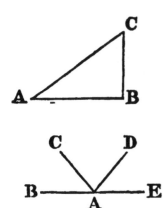

54. In a right-angled triangle, the side opposite to the right angle, is called the hypothenuse; and the other two sides are called the legs, and sometimes the base and perpendicular; thus, A, B is the base, B, C perpendicular, and A, C hypothenuse.

55. When an angle is denoted by three letters, of which, one stands at the angular point, and the other two on the two sides, that which stands at the angular point, is read in the middle. Thus, the angle contained by the lines B, A and A, D is called the angle B, A, D, or D, A, B.

56. The circumference of every circle is supposed to be divided into 360 equal parts, called degrees; and each degree into 60 minutes, each minute into 60 seconds; and so on. Hence a semicircle contains 180 degrees, and a quadrant, 90 degrees.

GEOMETRICAL PROBLEMS.

PROBLEM I.

To divide a line A, B into two equal parts.

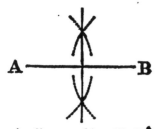

Set one foot of the compasses in A, and opening them beyond the middle of the line, describe arches above and below the line; with the same extent of the compasses, set one foot in the point B, and describe arches crossing the former; draw a line from the intersection above the line to the intersection below the line, and it will divide the line A, B into two equal parts.

PROBLEM II.

To erect a perpendicular on the point C, in a given line.

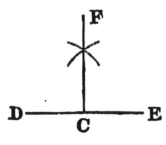

Set one foot of the compasses in the given point C, extend the other foot to any distance at pleasure, as to D, and with that extent make the marks D and E. With the compasses, one foot in D, at any extent above half the distance D, E, describe an arch above the line, and with the same extent, and one foot in E, describe an arch crossing the former; draw a line from the intersection of the arches to the given point C, which will be perpendicular to the given line in the point C.

PROBLEM III.

To erect a perpendicular upon the end of a line.

Set one foot of the compasses in the given point B, open them to any convenient distance, and describe the arch C, D, E; set one foot in C, and with the same extent, cross the arch at D; with the same extent, cross the arch again from D to E; then with one foot of the compasses in D,

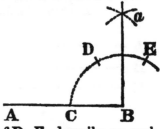

and with any extent above the half of D, E, describe an arch *a*; take the compasses from D, and keeping them at the same extent, with one foot in E, intersect the former arch *a* in *a* from thence draw a line to the point B, which will be a perpendicular to A, B.

PROBLEM IV.

From a given point a, to let fall a perpendicular to a given line A, B.

Set one foot of the compasses in the point *a*, extend the other so as to reach beyond the line A, B, and describe an arch to cut the line A, B in C and D; put one foot of the compasses in C, and with any extent above half C, D, describe an arch *b*, keeping the compasses at the same extent, put one foot in D, and intersect the arch *b* in *b*; through which intersection, and the point *a*, draw *a*, E, the perpendicular required.

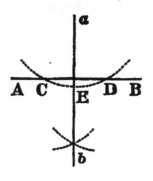

PROBLEM V.

To draw a line parallel to a given line A, B.

Set one foot of the compasses in any part of the line, as at *c*; extend the compasses at pleasure, unless a distance be assigned, and describe an arch *b*; with the same extent in some other part

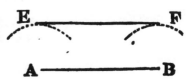

of the line A, B, as at *e*, describe the arch *a ;* lay a rule to the extremities of the arches, and draw the line E, F, which will be parallel to the line A, B.

PROBLEM VI.

To make a triangle, whose sides shall be equal to three given lines, any two of which are longer than the third.

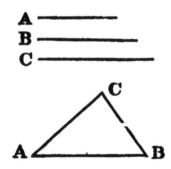

Let A, B, C, be the three given lines ; draw a line, A, B, at pleasure ; take the line C, in the compasses, set one foot in A, and with the other make a mark at B ; then take the given line B, in the compasses, and setting one foot in A, intersect the arch C in C ; lastly, draw the lines A, C and B, C, and the triangle will be completed.

PROBLEM VII.

To make a square, whose sides shall be equal to a given line.

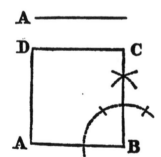

Let A be the given line ; draw a line, A, B, equal to the given line ; from B raise a perpendicular to C, equal to A, B ; with the same extent, set one foot in C, and describe the arch D ; also, with the same extent, set one foot in A, and intersect the arch D ; lastly, draw the line A, D, and C, D, and the square will be completed.

In like manner, may a parallelogram be constructed, only attending to the difference between the length and breadth.

PROBLEM VIII.

To describe a circle, which shall pass through any three given points, not in a straight line.

Let the three given points be A, B, C, through which the circle is to pass. Join the points A, B and B, C, with right lines, and bisect these lines; the point D, where the bisecting lines cross each other, will be the centre of the circle required. Therefore, place one foot of the compasses in D, extending the other to either of the given points, and the circle, described by that radius, will pass through all the points.

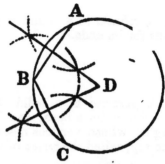

Hence it will be easy to find the centre of any given circle; for, if any three points are taken in the circumference of the given circle, the centre will be found as above. The same may also be observed, when only a part of the circumference is given.

MENSURATION OF SOLIDS.

DEFINITIONS.

1. Solids are figures, having length, breadth, and thickness.

2. A prism is a solid, whose ends are any plane figures, which are equal and similar, and its sides are parallelograms.

Note.—A prism is called a *triangular prism*, when its ends are triangles; *a square prism*, when its ends are squares; *a pentagonal prism*, when its ends are pentagons; and so on.

3. A *cube* is a square prism, having six sides, which are all squares.

4. A *parallelopiped* is a solid, having six rectangular sides, every opposite pair of which are equal, and parallel.

5. A *cylinder* is a round prism, having circles for its ends.

6. A *pyramid* is a solid, having any plane figure for a base, and its sides are triangles, whose vertices meet in a point at the top, called the *vertex* of the pyramid.

7. A *cone* is a round pyramid, having a circular base.

8. A *sphere* is a solid, bounded by one continued convex surface, every point of which is equally distant from a point within, called the *centre*. The sphere may be conceived to be formed by the revolution of a semicircle about its diameter, which remains fixed.

A *hemisphere* is half a sphere.

9. The *segment* of a pyramid, sphere, or any other solid, is a part cut off the top by a plane parallel to the base of that figure.

10. A *frustum* is the part that remains at the bottom, after the segment is cut off.

11. The *sector of a sphere* is composed of a segment less than a hemisphere, and of a cone, having the same base with the segment, and its vertex in the centre of the sphere.

12. The *axis of a solid* is a line drawn from the middle of one end to the middle of the opposite end ; as between the opposite ends of a prism. The axis of a sphere is the same as a diameter, or a line passing through the centre, and terminating at the surface on both sides.

13. The *height* or altitude of a solid, is a line drawn from its vertex, or top, perpendicular to its base.

MENSURATION OF SUPERFICIES AND SOLIDS.

PROBLEM I.

To find the area of a square or parallelogram.

RULE.

Multiply the length by the breadth and the product is the superficial contents.

1. What are the contents of a board 15 feet long, and 2 feet wide ?   Ans. 30 feet.

2. The State of Massachusetts is about 128 miles long and 48 miles wide. How many square miles does it contain ?
                                        Ans. 6144 miles.

3. The largest of the Egyptian pyramids is square at its base, and measures 693 feet on a side. How much ground does it cover ?   Ans. 11 acres, 4 poles.

4. What is the difference between a floor 40 feet square, and 2 others, each 20 feet square ?   Ans. 800 feet.

5. There is a square of 3600 yards area ; what is the side of a square, and the breadth of a walk along each side of the square, and each end, which may take up just one half of the square ?
Ans. $\begin{cases} 42.42 + \text{yards side of the square.} \\ 8.78 + \text{yards breadth of the walk.} \end{cases}$

PROBLEM II.

To find the area of a rhombus or rhomboid.

RULE. — *Multiply the length of the base by the perpendicular height.*

6. The base of a rhombus being 12 feet, and its height 8 feet. Required the area.   Ans. 96 feet.

PROBLEM III.

To find the area of a triangle.

RULE. — *Multiply the base by half the perpendicular height; or add the three sides together; then take half of that sum, and out of it subtract each side severally; multiply the half of the sum, and these remainder together, and the square root of this product will be the area of the triangle.*
22*

7. What are the contents of a triangle, whose perpendicular height is 12 feet, and the base 18 feet ?　　　　Ans. 108 feet.

8. There is a triangle, the longest side of which is 15.6 feet, the shortest side 9.2 feet, and the other side 10.4 feet. What are the contents ?　　　　Ans. 46.139+feet.

PROBLEM IV.

Having the diameter of a circle given, to find the circumference.

RULE. — *Multiply the diameter by* 3.141592, *and the product is the circumference.*

NOTE.—The exact proportion, which the diameter of a circle bears to the circumference has never been discovered, although some mathematicians have carried it to 200 places of decimals. If the diameter of a circle be 1 inch, the circumference will be 3.1415926535897932384626433832795028841971693993751 0582097494459230781640628620899862803483253421170679821480865132823 0664709384464609550582231725359408128480 inches nearly.

9. If the diameter of a circle is 144 feet, what is the circumference ?　　　　Ans. 452.389248 feet.

10. If the diameter of the earth be 7964 miles, what is its circumference ?　　　　Ans. 25019.638688+ miles.

PROBLEM V.

Having the diameter of a circle given, to find the side of an equal square.

RULE. — *Multiply the diameter by* .886227, *and the product is the side of an equal square.*

11. I have a round field 50 rods in diameter ; what is the side of a square field, that shall contain the same area ?　　　　Ans. 44.31135 + rods.

PROBLEM VI.

Having the diameter of a circle given, to find the side of an equilateral triangle inscribed.

RULE. — *Multiply the diameter by* .707016, *and the product is the side of a triangle inscribed.*

12. Required the side of a triangle, that may be inscribed in a cle 80 feet diameter.　　　　Ans. 56.56128 + feet.

PROBLEM VII.

Having the diameter of a circle given, to find the area.

RULE. — *Multiply the square of the diameter by .785398, and the product is the area.   Or multiply half the diameter by half the circumference, and the product is the area.*

13. If the diameter of a circle be 761 feet, what is the area ?
                                      Ans. 454840.475158 feet.
14. There is a circular island three miles in diameter, how many acres does it contain ?   Ans. 4523.89 + acres.

PROBLEM VIII.

Having the circumference of a circle given, to find the diameter.

RULE. — *Multiply the circumference by 31831, and the product is the diameter.*

15. If the circumference of a circle be 25000 miles, what is the diameter ?   Ans. 7957¼ miles.

PROBLEM IX.

Having the circumference of a circle given, to find the side of an equal square.

RULE. — *Multiply the circumference by .282094, and the product is the side of an equal square.*

16. I have a circular field, 360 rods in circumference, what must be the side of a square field, that shall contain the same area ?   Ans. 101.55 + rods.

PROBLEM X.

Having the circumference of a circle given, to find the side of an equilateral triangle inscribed.

RULE. — *Multiply the circumference by .2756646, and the product is the side of an equilateral triangle inscribed.*

17. How large a triangle may be inscribed in a circle, whose circumference is 5000 feet ?   Ans. 1378.323 feet.

PROBLEM XI.

Having the circumference of a circle given, to find the side of an inscribed square.

RULE. — *Multiply the circumference by .225079; and the product is the side of a square inscribed.*

18. I have a circular field whose circumference is 5000 rods; what is the side of a square field that may be made in it ?
Ans. 1125.395 + rods.

PROBLEM XII.

To find the contents of a cube or parallelopipedon.

RULE. — *Multiply the length, height, and breadth, continually together, and the product is the contents.*

19. How many cubic feet are there in a cube, whose side is 18 inches ?   Ans. 3¾ feet.
20. What are the contents of a parallelopipedon, whose length is 6 feet, breadth 2½ feet, and altitude 1¾ feet ?   Ans. 26¼ ft.
21. How many cubic feet in a block of marble, whose length is 3 feet 2 inches, breadth 2 feet 8 inches, and depth 2 feet 6 inches ?   Ans. 21½ feet.

PROBLEM XIII.

To find the solidity of a prism.

RULE. — *Multiply the area of the base, or end, by the height*

22. What are the contents of a triangular prism, whose length is 12 feet, and each side of its base 2½ feet ?   Ans. 32.47 + ft.
23. Required the solidity of a triangular prism, whose length is 10 feet, and the three sides of its triangular end, or base, are 5, 4, and 3 feet ?   Ans. 60 feet.

PROBLEM XIV.

To find the solidity of a cone or pyramid.

RULE. — *Multiply the area of the base by ⅓ of its height.*

24. What is the solidity of a cone, whose height is 12½ feet, and the diameter of the base 2½ feet ?   Ans. 20.45 + feet.
25. What are the contents of a triangular pyramid, whose height is 14 feet 6 inches, and the sides of its base 5, 6, and 7 feet ?   Ans. 71.035 + feet.

QUESTIONS TO EXERCISE THE ABOVE PROBLEMS.

26. I have a round garden, containing 75 square rods ; how large a square garden can be made in it ?

Ans. 47.7464 + square rods.

27. I have a circular garden containing 75 square rods ; what must be the side of a square field, that would contain it ?

Ans. 9.772 + rods.

28. There is a small circular field, 25 rods in diameter; what is the difference of the areas of the inscribed and circumscribed squares, and how much do they differ from the areas of the field ?

Ans. 312.5 rods, the difference of the squares ; 134.12625 + rods, the circumscribed square, more than the area ; 178.373 rods, inscribed square less than the area.

PROBLEM XV.

To find the surface of a cone.

RULE. — *Multiply the circumference of the base by half its slant height.*

29. What is the convex surface of a cone, whose side is 20 feet, and the circumference of its base 9 feet ?　　Ans. 90 feet.

PROBLEM XVI.

To find the solidity of the frustum of a cone, or pyramid.

RULE. — *Multiply the diameters of the two bases together, and to the product add ⅓ of the square of the difference of the diameters ; then multiply this sum by .785398, and the product will be the mean area, between the two bases ; lastly, multiply the mean area by the length of the frustum, and the product will be the solid contents.*

NOTE. — This is the exact rule for measuring round timber.

30. What is the content of a stick of timber, whose length is 40 feet, the diameter of the larger end 24 inches, and the smaller end 12 inches ?　　Ans. 73½ feet, nearly.

PROBLEM XVII.

To find the solidity of a sphere or globe.

RULE. — *Multiply the cube of the diameter by .5236.*

31. What is the solidity of a sphere, whose axis or diameter is 12 inches ?   Ans. 904.78 + inches.

32. Required the content of the earth, supposing its circumference to be 25000 miles. Ans. 263858149120006886875 miles.

PROBLEM XVIII.

To find the convex surface of a sphere, or globe.

RULE. — *Multiply its diameter by its circumference.*

33. Required the convex surface of a globe, whose diameter or axis is 24 inches.   ·   ·   Ans. 1809.55 + inches.

34. Required the surface of the earth, its diameter being 7957¾ miles, and its circumference 25000 miles.
                    Ans. 198943750 square miles.

PROBLEM XIX.

To find how large a cube may be cut from any given sphere, or be inscribed in it.

* RULE. — *Square the diameter of the sphere, divide that product by 3, and extract the square root of the quotient for the answer.*

DEMONSTRATION. — It is evident, that if a cube be inscribed in a sphere, its corners or angles will be in contact with the surface of the sphere, and that a line passing from the lower corner of the cube to its opposite upper corner, will be the diameter of the sphere ; and that the square of this oblique line is equal to the sum of the squares of three sides of the inscribed cube, is evident from the fact, that the square of any two sides of the cube (suppose two sides at the base) is equal to the square of the diagonal across the base ; and that the square of this diagonal, (which we have just proved to be equal to the square of two sides at the base,) and the square of the height of the cube are equal to the square of the diagonal line, which passes from

---

* The author is not aware, that any other person but himself has ever discovered the truth of this rule; and although it has appeared in some recent publications, yet it was at *his* request.

the lower corner of the square, to the opposite upper corner,
which line is the diameter of the sphere.   Therefore, the square
of the diameter of any sphere is equal to the sum of the squares
of any three sides of an inscribed cube ; or ⅓ of the square of
the diameter of any sphere is equal to the square of one of the
sides of an inscribed cube.   Q. E. D.

35. How large a cube may be inscribed in a globe 12 inches
in diameter ?   Ans. 6.928 + in. in the side of the cube.

$$\frac{12 \times 12}{3} = 48, \; \sqrt{48} = 6.928 + \text{Answer.}$$

36. How large a cube may be inscribed in a sphere 40 inches
in diameter ?   Ans. 23.09 + inches.

37. How many cubic inches are contained in a cube, that may
be inscribed in a sphere 20 inches in diameter ?
                                        Ans. 1539.6 + inches.

GAUGING.

GAUGING is the art of finding the contents of any regular vessel, in gallons, bushels, &c.

PROBLEM I.

To find the number of gallons, &c., in a square vessel.

- RULE.

*Take the dimensions in inches ; then multiply the length,
breadth, and height together ; divide the product by 282 for ale
gallons, 231 for wine gallons, and 2150.42 for bushels.*

38. How many wine gallons will a cubical box contain, that
is 10 feet long, 5 feet wide, and 4 feet high ?
                                        Ans. 1496⅘ galls.

39. How many ale gallons will a trough contain, that is 12
feet long, 6 feet wide, and 2 feet high ?   Ans. 882¹⁵⁄₄₇ galls.

40. How many bushels of grain will a box contain, that is 15
feet long, 5 feet wide, and 7 feet high ?   Ans. 421.8 bushels.

PROBLEM II.

To find the contents of a cask.

RULE. — *Take the dimensions of the cask in inches ; viz. the
diameter of the bung and head, and the length of the cask. Note
the difference between the bung diameter and head diameter.*

If the staves of the cask be much curved between the bung and the head, multiply the difference by .7; if not quite so much curved, by .65; if they bulge yet less, by .6; and if they are almost straight, by .55; add the product to the head diameter; the sum will be a mean diameter by which the cask is reduced to a cylinder.

Square the mean diameter thus found, then multiply it by the length; divide the product by 359 for ale or beer gallons, and by 294 for wine gallons.

41. Required the contents in wine gallons, of a cask, whose bung diameter is 35 inches, head diameter 27 inches, and length 45 inches.

$$35 - 27 \times .7 = 5.6 \qquad 32.6 \times 32.6 \times 45 = 47824.20$$
$$27 + 5.6 \quad = 32.6 \qquad 47824.20$$
$$\frac{47824.20}{294} = 162.66 \text{ wine gallons.}$$

42. What are the contents of a cask, in ale gallons, whose bung diameter is 40 inches, head diameter 30 inches, and length 50 inches ? **Ans. 185.55 + ale gallons.**

PROBLEM III.

To find the contents of a round vessel, wider at one end than the other.

RULE. — *Multiply the greater diameter by the less; to this product, add ⅓ of the square of their difference, then multiply by the height, and divide as in the last rule.*

43. What are the contents in wine measure, of a tub 40 inches in diameter at the top, 30 inches at the bottom, and whose height is 50 inches ? **Ans. 209.75 wine galls**

SECTION LXVI.

TONNAGE OF VESSELS.

CARPENTER's RULE. — For single-decked vessels, multiply the length, breadth, at the main beam, and depth in the hold together, and divide the product by 95, and the quotient is the tons.

But for a double-decked vessel, take half of the breadth of the main beam for the depth of the hold, and proceed as before.

1. What is the tonnage of a single-decked vessel, whose length is 65 feet, breadth 20 feet, and depth 10 feet ?

Ans. $136\frac{18}{15}$ tons.

2. What is the tonnage of a double-decked vessel, whose length is 70 feet, and breadth 24 feet ?   Ans. $212\frac{4}{15}$ tons.

GOVERNMENT RULE.— 'If the vessel be double-decked, take the length thereof from the fore part of the main stem to the after part of the stern-post above the upper deck ; the breadth thereof at the broadest part above the main wales, half of which breadth shall be accounted the depth of such vessel, and then deduct from the length $\frac{3}{5}$ of the breadth ; multiply the remainder by the breadth, and the product by the depth, and divide this last product by 95, the quotient whereof shall be deemed the true contents or tonnage of such ship or vessel ; and if such ship or vessel be single-decked, take the length and breadth, as above directed, deduct from the said length $\frac{3}{5}$ of the breadth, and take the depth from the under side of the deck-plank to the ceiling in the hold, and then multiply and divide as aforesaid, and the quotient shall be deemed the tonnage.'

3. What is the government tonnage of a single-decked vessel, whose length is 70 feet, breadth 30 feet, and depth in the hold 9 feet ?   Ans. $147\frac{18}{15}$ tons.

4. What is the government tonnage of a single-decked vessel, whose length is, 75 feet, breadth 22 feet, and depth in the hold 12 feet ?   Ans. $171\frac{381}{475}$ tons.

5. What is the government tonnage of a double-decked vessel, which has the following dimensions ; length 98 feet, breadth 35 feet ?   Ans. $496\frac{17}{38}$ tons.

6. Required the government tonnage of a double-decked vessel, whose length is 180 feet, and breadth 40 feet.

Ans. $1313\frac{43}{48}$ tons.

7. Required the government tonnage of a single-decked vessel, whose length is 78 feet, width 21 feet, and depth 9 feet.

Ans. $130\frac{53}{475}$ tons.

8. What is the government tonnage of a double-decked vessel, whose length is 159 feet, and width 30 feet ?

Ans. $667\frac{17}{19}$ tons.

9. What is the government tonnage of Noah's ark, admitting its length to have been 479 feet, its breadth 80 feet, and its depth 48 feet.   Ans. $17421\frac{3}{15}$ tons.

23

SECTION LXVII.

MENSURATION OF LUMBER.

PROBLEM I.

To find the contents of a board.

RULE.

Multiply the length of the board, taken in feet, by its breadth taken in inches, divide this product by 12, and the quotient is the contents in square feet.

1. What are the contents of a board 24 feet long, and 8 inches wide ?   Ans. 16 feet.
2. What are the contents of a board 30 feet long, and 16 inches wide ?   Ans. 40 feet.

PROBLEM II.

To find the contents of joist.

RULE. — *Multiply the depth and width, taken in inches, by the length in feet; divide this product by 12, and the quotient is the contents in feet.*

3. How many feet are there in 3 joist, which are 15 feet long, 5 inches wide, and 3 inches thick ?   Ans. 56¼ feet.
4. How many feet in 20 joist, 10 feet long, 6 inches wide, and 2 inches thick ?   Ans. 200 feet.

PROBLEM III.

To measure round timber.

RULE. — *Multiply the length of the stick, taken in feet, by the square of ¼ the girt, taken in inches; divide this product by 144, and the quotient is the contents in cubic feet.*

NOTE. — The girt is usually taken about ⅓ of the distance from the larger to the smaller end.

5. How many cubic feet in a stick of timber, which is 30 feet long, and whose girt is 40 inches ?   Ans. 20⅚ feet.
6. If a stick of timber is 50 feet long, and its girt is 56 inches, what number of cubic feet does it contain ?   Ans. 68¹⁄₁₈ feet.
7. What are the contents of a log 90 feet long and whose circumference is 120 inches ?   Ans. 562½ feet.

SECTION LXVIII.

PHILOSOPHICAL PROBLEMS. *

PROBLEM I.

To find the time in which pendulums of different lengths would vibrate; that which vibrates seconds being 39.2 inches.

The time of the vibrations of pendulums are to each other, as the square roots of their lengths; or their lengths are as the squares of their times of vibrations.

RULE. — *As the square of one second is to the square of the time in seconds in which a pendulum would vibrate, so is 39.2 inches to the length of the required pendulum.*

EXAMPLES.

1. Required the length of a pendulum, that vibrates once in 8 seconds.

$$1^2 : 8^2 :: 39.2 \text{ in} : 2508.8 \text{ in.} = 209\tfrac{1}{15} \text{ ft. Ans.}$$

2. Required the length of the pendulum, that shall vibrate 4 times a second.   .   Ans. $2\tfrac{9}{50}$ in.

3. Required the length of a pendulum, that shall vibrate once a minute.   Ans. 3920 yards.

4. How often will a pendulum vibrate, whose length is 100 feet   `.   Ans. 5.53 + seconds.

PROBLEM II.

To find the weight of any body, at any assignable distance above the earth's surface.

The gravity of any body above the earth's surface *decreases*, as the squares of its distance in semi-diameters of the earth from its centre *increases*.

RULE. — *As the square of the distance from the earth's centre is to the square of the earth's semi-diameter, so is the weight of the body on the earth's surface, to its weight at any assignable distance above the surface of the earth, and vice versa.*

5. If a body weigh 900 pounds at the earth's surface, what would it weigh 2000 miles above its surface ?   Ans. 400 lbs.

* For demonstrations of the following problems, the student is referred to Enfield's Philosophy, or to the Cambridge Mathematics.

6. Admitting the semi-diameter of the earth to be 4000 miles, what would be the weight of a body 20000 miles above its surface, that on its surface weighed 144 pounds ?   Ans. 4 lbs.

7. How far must a body be raised to lose half its weight ?
                                                Ans. 1656.85 + miles.

8. If a man at the earth's surface could carry 150 pounds, how much would that burden weigh at the earth, which he could sustain at the distance of the moon, whose centre is 240000 miles from the earth's centre ?   Ans. 540000 lbs.

9. If a body at the surface of the earth weigh 900 pounds, but being carried to a certain height, weighs only 400 pounds ; what is that height ?   Ans. 2000 miles.

PROBLEM III.

By having the height of a tide on the earth given, to find the height of one at the moon.

RULE. — *As the cube of the moon's diameter multiplied by its density, is to the cube of the earth's diameter multiplied by its density, so is the height of a tide on the earth, to the height of one at the moon.*

10. The moon's diameter is 2180 miles, and its density 494 ; the earth's diameter is 7964 miles, and its density 400. If then, by the attraction of the moon, a tide of 6 feet is raised at the earth, what will be the height of a tide raised by the attraction of the earth at the moon ?   Ans. 236.8 + feet.

PROBLEM IV.

To find the weight of a body at the sun and planets, having its weight given at the earth.

If the diameters of two globes be equal, and their densities different, the weight of a body on their surfaces will be as their densities.

If their densities be equal, and their diameters different, the weight of a body will be as ⅓ of their circumference.

If their diameters and densities be both different, the weight of a body will be as ⅓ of their semi-diameters, multiplied by their densities.

Therefore, having the weight of a body on the surface of the earth given, to find its weight at the surface of the sun, and the several planets, we adopt the following

RULE. — *As ⅓ of the earth's semi-diameter, multiplied by its ... is to ⅓ of the sun's, or planet's semi-diameter, multipli-*

ed by its density, so is the weight of a body at the surface of the earth, to the weight of a body at the surface of the sun, or planet.

11. If the weight of a man at the surface of the earth be 170 pounds, what will be his weight at the surface of the sun, and the several planets, whose densities, &c. are as in the following

TABLE.

	Density.	Diameter.	Semi-diameter.	⅓ Semi-diameter.
Sun,	100	883246	441623	294415
Jupiter,	94.5	89170	44585	29723
Saturn,	67	79042	39521	26347
Earth,	400	7964	3982	2654
Moon,	494	2180	1090	726

Ans. Weight at the Sun 4714.6 + lbs., at Jupiter 449.7 + lbs., at Saturn 282.6 + lbs., at the Moon 57.4 + lbs.

PROBLEM V.

To find how far a heavy body will fall in a given time, near the surface of the earth.

Heavy bodies near the surface of the earth fall 16 feet in one second of time ; and the velocities they acquire in falling, are as the squares of the times ; therefore, to find the distance any body will fall in a given time, we adopt the following

RULE. — *As 1 second is to the square of the time in seconds, that the body is falling, so is 16 feet to the distance in feet, that the body will fall in the given time.*

12. How far will a leaden bullet fall in 8 seconds ?

$$1^2 : 8^2 : : 16\,\text{ft.} : 1024\,\text{ft.} : = \text{Answer.}$$

13. How far would a body fall in 1 minute ?

Ans. 10 miles, 1600 yards.

14. How far would a body fall in 1 hour ?

Ans. 39272 miles, 1280 yards.

15. How far did the lost spirits fall in 9 days, in Milton's Paradise Lost ?   Ans. 1832308363 miles, 1120 yards.

PROBLEM VI.

The velocity given to find the space fallen through, to acquire that velocity.

RULE. — *Divide the velocity by 8, and the square of the quotient will be the distance fallen through to acquire that velocity.*

23*

16. The velocity of a cannon ball is 660 feet per second.
From what height must it fall to acquire that velocity?
                                    Ans. 6806¼ feet.
17. At what distance must a body have fallen to acquire the
velocity of 1000 feet per second?   Ans. 2 miles, 5065 feet.

PROBLEM VII.

The velocity given per second to find the time.

Rule. — *Divide the velocity by 8, and a fourth part of the
quotient will be the time in seconds.*

18. How long must a body be falling to acquire a velocity of
200 feet per second ?   Ans. 6¼ seconds.
19. How long must a body be falling to acquire a velocity of
320 feet per second ?   Ans. 10 seconds.

PROBLEM VIII.

The space through which a body has fallen, given to find the
time it has been falling.

Rule. — *Divide the square root of the space fallen through
by 4, and the quotient will be the time in which it was falling.*

20. How long would a body be falling through the space of
40000 feet ?   Ans. 50 seconds.
21. How long would a ball be falling from the top of a tower,
that was 400 feet high, to the earth ?   Ans. 5 seconds.

PROBLEM IX.

The weight of a body and the space fallen through, given to
find the force with which it will strike.

Rule. — *Multiply the space fallen through by 64, then multiply the square root of this product by the weight, and the product is the momentum, or force with which it will strike.*

22. If the rammer for driving the piles of Warren Bridge
weighed 1000 pounds, and fell through a space of 16 feet, with
what force did it strike the pile ?

$\sqrt{16 \times 64} = 32$ $32 \times 1000 = 32000$ lbs. Answer.

23. Bunker Hill Monument at this time, October 13, is about 64 feet high ; supposing a stone weighing 4 tons, should fall from its top to the earth, what would be its force, or momentum ?

Ans. 573440 lbs.

That is, it would strike the earth with more force, than the weight of two hundred and fifty tons.

SECTION LXIX.

MECHANICAL POWERS.

THAT body which communicates motion to another, is called the *power*.

The body which receives motion from another, is called the *weight*.

The mechanical powers are six, the Lever, the Wheel and Axle, the Pulley, the Inclined Plane, the Screw and the Wedge.

LEVER.

The *lever* is a bar, movable about a fixed point, called its *fulcrum* or *prop*. It is in theory considered as an inflexible line, without weight. It is of three kinds ; the first, when 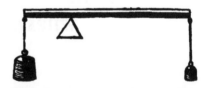 the prop is between the weight and the power ; the second, when the weight is between the prop and the power ; the third, when the power is between the prop and the weight.

A power and weight acting upon the arms of a lever, will balance each other, when the distance of the point, at which power is applied to the lever from the prop, is to the distance of the point at which the weight is applied, as the weight is to the power.

Therefore, to find what weight may be raised by a given power, we adopt the following

RULE.—*As the distance between the body to be raised, or balanced, and the fulcrum or prop, is to the distance between the prop and the point, where the power is applied; so is the power to the weight which it will balance.*

1. If a man weighing 170 pounds be resting upon a lever 10 feet long, what weight will he balance on the other end, supposing the prop one foot from the weight ? Ans. 1530 lbs.

2 If a weight of 1530 pounds were to be raised by a lever 10 feet long, and the prop fixed one foot from the weight, what power applied to the other end of the lever would balance it ?
Ans. 170 lbs.

3. If a weight of 1530 pounds be placed one foot from the prop, at what distance from the prop must a power of 170 pounds be applied to balance it ? Ans. 9 feet.

4. At what distance from a weight of 1530 pounds must a prop be placed, so that as a power of 170 pounds, applied 9 feet from the prop, may balance it ? Ans. 1 foot.

5. Supposing the earth to contain 4.000.000.000.000.000.000.000 cubic feet, and each foot to weigh 100 pounds, and that the earth was suspended at one end of a lever, its centre being 6000 miles from the fulcrum or prop ; and that a man was able, at the other end of the lever, to pull, or press with a force of 200 pounds ; what must be the distance between the man and the fulcrum, that he might be able *to move the earth?*
Ans. 12.000.000.000.000.000.000.000.000 miles.

6. Supposing the man in the last question to be able to move *his* end of the lever 100 feet per second, how long would it take him to raise the earth *one inch?*
Ans. 52.813.479.690 years, 17 days, 14 hours, 57 min. 46¾ sec.

WHEEL AND AXLE.

The *wheel* or *axle*, is a wheel turning round together with its axis ; the power is applied to the circumference of the wheel, and the weight to that of the axis by means of cords.

An equilibrium is produced in the wheel and axis, when the wheel is to the power, as the diameter of the wheel to the diameter of its axis.

To find, therefore, how large a power must be applied to the wheel to raise a given weight on the axis, we adopt the following

RULE. — *As the diameter of the wheel is to the diameter of the axle, so is the weight to be raised by the axle, to the power that must be applied to the wheel.*

7. If the diameter of the axle be 6 inches, and the diameter of the wheel 4 feet, what power must be applied to the wheel to raise 960 pounds at the axle ?   Ans. 120 lbs.

8. If the diameter of the axle be 6 inches, and the diameter of the wheel 4 feet, what power must be applied to the axle to raise 120 pounds at the wheel ?   Ans. 960 lbs.

9. If the diameter of the axle be 6 inches, and 120 pounds applied to the wheel, raise 960 pounds at the axle, what is the diameter of the wheel ?   Ans. 4 feet.

10. If the diameter of the wheel be 4 feet, and 120 pounds applied to the wheel, raise 960 pounds at the axle, what is the diameter of the axle ?   Ans. 6 inches.

PULLEY.

The *pulley* is a small wheel, movable about its axis by means of a cord, which passes over it.

When the axis of a pulley is fixed, the pulley only changes the direction of the power; if movable pulleys are used, an equilibrium is produced, when the power is to the weight, as one to the number of ropes applied to them. If each movable pulley has its own rope, each pulley will be double the power.

To find the weight, that may be raised by a given power.

RULE. — *Multiply the power by twice the number of movable pulleys, and the product is the weight.*

11. What power must be applied to a rope, that passes over one movable pulley, to balance a weight of 400 pounds ?
      Ans. 200 lbs

12. What weight will be balanced by a power of 10 pounds, attached to a cord that passes over 3 movable pulleys ?
                                                    Ans. 60 lbs.
13. What weight will be balanced by a power of 144 pounds, attached to a cord, that passes over 2 movable pulleys ?
                                                    Ans. 576 lbs.
14. If a cord, that passes over two movable pulleys be attached to an axle 6 inches in diameter ; and if the wheel be 60 inches in diameter, what weight may be raised by the pulley, by applying 144 pounds to the wheel ?   Ans. 5760 lbs.

INCLINED PLANE.

An *inclined plane* is a plane, which makes an acute angle with the horizon.

The motion of a body, descending down an inclined plane, is uniformly accelerated.

The force with which a body descends, by the force of attraction, down an inclined plane, is to that, with which it would descend freely, as the elevation of the plane to its length ; or, as the size of its angle of inclination to radius.

To find the power, that will draw a weight up an inclined plane.

RULE. — *Multiply the weight by the perpendicular height of the plane, and divide this product by the length.*

15. An inclined plane is 50 feet in length, and 10 feet in perpendicular height ; what power is sufficient to draw up a weight of 1000 pounds ?   Ans. 200 lbs.
16. What weight applied to a cord passing over a single pulley at the elevated part of an inclined plane, will be able to sustain a weight of 1728 pounds, provided the plane was 600 feet long, and its perpendicular height 5 feet ?   Ans. 14⅖ lbs.
17. A certain railroad, one mile in length, has a perpendicular elevation of 50 feet ; what power is sufficient to draw a train of cars weighing 20000 pounds, up this elevation ?
                                                    Ans. 189¹³⁄₁₃ lbs.

THE SCREW.

The *screw* is a cylinder, which has either a prominent part, or a hollow line passing round it in a spiral form, so inserted in one of the opposite kind, that it may be raised or depressed at pleasure, with the weight upon its upper, or suspended beneath its lower surface.

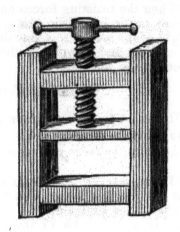

In the *screw*, the equilibrium will be produced, when the power is to the weight, as the distance between the two contiguous threads, in a direction parallel to the axis of the screw, to the circumference of the circle described by the power in one revolution.

To find the power that should be applied to raise a given weight.

RULE. — *As the distance between the threads of the screw is to the circumference of the circle described by the power, so is the power to the weight to be raised.*

NOTE. — One third of the power is lost in overcoming friction.

18. If the threads of a screw be 1 inch apart, and a power of 100 pounds be applied to the end of a lever 10 feet long, what force will be exerted at the end of the screw ?
Ans. 75398.20 + lbs.

19. If the threads of a screw be ½ an inch apart, what power must be applied to the end of a lever 100 inches in length to raise 100000 pounds ? Ans. 79.5774 + lbs.

20. If the threads of a screw be ½ an inch apart, and a power of 79.5774 + pounds be applied to the end of a lever 100 inches in length, what will be raised ? Ans. 100.000 lbs.

21. If a power of 79.5774 + pounds be applied to the end of a lever 100 inches long ; and if by this force a weight of 100.000 pounds be raised, what was the distance of the threads of the screw ? Ans. ½ an inch.

22. If a power of 79.5774 + pounds be applied to the end of a lever ; and if by this force a weight of 100.000 pounds be raised, what must be the length of the lever, if the threads of the screw be ½ an inch apart ? Ans. 100 inches.

WEDGE.

The *wedge* is composed of two inclined planes, whose bases are joined.

When the resisting forces and the power which acts on the wedge, are in equilibrio, the weight will be to the power, as the height of the wedge to a line drawn from the middle of the base to one side, and parallel to the direction in which the resisting force acts on that side.

To find the force of the wedge.

RULE. — *As the breadth or thickness of the head of the wedge is to one of its slanting sides, so is the power which acts against its head, to the force produced at its side.*

23. Suppose 100 pounds to be applied to the head of a wedge, that was 2 inches broad, and 20 inches long, what force would be affected on each side ?   Ans. 1000 lbs.

24. If a wedge be 12 inches long, and its head 1½ inches broad ; and if a screw whose threads are ¼ of an inch asunder, be applied to the head of this wedge ; and if a power of 200 pounds were applied to the end of the lever, 16 feet long, what would be the force exerted on the sides of the wedge ?
Ans. 2573592.1 + lbs.

SECTION LXX.

SPECIFIC GRAVITY.*

To find the specific gravity of a body.

RULE.

Weigh the body both in water and out of the water, and note the difference, which will be the weight lost in water; then, as the weight lost in water, is to the whole weight, so is the specific gravity of water to the specific gravity of the body. But if the body, whose specific gravity is required, is lighter than water, affix to it another body heavier than water, so that the mass, compounded of the two, may sink together. Weigh the dense

* The specific gravity of a body is its weight compared with water; the water being considered 1000.

body and the compound mass separately, both in water, and out of it; then find how much each loses in water by subtracting its weight in water from its weight in air; and subtract the less of these remainders from the greater; then say, as the last remainder is to the weight of the body in air, so is the specific gravity of water to the specific gravity of the body.

NOTE. — A cubic foot of water weighs 1000 ounces.

1. A stone weighed 10 pounds, but in water only 6⅜ pounds Required the specific gravity. Ans. 2608.6+.
2. Suppose a piece of elm weigh 15 pounds in air, and that a piece of copper, which weighs 18 pounds in air, and 16 pounds in water is affixed to it, and that the compound weighs 6 pounds in water. Required the specific gravity of the elm.

Ans. 600.

SECTION LXXI.

ASTRONOMICAL PROBLEMS.

PROBLEM I.

To find the dominical letter for any year in the present century, and also to find on what day of the week, January will begin.

RULE.

To the given year add its fourth part, rejecting the fractions; divide this sum by 7; if nothing remains, the dominical letter is A; but, if there be a remainder, subtract it from 8, and the residue will show the dominical letter, reckoning 0 = A, 2 = B, 3 = C, 4 = D, 5 = E, 6 = F, 7 = G. These letters will also show on what day of the week January begins. For when A is the dominical letter, January begins on the Sabbath; when B is the dominical letter, January begins on Saturday; C begins it on Friday; D on Thursday; E on Wednesday; F on Tuesday; G on Monday.

NOTE. — If it be required to find the dominical letter for the last century, proceed as above; only, if there be a remainder after division, subtract it from 7, and the remainder shows the dominical letter, reckoning 1 = A, 2 = B, 3 = C, 4 = D, 5 = E, 6 = F, 0 = G.

24

1. Required the dominical letter for 1835.

OPERATION.

4)1835   8 — 4 = 4 = D = dominical letter.
  458

7)2293   `   As D is the dominical letter, January will
———   begin on Thursday, and the fourth day will
327–4   be the Sabbath.

2. Required the dominical letter for 1836.

OPERATION.

4)1836   8 — 6 = 2 = B and C = dominical letters.
  459

  ———   In leap years, there are two dominical let-
7)2295   ters. The last letter, C, is for January and
———   February, and B, for the remainder of the
327–6   year. As C is the dominical letter, January
                   will begin on Friday, and the third day will
                   be the Sabbath.

3. Required the dominical letter for 1841.   Ans. C.
4. Required the dominical letter for 1899.   Ans. A.
5. Required the dominical letters for 1896.   Ans. D & E.
6. What is the dominical letter for 1786 ?   Ans. A.
7. What is the dominical letter for 1837 ?   Ans. A.

PROBLEM 11.

To find on what day of the week, any given day of the month
will happen.

RULE. — *Find by the last problem the dominical letter for the
given year, and on what day in January will be the first Sabbath; and the corresponding days in the succeeding months
will be as follows; Wednesday for February; Wednesday for
March; Saturday for April; Monday for May; Thursday
for June; Saturday for July; Tuesday for August; Friday
for September; Sabbath for October; Wednesday for November; Friday for December. Having found the day of the
week for any day in the month, any other day may be easily
obtained, as may be seen in the following example.*

8. Let it be required to ascertain on what day of the week
will be the 25th day of September, 1838.

The dominical letter for 1838 is G ; therefore, the 7th of January will be the Sabbath, and by the above rule, the 7th of February will be Wednesday, the 7th of March will be Wednesday, the
7th of April will be Saturday, the 7th of May will be Monday, the
7th of June will be Thursday, the 7th of July will be Saturday,

the 7th of August will be Tuesday, the 7th of September will be Friday. If the 7th be Friday, the 14th, the 21st, and the 28th will be Fridays. And if the 21st be Friday, the 22d will be Saturday, the 23d will be the Sabbath, the 24th will be Monday, and the 25th, the day required, will be Tuesday.

The following distich will assist the memory in finding the corresponding days of the month.

At Dover Dwells George Brown Esquire,
Good Christian Friend, And David Friar.

It will be recollected, that the initial A is for the Sabbath, B for Monday, C for Tuesday, D for Wednesday, E for Thursday, F for Friday, and G for Saturday.

NOTE.— When it is leap year, the days of the week, after February, will be one day later than on other years.

9. Required the day of the week for the 4th of July, 1836.

We find the dominical letters to be B and C, the 3d day of January therefore will be on the Sabbath, and by the above rule the 3d day of July would be on Saturday ; but as it is leap year, the third of July will be one day later ; that is, it will be on the Sabbath ; and if the 3d be the Sabbath, the 4th of July will be on Monday.

10. On what day of the week will be Dec. 8, 1849 ?
Ans. Saturday.

11. On what day will happen July 4, 1857 ?   Ans. Saturday.

12. On what day will March begin in the year 1890 ?
Ans. Saturday.

13. On what day of the week was our Independence declared ?
Ans. Thursday.

14. There will be a "Transit of Venus," December 8, 1874; on what day of the week will it happen ?   Ans. Tuesday.

15. On what day of the week were you born ?

MISCELLANEOUS QUESTIONS.

1. What number must $7\frac{3}{4}$ be multiplied by, that the product may be $6\frac{2}{3}$?   Ans. $\frac{8}{9}$.

2. What number is that, which being multiplied by half itself, the product shall be $4\frac{1}{2}$?   Ans. 3.

3. What fraction is that, which being divided by $11\frac{3}{7}$, the quotient shall be 5 ?   Ans. $57\frac{1}{7}$.

4. What part of $7\frac{3}{4}$ is $9\frac{4}{7}$?   Ans. $\frac{261}{217}$ or $1\frac{51}{217}$.

5. Reduce $\dfrac{7}{19\frac{2}{3}}$ to a simple fraction.   Ans. $\frac{3}{14}$.

6. Add $\frac{4}{7}$ of a ton to $\frac{5}{10}$ of a cwt.   Ans. 12cwt. 1qr. $8\frac{4}{7}$ lbs.

7. If the earth make one complete revolution in 23 hours, 56 minutes, 3 seconds, in what time does it move one degree?
Ans 3m. 59″. 20‴

8. What number is that to which, if $\frac{2}{7}$ of $\frac{5}{9}$ be added, the sum will be 1? Ans. $\frac{53}{63}$.

9. A, B, and C are to share $100.000 in the proportion of $\frac{1}{3}$, $\frac{1}{4}$, and $\frac{1}{5}$ respectively; but C's part being lost by his death, it is required to divide the whole sum, properly, between the other two. Ans. A's part is 57142\frac{6}{7}$, and B's 42857\frac{1}{7}$.

10. A gentleman being asked what o'clock it was, said that it was between 5 and 6; but to be more particular, he said that the minute hand had past as far beyond the 6, as the hour hand wanted of being to the 6; that is, that the hour and minute hands made equal acute angles, with a line passing from the 12 through the 6. Required the time of day.

Ans. 32m. 18$\frac{6}{13}$ sec. past 5.

11. A certain gentleman, at the time of marriage, agreed to give his wife $\frac{2}{3}$ of his estate, if, at the time of his death, he left only a daughter, and if he left only a son, she should have $\frac{1}{3}$ of his property; but, as it happened, he left a son and a daughter, by which the widow lost in equity $2400 more than if there had been only a daughter. What would have been his wife's dowry if he had left only a son?

Ans. $2100.

12. A father devised $\frac{7}{18}$ of his estate to one of his sons and $\frac{7}{18}$ of the residue to the other, and the remainder to his wife. The difference of his sons' legacies was found to be £257. 3s. 4d. What money did he leave for his widow?

Ans. £635. 0s. 10$\frac{30}{49}$d.

13. In the walls of Balbeck, in Turkey, the ancient Heliopolis, there are three stones laid end to end, now in sight, that measure in length 61 yards; one of which is 63 feet long, 12 feet thick, and 12 feet broad; what is the weight, supposing its specific gravity to be 3 times that of water? Ans. 759$\frac{3}{8}$ tons.

14. Two men carrying a burden of 200 pounds, hung on a pole, the ends of which rest on their shoulders; how much of this load is borne by each man, the weight hanging 6 inches from the middle, and the whole length of the pole being 4 feet?

Ans. 125lbs., and 75lbs.

15. The new Court House in Boston, has 8 pillars of granite, that are 25 feet 4 inches in length, 4 feet 5 inches in diameter at one end, and 3 feet 5 inches in diameter at the other end. How many cubic feet do they contain, and what is their weight, allowing a cubic foot to weigh 3000 ounces?

Ans. 2455.03 cubic feet, 205.4 tons.

16. A father dying, left his son a fortune, $\frac{1}{4}$ of which he spent in 8 months; $\frac{3}{7}$ of the remainder lasted him 12 months longer; after which he had only $410 left. What did his father bequeath him? Ans. 956.66\frac{2}{3}$.

17. A merchant has sold goods to a certain amount, on a commission of 4 per cent., and having remitted the net proceeds to the owner, he received ¼ per cent. for prompt payment, which amounted to $15.60. What was the amount of his commission ? Ans. $260.00.

18. A, of Boston, remits to B, of New York, a bill of Exchange on London, the avails of which he wishes to be invested in goods on his account. B having disposed of the bill at 7½ per cent. advance, he received $9675.00, and having reserved for himself ¼ per cent. on the sale of the bill, and 2 per cent. for commission, what will remain for investment, and for how much was the bill drawn ? Ans. For investment, $9461.58₇₁³.

The bill was £2025.

19. Bunker Hill Monument is 30 feet square at its base, and *is to be* 15 feet square at its top ; its height is to be 220. From the bottom to the top, through its centre, is a cylindrical avenue 15 feet in diameter at the bottom, and about 11 feet at the top. How many cubic feet will there be in the monument ?
Ans. 86068.518 + feet.

20. A hired a house for one year for $300 ; at the end of 4 months he takes in M as a partner, and at the end of 8 months he takes in P. At the end of the year, what rent must each pay ? Ans. A pays $183⅓, M pays $83⅓, P pays $33⅓.

21. A merchant receives on commission, three kinds of flour ; from A he receives 20 barrels, from P 25 barrels, and from C 40 barrels. He finds that A's flour is 10 per cent. better than B's, and that B's is 20 per cent. better than C's. He sells the whole at $6 per barrel. What in justice should each man receive ? Ans. A receives $139₄₁¹¹, B $158₄₁⁷⁷, C $211₄₁¹⁴.

22. Bought 100 barrels of flour, at $5.00 per barrel, and immediately sold it on a credit of 6 months. The note which I received for pay, I got discounted at the Suffolk Bank ; and, on examining my money, I found that I had gained 20 per cent. on my purchase. What did I receive per barrel for the flour ?
Ans. $6.18₁₃₃⁹⁹.

23. Purchased for a cloak 5¼ yards of broadcloth, that was 1⅜ yards wide ; to line this, I purchased flannel that was ⅞ yards wide, but on being wet it shrunk 1 nail in width, and 1 yard in every 20 yards in length. How many yards of flannel was it necessary to buy ? Ans. 12₂₄₇¹⁶ yds.

24. How many bricks would it require to build the walls of a house 40 feet long, 30 feet wide, and 20 feet high, admitting the walls to be a foot thick, and that each brick was 8 inches long, 4 inches wide, and 2 inches thick ? Ans. 73440 bricks.

25. How many feet of boards would it require to cover a house, that was 40 feet square, and whose height to the top of the plate was 20 feet. The roof projected 1 foot over the plate, and came to a point over the centre of the house, 15 feet above the garret floor ? Ans. 5367.7 + feet.

24*

26. The dimensions of a bushel measure are 18½ inches wide, and 8 inches deep ; what should be the dimensions of a similar measure, that would contain 8 bushels ?

Ans. 37in. wide, 16in. deep.

27. What is the weight of a hollow spherical iron shell 5 inches in diameter, the thickness of the metal being 1 inch, and a cubic inch of iron, weighing ¼¾ of a pound ? Ans. 13.2387lbs.

28. At a certain time between two and three o'clock, the minute hand of the clock was between three and four. Within an hour after, the hour hand and minute hand had exactly changed places with each other. What was the precise time, when the hands were in the first position ? Ans. 2 h. 15 m. 56$\frac{52}{143}$ sec.

29. How much larger is a cube, that will contain a globe of 20 inches in diameter, than a cube inscribed within such globe?

Ans. Larger cube 8000in., smaller 1539in.

30. If in a pair of scales, a body weigh 90 pounds in one scale, and only 40 pounds in the other. Required the true weight, and the proportion of the lengths of the two arms of the balance beam on each side of the point of suspension.

Ans. The weight 60lbs., and the proportions 3 to 2.

31. In turning a one-horse chaise within a ring of a certain diameter, it was observed that the outer wheel made two turns, while the inner wheel made but one ; the wheels were both four feet high ; and supposing them both fixed, at the distance of five feet asunder on the axletree, what was the circumference of the track described by the outer wheel ? Ans. 62.83 + feet.

32. The ball on the top of St. Paul's Church is 6 feet in diameter. What did the gilding of it cost at 3½d. per square inch ?

Ans. £237. 10s. 1d.

33. There is a conical glass, 6 inches high, 5 inches wide at the top, and is ⅓ part filled with water. What must be the diameter of a ball, let fall into the water, that shall be immersed by it ? Ans. 2.445 + in.

34. A certain lady, the mother of three daughters, had a farm of 500 acres, in a circular form, with her dwelling-house in the centre. Being desirous of having her daughters settled near her, she gave to them three equal parcels, as large as could be made in three equal circles, included within the periphery of her farm, one to each, with a dwelling-house in the centre of each ; that is, there were to be three equal circles, as large as could be made within a circle, that contained 500 acres. How many acres did the farm of each daughter contain ; how many acres did the mother retain ; how far apart were the dwelling-houses of the daughters ; and how far was the dwelling-house of each daughter from that of the mother ?

Ans. Each daughter's farm contained 107 acres, 2 rods, 31.22 + rods. The mother retained 176 acres, 3 rods, 26.34 + rods. The distance from one daughter's house to the other, was 148.-119817 + rods. The mother's dwelling-house was distant from her daughters' 85.51+rods.

A

NEW AND CONCISE SYSTEM

OF

BOOK-KEEPING,

BY

SINGLE AND DOUBLE ENTRY:

CALCULATED FOR THE VARIOUS BRANCHES

OF

FOREIGN AND DOMESTIC TRADE.

DESIGNED FOR THE USE OF SCHOOLS AND ACADEMIES.

BY BENJAMIN GREENLEAF, A. M.,
Principal of Bradford Teachers' Seminary.

BOSTON:
PUBLISHED BY ROBERT S. DAVIS.
....
1843.

BOOK-KEEPING.

BOOK-KEEPING is the art of recording mercantile transactions. Two methods have been usually adopted; *Single* and *Double Entry*.

In Single Entry, two books are necessary, the Day or Waste-Book, and Leger; others are occasionally used.

DAY BOOK.

This book commences with an inventory of all the property invested in trade; viz. money, notes, goods, &c.; and also, an account of all the debts, which the merchant owes.

After the inventory, the daily occurrences of trade should be narrated one after the other, according to their dates; such as buying, selling, receiving, paying, &c.

In entering which, every circumstance, that respects the transaction, should be clearly and distinctly expressed; such as the names of the parties, the conditions of bargains, the terms of payment, the quantity, quality and price of goods, with every thing else, that is material or expressive of the nature and design of the transaction.

Dr. and Cr. may be easily distinguished by the following rules.

1. The person to whom goods are sold on credit is Dr.
2. The person of whom goods are bought on credit is Cr.
3. The person to whom money is paid is Dr.
4. The person from whom money is received is Cr.
5. The person to whom the merchant, in any way becomes indebted, is entered Cr.
6. The person, who in any way becomes indebted to the merchant, is entered Dr.
7. The receiver is Dr. and the giver is Cr.

THE LEGER.

The Leger is the merchant's principal book, as in it are collected the scattered accounts of the Waste Book, and disposed in spaces assigned for them.

RULES FOR POSTING THE LEGER.

Enter the several transactions on the debtor and creditor side in the Leger, as they stand debited or credited in the Day Book.

When all the transactions are correctly posted into the Leger, each account is balanced by subtracting the less side from the greater, entering the balance on the less side, by which both sides will be made equal.

NOTE.—In Single Entry, the Day Book and Journal are considered the same thing

Boston, January 1, 1835.

L. F.		$	C.
	INVENTORY.		
	Inventory of goods, money, and debts due to James Williams, merchant, Boston.		
	Money on hand - - - - - $1000		
	Various merchandise - - - - $000		
1	Daniel Snow owes me - - - 100		
3	James George " " - - - - 200		
		4300	00
	"		
	DEBTS OWED.		
	To John Dole, on account - - $150.00		
	James Ames - - - - - 200.00		
		350	00
	——————— 2 ———————		
1	John True Dr.		
	To 7 lbs. coffee - - - - a .14		
	9 lbs. tea - - - - - a .38		
	50 lbs. fish - - - - - a .05		
		6	90
	——————— 3 ———————		
1	Daniel Snow Dr.		
	To 9 yards white flannel - - - a $ .66		
	7 lbs. colored thread - - - - a .98		
	4 silk vests - - - - - a 3.75		
	6 ivory combs - - - - - a .30		
		29	60
	——————— 5 ———————		
1	Samuel Brown Dr.		
	To 10 lbs. hyson tea - - - - a .38		
	17 " coffee - - - - a .14		
	16 yards durant - - - - a .40		
	15 " white flannel - - - a .67		
		22	63
	——————— 6 ———————		
1	James Ames Dr.		
	To 7 galls. wine - - - - - a $1.25		
	1 barrel oil - - - - - a 23.82		
		32	57

Boston, January 7, 1835.

L. F.		$	C.
1	**Samuel Brown** Cr.		
	By 19 pair shoes - - - - - a $1.75		
	12 " boots - - - - a 2.25		
	6 silk vests - - - - - a 3.18		
	7 cords wood - - - - - a 5.17		
		115	52
	———————— 8 ————————		
1	**Enoch Carey** Dr.		
	To 16 lbs. coffee - - - - - a .25		
	8 " sugar - - - - - a .11		
	15 " hyson tea - - - - - a .42		
	24 " loaf sugar - - - - - a .17		
		15	26
	———————— 9 ————————		
1	**Daniel Snow** Cr.		
	By 16 bushels corn - - - - a $ .83		
	7 " rye - - - - a 1.13		
		21	19
	———————— 10 ————————		
2	**John Day** Dr.		
	To 1 bale cotton, 300 yards - - - a $ .09		
	1 " shirting, 400 " - - - a .15		
	1 ps. cassimere, 20 " - - - a 1.25		
		112	00
	———————— 12 ————————		
2	**Luke Frye** Dr.		
	To 2 yards broadcloth - - - a $4.25		
	silk - - - - a .87		
	2 pair gloves - - - - a .75		
		10	87
	———————— 13 ————————		
1	**John True** Cr.		
	By cash - - - - - - - - -	3	00
	———————— 14 ————————		
2	**Mark George** Dr.		
	To 14 lbs. sugar - - - - - a .12		
	16 " coffee - - - - - a .15		
	12 " butter - - - - - a .16		
		6	00
	———————— 15 ————————		
2	**John How** Dr.		
	To 275 lbs. fish - - - - - a .03		
	50 " ginger - - - - - a .16		
	23 " pepper - - - - a .24		
		21	72

Boston, January 16, 1835.

L. F.			$	C.
2	John Day　　　　　　　　　　　　　　　Cr.			
	By cash on account　-　-　-　-　-			
			85	00
	———— 17 ————			
2	Samuel Ives　　　　　　　　　　　　　Dr.			
	To　4 umbrellas　-　-　-　-　-　a $1.25			
	9 yards bombazine　-　-　-　a　.45			
	5 " 　satinett　-　-　-　-　a　.50			
			11	55
	———— 19 ————			
2	Daniel Jones　　　　　　　　　　　　Dr.			
	To 4 yards broadcloth　-　-　-　a $5.25			
	14 " 　linen -　-　-　-　-　a　.15			
	trimmings　-　-　-　-　a　1.50			
			24	60
	———— 20 ————			
3	James George　　　　　　　　　　　Dr.			
	To　6 galls. molasses -　-　-　-　a .30			
	25 lbs. flour　-　-　-　-　-　a .05			
	10 " 　raisins　-　-　-　-　a .10			
			4	05
	———— 21 ————			
1	Enoch Carey　　　　　　　　　　　Cr.			
	By 16 sides leather, 24 lbs. -　-　-　a .20			
	20 lbs. calfskin -　-　-　-　a .45			
			85	80
	———— 22 ————			
1	Daniel Snow　　　　　　　　　　　Dr.			
	To 17 yards cambric　-　-　-　a $ .44			
	5 worsted shawls -　-　-　a　1.50			
			14	98
	———— 23 ————			
2	John Day　　　　　　　　　　　　Cr.			
	By order on Jacob Flint　-　-　-　-		20	00
	———— 24 ————			
1	James Ames　　　　　　　　　　　Dr.			
	To 5 yards brown linen -　-　-　a $ .30			
	9 " 　plaid print　-　-　-　a　.29			
	3 " 　silk velvet　-　-　-　a　2.50			
			11	61
	———— 26 ————			
2	John Day　　　　　　　　　　　　Dr.			
	To 1 bale shirting, 500 yards　-　-　a .10		50	00

Boston, January 27, 1835.

L. F.		$	C.
2	John How Cr.		
	By cash on account - - - - - -		
		20	00
	————— 28 —————		
1	Samuel Brown Dr.		
	To 7 yards broadcloth - - - - a $5.50		
	5 " cassimere - - - a 1.50		
	10 " jean - - - - - a .20		
		48	00
	————— 29 —————		
2	Luke Frye Dr.		
	To 3 yards kersey - - - - - a $1.00		
	8 " flannel - - - - a .37½		
		6	00
	————— 30 —————		
2	Mark George Dr.		
	To 1 keg molasses, 10 galls. - - - a $ .35		
	1 bbl. flour - - - - - a 6.75		
		10	25
	————— 31 —————		
1	John True Cr.		
	By 1 load wood, 1½ cords - - - $4.50	6	75
	————— February 2 —————		
	Samuel Ives Dr.		
	To 2 pair boots - - - - - a $3.00		
	1 " shoes - - - - - a 1.25		
		7	25
	————— 3 —————		
2	Daniel Jones Dr.		
	To 14 yards tow cloth - - - a $ .08		
	40 " brown linen - - - a .25		
	4 pieces nankin - - - - a 1.87		
	9 yards striped jean - - - a .20		
		20	40
	————— 4 —————		
3	James George Dr.		
	To 3 bushels corn - - - - - a .85		
	2 " rye - - - - - a .92		
	8 " oats - - - - - a .33		
		7	03
	————— 5 —————		
2	John Day Cr.		
	By cash on account - - - - -	50	00

Boston, February 7, 1835.

L. F.			$	
1	John True Dr.			
	To 3 bbls. flour - - - - - a $5.50			
	15 lbs. cheese - - - - - a .13			
	18 " raisins - - - - - a .08			
	15 " pepper - - - - - a .17			
			22	44
	————————— 9 —————————			
1	Samuel Brown Dr.			
	To 1 bbl. flour - - - - - a $5.60			
	7 lbs. green tea - - - - a .38			
	5 " sugar - - - - - a .12			
			8	36
	————————— 10 —————————			
1	John True Cr.			
	By cash on account - - - - - -		20	00
	————————— 11 —————————			
1	Daniel Snow Dr.			
	To 30 lbs. sugar - - - - - a .12			
	70 " flour - - - - - a .06			
	5 " raisins - - - - - a .10			
			8	30
	————————— 12 —————————			
1	Samuel Brown Cr.			
	By cash on account - - - - - -		30	00
	————————— 13 —————————			
1	Daniel Snow Cr.			
	By 40 bushels salt - - - - - a $ .45			
	37 " rye - - - - - a 1.75			
	15 " corn - - - - - a 1.12			
			99	55
	————————— 14 —————————			
1	John True Cr.			
	By cash on account - - - - - -		300	39
	————————— 16 —————————			
2	John Day Dr.			
	To 7 yards broadcloth - - - - a $3.75			
	6 " calico - - - - - a .17			
	12 lbs. sugar - - - - - a .17			
			29	31
	————————— 17 —————————			
1	Enoch Carey Dr.			
	To 15 lbs. sugar - - - - - - a .09			
	16 " tea - - - - - - a .65			
	3 bushels salt - - - - - a .47			
			13	16

Boston, February 18, 1835

L. F.			$	C
2	**John Day**	**Cr.**		
	By cash on account		36	31
	—————— 19 ——————			
. 1	**James Ames**	**Dr.**		
	To 7 bbls. flour	a $5.60		
	8 bushels salt	a .81		
	81 " rye	a 1.25		
			146	93
	—————— 20 ——————			
1	**Enoch Carey**	**Dr.**		
	To 3 cords wood	a $5.50		
	18 tons timber	a 7.80		
			156	90
	—————— 21 ——————			
2	**Luke Frye**	**Dr.**		
	To 8 yards broadcloth	a $5.75	46	00
	—————— 23 ——————			
2	**Mark George**	**Cr.**		
	By 7 cords oak wood	a $6.75	47	25
	—————— 24 ——————			
2	**Mark George**	**Dr.**		
	To 17 lbs. butter	a .25		
	18 " coffee	a .16		
	30 " cheese	a .12		
			10	73
	—————— 25 ——————			
2	**Mark George**	**Cr.**		
	By 15 bushels salt	a $ .80		
	14 " rye	a 1.67		
			35	38
	—————— 26 ——————			
2	**John How**	**Dr.**		
	To 10 yards ticking	a .25		
	16 " shirting	a .18		
			5	38
	—————— 27 ——————			
2	**Luke Frye**	**Dr.**		
	To 5 yards linen	a $ .80		
	8 " full cloth	a 1.65		
	7 " bombazine	a .45		
			20	35
	—————— 28 ——————			
2	**John How**	**Cr.**		
	By cash		45	00

Boston, February 28, 1835.

L. F.		$	C.
2	Luke Frye Cr.		
	By 5 bushels apples - - - - - - a $1.45		
	8 " walnuts - - - - a 1.75		
		12	50
	——————— March 2 ———————		
1	John True Dr.		
	To 28 yards broadcloth - - - - a $5.60		
	15 bbls. flour - - - - - a 6.70		
	90 galls. molasses - - - - a .46		
	14 lbs coffee - - - - - a .15		
		800	80
	————————— 3 —————————		
1	James Ames Cr.		
	By 46 bushels corn - - - - - a $1.12		
	17 " rye - - - - a 1.25		
	30 " wheat - - - - a 1.50		
		117	77
	————————— 4 —————————		
1	Enoch Carey Cr.		
	By 7 cords oak wood - - - - a $5.50		
	18 lbs. clover seed - - - - a .10		
		40	30
	————————— 5 —————————		
1	James Ames Cr.		
	By cash - - - - - - -	73	34
	————————— 6 —————————		
2	Luke Frye Cr.		
	By cash - - - - - - -	70	72
	————————— 7 —————————		
2	Mark George Cr.		
	By 1 cord wood - - - - - a $5.67	5	67
	————————— 9 —————————		
2	Samuel Ives Cr.		
	By 17 bushels beans - - - - a $1.50		
	13 " corn - - - - a 1.12		
		45	66
	———————— 10 ————————		
2	Daniel Jones Cr.		
	By an order on John Wait - - - -	30	00
	———————— 11 ————————		
2	Samuel Ives Cr.		
	By an order on John Ames - - - -	5	00
	———————— 12 ————————		
3	James George Cr.		
	By cash on account - - - - -	201	08

Boston, March 13, 1835.

				$	C.
2	Daniel Jones	Cr.			
	By 1 ton hay - - - - - - - - -			15	00
	——————— 14 ———————				
3	James George	Cr.			
	By 7 day's labor - - - - - a $1.25				
	1 bushel corn - - - - - a 1.25				
				10	00

INDEX TO THE LEGER.

NOTE. — The first column in the Day Book refers to the page in the Ledger, to which the account is transferred; and the third column in the Ledger, under J. F., refers to the page in the Journal or Waste Book, from which the account has been transferred

Dr.   John True.

1835.		JF		$	C.
Jan.	2	1	To sundries as per day book - -	6	90
Feb.	7	5	Sundries - - - - -	22	44
March	2	7	Sundries - - - - -	300	80
				330	14

Dr.   Daniel Snow.

1835.		JF		$	C.
Jan.	1	1	To stock - - - - - -	100	00
	3	1	Sundries - - - - -	29	60
	22	3	Sundries - - - - -	14	98
Feb.	11	5	Sundries - - - - -	8	30
				152	88

Dr.   Samuel Brown.

1835.		JF		$	C.
Jan.	5	1	To sundries - - - - - -	22	63
	28	4	Sundries - - - -	48	00
Feb.	9	5	Sundries - - - - - -	8	86
			Balance transferred to folio 1, Leger B.	66	03
				145	52

Dr.   James Ames.

1835.		JF		$	C.
Jan.	6	1	To Sundries - - - - - -	32	57
	24	3	Sundries - - - - -	11	61
Feb.	19	6	Sundries - - - - - -	146	93
			Balance transferred to folio 1, Leger B.	200	00
				391	11

Dr.   Enoch Carey.

1835.		JF		$	C.
Jan.	8	2	To sundries - - - - - -	15	26
Feb.	17	5	Sundries - - - - -	13	16
	20	6	Sundries - - - - -	156	90
				185	32

Cr.

1835.				JF	$	C.
Jan.	13	By cash - - - - - - - -		2	3	00
	31	A load of wood - - - -		4	6	75
Feb.	10	Cash on account - - - - -		5	20	00
	14	Cash on account - - - -		5	300	39
					330	14

Cr.

1835.				JF	$	C.
Jan.	9	By sundries - - - - - -		2	21	19
Feb.	13	Sundries - - - - -		5	99	55
		Balance transferred to folio 1, Leger B.			32	14
					152	88

Cr.

1835.				JF	$	C.
Jan.	7	By sundries - - - - -		2	115	52
Feb.	12	Cash on account - - - -		5	30	00
					145	52

Cr.

1835.				JF	$	C.
Jan.	1	By stock - - - - - - -		1	200	00
March	3	Sundries - - - - -		7	117	77
	5	Cash - - - - - -		7	73	34
					391	11

Cr.

1835.				JF	$	C.
Jan.	21	By sundries - - - - -		3	85	80
March	4	Sundries - - - - -		7	40	30
		Balance transferred to folio 1, Leger B.			59	22
					185	32

Dr.   John Day.

1835.		JF		$	C.
Jan.	10	2	To sundries - - - - - -	112	00
	26	3	Bale of shirting - - - -	50	00
Feb.	16	5	Sundries - - - - - -	29	31
				191	31

Dr.   Luke Frye.

1835.		JF		$	C.
Jan.	12	2	To sundries - - - - -	10	87
	29	3	Sundries - - - - -	6	00
Feb.	21	6	8 yards broadcloth - - -	46	00
	27	6	Sundries - - - -	20	35
				83	22

Dr.   Mark George.

1835.		JF		$	C.
Jan.	14	2	To sundries - - - - -	6	00
	30	4	Sundries - - - - -	10	25
Feb.	24	6	Sundries - - - - - -	10	73
			Balance transferred to folio 1, Leger B.	61	32
				88	30

Dr. -   John How.

1835.		JF		$	C.
Jan.	15	2	To sundries - - - - -	21	77
Feb.	26	6	Sundries - - - - -	5	38
			Balance transferred to folio 1, Leger B.	37	85
				65	00

Dr.   Samuel Ives.

1835.		JF		$	C.
Jan.	17	3	To sundries - - - - -	11	55
Feb.	2	4	Sundries - - - - -	7	25
			Balance transferred to folio 1, Leger B.	31	86
				50	66

Dr.   Daniel Jones.

1835.		JF		$	C
Jan.	19	3	To sundries - - - - -	24	60
Feb.	3	4	Sundries - - - - -	20	40
				45	00

Cr.

1839.			JF	$	C.
Jan.	16	By Cash on account - - - -	3	85	00
	23	Order on Jacob Flint - - - -	3	20	00
Feb.	5	Cash on account - - - -	4	50	00
	18	Cash on account - - - -	6	36	31
				191	31

Cr.

1835.			JF	$	C.
Feb.	28	By sundries - - - - -	7	12	50
March	6	Cash - - - - - -	7	70	72
				83	22

Cr.

1835.			JF	$	C.
Feb.	23	By 7 cords wood - - - - -	6	47	25
	25	Sundries - - - - -	6	35	38
March	7	1 cord wood - - - - -	7	5	67
				88	30

Cr.

1835.			JF	$	C.
Jan.	27	By cash on account - - - -	4	20	00
Feb.	28	Cash - - - - - -	6	45	00
				65	00

Cr.

1835.			JF	$	C.
March	9	To sundries - - - - -	7	45	66
	11	Order on John Ames - - -	7	5	00
				50	66

Cr.

1835.			JF	$	C.
March	10	By order on John Wait - - -	7	30	00
	13	1 ton of hay - - - - -	8	15	00
				45	00

Dr.   James George.

1835.		JF		$	C.
Jan.	1	1	To stock - - - - - - -	200	00
	20	3	Sundries - - - - -	4	05
Feb.	4	4	Sundries - - - - - -	7	03
				211	08

Dr.   John Dole.

1835.		JF		$	C.
Jan.	1		To balance transferred to Leger B. -	150	00

LEGER B.   [1

Dr.   Daniel Snow.

1835.		JF		$	C.
April	1		To Balance at folio 1, Leger A. - -	32	14

Dr.   Samuel Brown.

1835.		JF		$	C.

Dr.   James Ames.

1825.		JF		$	C.

Dr.   Enoch Carey.

1835.		JF		$	C.
April.	1		To balance at folio 1, Leger A. - -	59	22

Dr.   Mark George.

1835.		JF		$	C.

Dr.   John How.

1835.		JF		$	C.

Dr.   Samuel Ives.

1835,		JF		$	C.

Dr.   John Dole.

1835.		JF		$	C.

Cr. .

1835.										JF	$	C.
March	12	By cash on account	-	-	-	-		7	201	08
	14	Sundries	-	-	-	-	-		8	10	00
											211	08

Cr.

1835.								JF	$	C.
Jan.	1	By stock	-	-	-	-			150	00

1]　　　　　　LEGER B.

Cr.

1835.			JF	$	C

Cr.

1835.			JF	$	C.
April	1	By balance at folio 1, Leger A. - -		66	03

Cr.

1835.			JF	$	C.
April	1	By balance at folio 1, Leger A. - -		200	00

Cr.

1835.			JF	$	C.

Cr.

1835.			JF	$	C.
April	1	By balance at folio 2, Leger A. - -		61	32

Cr.

1835.			JF	$	C.
April	1	By balance at folio 2, Leger A. - -		37	85

Cr.

1835.			JF	$	C.
April	1	By balance at folio 3, Leger A. -		31	86

Cr.

1835.			JF	$	C.
April	1	By balance at folio 3, Leger A. - -		150	00

BOOK-KEEPING BY DOUBLE ENTRY.

THE art of book-keeping by *Double Entry* consists in recording correctly and intelligibly every transaction, in relation to buying and selling each kind of goods employed in trade.

The three principal books used, are the Waste or Day Book, the Journal, and the Leger.

The auxiliary books, are the Cash, Bill, Invoice, Sales, Commission, Letter, Postage, Expense, Receipt, and Memorandum books.

THE WASTE, OR DAY BOOK.

For the forms, &c. of this book, see Single Entry.

JOURNAL.

The *Journal* is a fair transcript of all the transactions in the Waste Book.

The art of book-keeping wholly depends on a correct discrimination of Dr. and Cr.

The following rules should be carefully observed.

DEBTOR AND CREDITOR.

Goods bought for ready money are Dr. to cash.

Goods bought on credit are Dr. to the seller.

Goods bought for part money, and part credit, are Dr. to sundries ; to cash paid in part, and to the seller for the remainder.

When money is paid, the receiver is Dr., and cash Cr.

NOTE.— When money is paid, it should be specified, whether in part or in full.

When goods are sold on credit, the *buyer* is Dr. to *goods*.

When goods are sold for cash, cash is Dr. to *goods*.

When goods are sold for part cash and part credit, sundries are Dr. to goods to *cash* for the sum received, and the *purchaser* for the rest.

The following *general rule*, if correctly understood, is sufficient for all the purposes of posting.

Whatever is received, as a consideration for goods sold, is Dr.; and whatever is paid for goods received, whether cash, notes, orders, bills, &c., is Cr.

LEGER.

The LEGER is the merchant's principal book, to which the several transactions, dispersed through the Waste Book, &c. are transferred each to its proper account.

In posting the Leger, the following circumstances should be carefully observed.

On the Dr. side should be written,

1. The date.
2. The Journal folio.
3. Each creditor title.
4. The transaction.
5. The quantity and quality of goods.
6. Leger folio of the creditor.
7. The amount in dollars and cents.

On the creditor side should be written,

1. The date.
2. The Journal folio.
3. Each debtor title.
4. The transaction.
5. The quantity and quality of goods.
6. Leger folio of the debtor.
7. Amount in dollars and cents.

The merchant should arrange his accounts in the Leger according to their importance.

Stock should be first, and cash second.

The column under J. F. refers to the page in the Journal, from which the account is transferred. The column under Cr. on the debtor side refers to the page in the Leger, where the same account is credited; and the column under Dr. on the credit side refers to the page in the Leger, where the same account is debited.

26

Boston, January 1, 1835.

J. F.		$	C.
	An Inventory of money, goods, debts, &c. belonging to me, James Hobart, and also of my debts.		
1	I have in ready money - - - - $100.00		
	34 barrels of flour, at $5 - - - - 170.00		
	40 pieces of linen, at $20 - - - 800.00		
	5 cwt. sugar, at $10 - - - - 50.00		
	20 quintals fish, at $3 - - - - 60.00		
	8 hhds. molasses, at $30 - - - 240.00		
		1420	00
1	I am indebted to Daniel Fox, on account $50.00		
	John Chase, on account - - - 20.00		
	John Jones, " - - - - 36.00		
		106	00
	——— 2 ———		
1	Sold John Dow 10 barrels flour, at $6 - -	60	00
	——— 3 ———		
1	Sold for cash, 12 barrels flour, at $6 -	72	00
	——— 5 ———		
1	Sold Daniel Fox 10 pieces of linen, at $25 -	250	00
	——— 6 ———		
1	Sold John Chase 5 cwt. of sugar, at $12 -	60	00
	——— 7 ———		
1	Sold for cash 8 barrels of flour, at $6.50 -	52	00
	——— 8 ———		
1	Bought for cash 10 cwt. of sugar, at $11 -	110	00
	——— 9 ———		
1	Sold John Jones 8 cwt. sugar, at $14 - -	112	00
	——— 10 ———		
1	Received of John Dow, on account - -	50	00
	——— 12 ———		
1	Sold Daniel Fox 2 cwt. sugar, at $14 - -	28	00
	——— 13 ———		
2	Received of Daniel Fox, on account - -	200	00
	——— 14 ———		
2	Received of John Dow, on account - -	10	00
	——— 15 ———		
2	*Sold John Jones 10 quintals of fish, at $4* -	40	00

Boston, January 16, 1835.

L. F.		$	C.
2	Shipped in the Patience, John Martin, master, for Hartford, consigned to my factor, Samuel Bond, on my account		
	2 barrels flour - - - - at $6.00		
	10 quintals fish - - - - - at 3.75		
	Charges at shipping - - - - 2.00	51	50
	———————— 17 ————————		
2	Sold John Chase 12 pieces of linen, at $25	300	00
	———————— 19 ————————		
2	Received of John Chase, on account - -	250	00
	———————— 20 ————————		
2	Samuel Bond advises me that he has received the goods sent him, and that, having disposed of them, the net proceeds are - - -	70	00
	———————— 21 ————————		
2	Sold John Dow 6 hhds. molasses, at $35 -	210	00
	———————— 22 ————————		
2	Sold John Jones 10 pieces linen, at $24 -	240	00
	———————— 23 ————————		
2	John Dow paid, on account - - -	130	00
	———————— 26 ————————		
2	John Jones paid, on account - - -	200	00
	———————— 27 ————————		
2	Paid my book-keeper one month's salary -	25	00

Boston, January 1, 1835.

L. F. Dr.	L. F. Cr.		$	C.
	1	Sundries Dr. to stock,		
1		Cash - - - - - - - $100.00		
1		Flour, 34 barrels, at - - - - 5.00		
1		Linen, 40 pieces, at - - - 20.00		
2		Sugar, 5 cwt., at - - - - 10.00		
2		Fish, 20 quintals, at - - - 3.00		
2		Molasses, 8 hhds., at - - - 30.00	1420	00
		"		
1		Stock Dr. to sundries,		
	2	To Daniel Fox, on account - - $50.00		
	2	John Chase - - - - 20.00		
	3	John Jones - - - - 36.00	106	00
		2		
3	1	John Dow Dr. to flour,		
		For 10 barrels, at $6 - - - - -	60	00
		3		
1	1	Cash Dr. to flour,		
		For 12 barrels, at $6 - - - - -	72	00
		5		
2	1	Daniel Fox Dr. to linen,		
		For 10 pieces, at $25 - - - -	250	00
		6		
2	2	John Chase Dr. to sugar,		
		For 5 cwt., at $12 - - - - -	60	00
		7		
1	1	Cash Dr. to flour		
		For 8 barrels, at $6.50 - - - -	52	00
		8		
2	1	Sugar Dr. to cash,		
		For 10 cwt., at $11 - - - - -	110	00
		9		
3	2	John Jones Dr. to sugar,		
		For 8 cwt., at $14 - - - - -	112	00
		10		
1	3	Cash Dr. to John Dow,		
		Received, on account - - - -	50	00
		12		
2	2	Daniel Fox Dr. to sugar,		
		For 2 cwt., at $14 - - - -	28	00

Boston, January 13, 1835.

L. F.	L. F.		$	C.
1	2	Cash Dr. to Daniel Fox, Received, on account - - - -	200	00
		——— 14 ———		
1	3	Cash Dr. to John Dow, Received, on account - - - -	10	00
		——— 15 ———		
3	2	John Jones Dr. to fish, For 10 quintals, at $4 - - - -	40	00
		——— 16 ———		
3		Adventure to Hartford Dr. to sundries,		
	2	For flour, 2 barrels, at - - $6.00		
	2	Fish, 10 quintals, at - - - 3.75		
	1	Charges - - - - - - 2.00	51	50
		——— 17 ———		
2	1	John Chase Dr. to linen, For 12 pieces, at $25 - - - -	300	00
		——— 19 ———		
1	2	Cash Dr. to John Chase, Received, on account - - - -	230	00
		——— 20 ———		
3	3	Samuel Bond Dr. to adventure to Hartford, For net proceeds consigned to him - -	70	00
		——— 21 ———		
3	2	John Dow Dr. to molasses, For 6 hhds., at $35 - - - - -	210	00
		——— 22 ———		
3	1	John Jones Dr. to linen, For 10 pieces, at $24 - - - -	240	00
		——— 23 ———		
1	3	Cash Dr. to John Dow, Received, on account - - - -	130	00
		——— 26 ———		
1	3	Cash Dr. to John Jones, Received, on account - - - -	200	00
		——— 27 ———		
3	1	Profit and loss Dr. to cash, For my book-keeper's salary - - -	25	00

NOTE. — The first column refers to the page in the Leger, where the account is debited; and the second column, where it is credited.

26*

Dr.   **Stock.**

1835.		JF		Cr. LF	$	C.
Jan.	1	1	To sundries - - - - - -		106	00
	31		Balance per net stock - -		1581	00
					1687	00

Dr.   **Cash**

1835.		JF		Cr. LF	$	C.
Jan.	1	1	To stock - - - - - -	1	100	00
	3	1	Flour - - - - -	1	72	00
	7	1	Flour - - - - -	1	52	00
	10	1	John Dow - - - -	3	50	00
	13	2	Daniel Fox - - - -	2	200	00
	14	2	John Dow - - - -	3	10	00
	19	2	John Chase - - - -	2	230	00
	23	2	John Dow - - - -	3	130	00
	26	2	John Jones - - - -	3	200	00
					1044	00

Dr.   **Flour.**

1835.		JF		Bbls.	Cr. LF	$	C.
Jan.	1	1	To stock, at - - -	34	1	170	00
	31		Profit and loss - -		3	36	00
				34		206	00

Dr.   **Linen.**

1835.		JF		Ps.	Cr. LF	$	C.
Jan.	1	1	To stock, at - - - -	40	1	800	00
	31		Profit and loss - -		3	150	00
				40		950	00

Contra.   Cr.

1835.		JF		Dr. L F	$	C.
Jan.	1	1	By sundries - - - -	3	1420	00
	31		Profit and loss - - - -		267	00
					1687	00

Contra.   Cr.

1835.		JF		Dr. L F	$	C.
Jan.	8	1	By sugar - - - - -	2	110	00
	16	2	Adventure to Hartford - -	3	2	00
	27	2	Cash - - - - -	3	25	00
	31		Balance remaining - -	4	907	00
					1044	00

Contra.   Cr.

1835.		JF		Dr. Bbls.	L F	$	C.
Jan.	2	1	By John Dow, at - - $6.00	10	3	60	00
	3	1	By cash, at - - - 6.00	12	1	72	00
	7	1	By cash, at - - 6.50	8	1	52	00
	16	2	Adventure to Hartford, at 6.00	2	3	12	00
	31		Balance, at - - 5.00	2	4	10	00
				34		206	00

Contra.   Cr.

1835.		JF		Dr. Ps.	L F	$	C.
Jan.	5	1	By Daniel Fox, at - $25.00	10	2	250	00
	17	2	John Chase, at - 25.00	12	2	300	00
	22	2	John Jones, at - - 24.00	10	3	240	00
	31		Balance - - - 20.00	8	4	160	00
				40		950	00

Dr.   Sugar.

1835.		JF				Cwt.	Cr. LF	$	C.
Jan.	1	1	To stock, at	- -	$10.00	5	1	50	00
	8	1	Cash, at	- - -	11.00	10	1	110	00
	31		Profit and loss	- -			3	40	00
						15		200	00

Dr.   Fish.

1835.		JF				Quint.	Cr. LF	$	C.
Jan.	1	1	To stock, at	- -	$3.00	20	1	60	00
	31		Profit and loss	- - -			3	17	50
						20		77	50

Dr.   Molasses.

1835.		JF				Hhds.	Cr. LF	$	C.
Jan.	1	1	To stock, at	- -	$30.00	8	1	240	00
	31		Profit and loss	- -			3	30	00
						8		270	00

Dr.   Daniel Fox.

1835.		JF					Cr. LF	$	C.
Jan.	5	1	To linen, 10 pieces	- -	$25.00		1	250	00
	12	1	Sugar, 2 cwt., at	- -	14.00		2	28	00
								278	00

Dr.   John Chase.

1835.		JF					Cr. LF	$	C.
Jan.	6	1	To sugar, 5 cwt., at	- -	$12.00		2	60	00
	17	2	Linen, 12 pieces, at	- -	25.00		1	300	00
								360	00

Contra.   Cr.

1835.		JF			Cwt.	Dr. L F	$	C.
Jan.	6	1	By John Chase, at - $12.00		5	2	60	00
	9	1	John Jones, at - - 14.00		8	3	112	00
	12	1	Daniel Fox, at - 14.00		2	2	28	00
					15		200	00

Contra.   Cr.

1835.		JF			Quint.	Dr. L F	$	C.
Jan.	15	2	By John Jones, at - $4.00		10	3	40	00
	16	2	Adventure to Hartford 3.75		10	3	37	50
					20		77	50

Contra.   Cr.

1835.		JF			Hhds.	Dr. L F	$	C.
Jan.	21	2	By John Dow, at - $35.00		6	3	210	00
	31		Balance - - - 30.00		2	4	60	00
					8		270	00

Contra.   Cr.

1835.		JF		Dr. L F	$	C.
Jan.	1	1	By stock - - - - - - -	1	50	00
	13	2	Cash - - - - - - -	1	200	00
	31		Balance - - - - -	4	28	00
					278	00

Contra.   Cr.

1835.		JF		Dr. L F	$	C.
Jan.	1	1	By stock - - - - - -	1	20	00
	19	2	By cash - - - - -	1	230	00
	31		Balance - - - - -	4	110	00
					360	00

Dr.　　　　　　John Jones.

1835.		JF			Cr. L F	$	C.
Jan.	9	1	To sugar, 8 cwt., at - -	$14.00	2	112	00
	15	2	Fish, 10 quintals, at - -	4.00	2	40	00
	31	2	Linen, 10 pieces, at - -	24.00	1	240	00
						392	00

Dr.　　　　　　John Dow.

1835.		JF			Cr. L F	$	C.
Jan.	2	1	To flour, 10 barrels, at - -	$6.00	1	60	00
	21	2	Molasses, 6 hhds., at - -	35.00	2	210	00
						270	00

Dr.　　　　Adventure to Hartford.

1835.		JF		Cr. L F	$	C.
Jan.	16	2	To sundries - - - - - -	3	51	50
	31		Profit and loss - - - -	3	18	50
					70	00

Dr.　　　　　Samuel Bond.

1835.		JF		Cr. L F	$	C.
Jan.	20	2	To adventure, per brig Patience -	3	70	00

Dr.　　　　　Profit and Loss.

1835.		JF		Cr. L F	$	C.
Jan.	27	2	To cash - - - - - -	1	25	00
	31		Stock, net gain - - - -	1	267	00
					292	00

Contra.   Cr.

1835.		JF		Dr. LF	$	C.
Jan.	1	1	By stock - - - - - -	1	36	00
	26	2	Cash - - . - - . -	1	00	00
	31		Balance - - - - -	4	156	00
					392	00

Contra.   Cr.

1835.		JF		Dr. LF	$	C.
Jan.	10	1	By cash - - - - - -	1	50	00
	14	2	Cash - - - - - -	1	10	00
	23	2	Cash - - - - - -	1	130	00
	31		Balance - - - - -	4	80	00
					270	00

Contra.   Cr.

1835.		JF		Dr. LF	$	C.
Jan.	20	2	By Samuel Bond - - - -	3	70	00

Contra.   Cr.

1835.		JF		Dr. LF	$	C.
Jan.	31	2	By balance - - - - -	4	70	00

Contra.   Cr.

1835.		JF		Dr. LF	$	C.
Jan.	1		By flour - - - - -	1	36	00
			Linen - - - - - -	1	150	00
			Sugar - - - - -	2	40	00
			Fish - - - - -	2	17	50
			Molasses - - - - -	2	30	00
			Adventure to Hartford - -	3	18	50
					292	00

Dr.   Balance.

1835.		JF									Cr. LF	$	C.
	31	1	To Cash - - - - - - -							1	907	00
			Flour - - - - -						1	10	00
			Linen - - - - -						1	160	00
			Molasses - - - - -						2	60	00
			Daniel Fox - - - - -						2	28	00
			John Chase - - - - -						2	110	00
			John Jones - - - - -						3	156	00
			John Dow - - - - -						3	80	00
			Samuel Bond - - - -						3	70	00
											1581	00

General Trial Balance.

Titles of Accounts.	LF	Dr.	Cr.	Dr.	Cr.
Stock - - -	1	$ 106.00	$1420.00	$	$1314.00
Cash - - - -	1	1044.00	137.00	907.00	
Flour - - -	1	170.00	196.00		26.00
Linen - - -	1	800.00	790.00	10.00	
Sugar - - -	2	160.00	200.00		40.00
Fish - - -	2	60.00	77.50		17.50
Molasses - -	2	240.00	210.00	30.00	
D. Fox - - -	2	278.00	250.00	28.00	
J. Chase - -	2	360.00	250.00	110.00	
J. Jones - - -	3	392.00	236.00	156.00	
J. Dow - - -	3	270.00	190.00	80.00	
Adventure to Hartford	3	51.50	70.00		18.50
S. Bond - - -	3	70.00	00.00	70.00	
Profit and Loss - -	3	25.00	00.00	25.00	
Proof.		$4026.50	$4026.50	$1416.00	$1416.00

Contra. Cr.

1835.		JF		Dr. L F	$	C
Jan.	31		Stock for net of my estate - -	1	1581	00

INDEX TO THE LEGER.

MERCANTILE FORMS.

General Receipt.

Boston, Nov. 7, 1835. Received of John Jacobs, ten dollars, in full of all accounts. JOHN JAY.

Receipt for Money on Account.

Boston, Sept. 25, 1835. Received of Samuel Eaton, twenty dollars, on account. SAMUEL SNOW.

Receipt for an Indorsement on a Note.

Salem, Nov. 1, 1835. Received of William Jones, seventy-five dollars, which is indorsed on his note, of June 7, 1834, for three hundred and fifty dollars. PHILIP QUARLES.

Form for Notes, see page 140.

Bill of Exchange, see page 232.

An Order for Merchandise.

Mr. Alfred Perkins,

Pay the bearer, fifty dollars, in merchandise, from your store, and charge your obt. servant,

Lowell, May 15th, 1835. THOMAS FREEMAN.

An Order for Money.

Messrs. Stanford & Davis,

Pay to the order of J. L. Fairbanks, thirty-five dollars and seventy cents, and this shall be your receipt for the same. JOHN MARSH.

Boston, Sept. 10, 1835.

NOTE. — A receipt given in full of *all accounts* cuts off accounts only. An order, when paid, should be receipted on the back of the same, by the person whom it is made payable, or his attorney; but when it is made payable to A., or bearer, it may be receipted by any one who presents it for payment.

CATALOGUE OF
APPROVED SCHOOL BOOKS,

PRINTED AND SOLD BY

ROBERT S. DAVIS,

Publisher and Bookseller,

No. 77 WASHINGTON STREET, (JOY'S BUILDING,)

BOSTON.

SOLD ALSO BY THE PRINCIPAL BOOKSELLERS IN THE UNITED STATES.

☞ *Also constantly on hand (in addition to his own publications) a general assortment of Books in every department of Education, which are offered to Booksellers, School Committees and Teachers, in large and small quantities, on very liberal terms; also, Stationery. Orders from the city and country respectfully solicited.*

THE NATIONAL ARITHMETIC, on the inductive system; combining the Analytic and Synthetic methods, in which the principles of Arithmetic are explained and illustrated in a perspicuous and familiar manner; containing, also, practical systems of Mensuration, Gauging, Geometry, Book-keeping, &c., and much practical information connected with Trade and Commerce—forming a complete Mercantile Arithmetic. Designed for Schools and Academies throughout the United States. By Benjamin Greenleaf, A. M., Principal of Bradford Seminary. Twelfth, Improved Stereotype Edition.

ADVERTISEMENT TO THE SECOND (STEREOTYPE) EDITION.

THE rapid and extensive sale of the first edition of the National Arithmetic, together with its flattering reception in various sections of our country, has induced the author thoroughly to revise and improve the work, which he trusts will give it additional merit.

The author believes that not an error or inaccuracy of essential importance will be found in the present (stereotype) edition, which could not be wholly avoided in the *first*.

It has been deemed expedient, that the new edition should embrace more of the inductive plan than the former, with the addition of much important and valuable matter.

The author has availed himself of the assistance of several experienced teachers, among whom he would acknowledge his obligations particularly to Mr. CHARLES H. ALLEN, one of the associate Principals of the Franklin Academy, Andover; and to Mr. DAVID P. PAGE, Principal of the English High School, Newburyport; also to several mercantile gentlemen, who have imparted valuable suggestions of a practical nature.

A KEY TO THE NATIONAL ARITHMETIC, exhibiting the operation of the more difficult questions in that work. By the Author. *Designed for the use of teachers only.* Fourth Stereotype Edition.

1

GREENLEAF'S NATIONAL ARITHMETIC.

RECOMMENDATIONS.

From the late Principal of the Young Ladies' High School, Boston.

Dear Sir: I have examined with great care Mr. Greenleaf's National Arithmetic, and have used it as a text-book for my pupils. In my view, the plan and execution of the work are quite perfect, the rules being deduced analytically from examples, and followed by copious questions for practice. The pupil can hardly fail to *understand* as he advances; nor can he go through the book, without being a master of the science of Arithmetic. This is not an old book with a new name, but the work of one who thoroughly understands the subject, and who has learned, from a long and successful experience in teaching, how to prepare one of the very best school books which has ever been issued from the American press. Very respectfully, E. BAILEY.
Mineral Spring School, Lynn, May 13th, 1839.

The undersigned, members of the General School Committee, of Haverhill, take the liberty of recommending to the purchasers of books for the use of schools in this town, Greenleaf's National Arithmetic, as a work comprehending most of the advantages of the various treatises on the subject now before the public, and as more directly adapted to the practical interests of the community than any which have fallen under their notice.

GEO. KEELY,	JOSEPH WHITTLESEY,
NATHL. GAGE,	SAMUEL H. PECKHAM,
HENRY PLUMMER,	ABIJAH CROSS,

Haverhill, 8th, 1st Month, 1836.   JOHN G. WHITTIER.

From the Principal of Merrimack Academy.

B. Greenleaf, Esq. Dear Sir: I have examined your National Arithmetic, and am happy to say, that it is truly a *practical* work. The *numerous* questions, both from their *nature* and *arrangement*, are well calculated to produce an increased interest, and to facilitate the acquisition of a thorough practical knowledge of this science. I have introduced it into my school, and the result has been, that the scholars have manifested a greater fondness for the study, and made more rapid progress, than when attending to books formerly in use.

I do most cheerfully recommend the work, believing it to be very happily adapted to the wants of our schools and academies. Very respectfully, yours,
Bradford, Feb. 8, 1836. SYLVANUS MORSE.

From Mr. Page, Principal of the English High School, Newburyport.

Benjamin Greenleaf, Esq. Dear Sir: I have with much care examined the NATIONAL ARITHMETIC, of which you are the author, and after having compared it, *article by article*, with the various other publications that have come to my hands, I hesitate not to say, that I think it contains a greater amount of matter, and a better arrangement of subjects, than any other book I have seen. Your rules and explanations are clear and definite, and your examples are well calculated to fix them in the mind. I congratulate the community on this valuable accession to our list of school books; and shall take pleasure in seeing your Arithmetic extensively introduced into all our schools, as also into that under my own care. Yours, with just respect,
Newburyport, Mass. March 5, 1836. DAVID P. PAGE.

To Benjamin Greenleaf, Esq., A. M., Preceptor of Bradford Academy.

Sir: The School Committee of the town of Bradford, having examined your National Arithmetic, are of opinion, that the various rules are well arranged, and the numerous operations judiciously selected; and that it is better adapted to the wants of our Academies and Schools than any Arithmetic now in common use, and do hereby recommend it to be used in the schools under our care.
JEREMIAH SPOFFORD, *Chairman of said Committee.*

2

GREENLEAF'S NATIONAL ARITHMETIC.

From Mr. J. P. Engles, A. M., Principal of the Classical Institute, Philadelphia.

I have examined, with considerable interest, Greenleaf's National Arithmetic, and have no hesitation in recommending it as an admirable system of Arithmetic, which contains all that is essential to a knowledge of the science, and nothing that is useless. The arrangement, too, is such as to make the contents easily available to the teacher and the pupil. Should it succeed in displacing the host of so called "Assistants," with which our schools are flooded, I conceive it would be equally to the comfort of teachers, and the profit of students. I shall cheerfully introduce it into my Academy.   J. P. ENGLES.

Philadelphia, Nov. 14, 1838.

I cheerfully concur in sentiment with Mr. Engles, respecting Mr. Greenleaf's Arithmetic ; it is the best work of the kind I have ever seen. With a great deal of pleasure, I shall introduce the same into my Seminary.

W. ALEXANDER, *Classical Teacher, Philadelphia.*

I have examined Greenleaf's National Arithmetic with a great deal of satisfaction, and have no hesitation in saying, that it is the most complete system of Mercantile Arithmetic with which I am acquainted ; and will cheerfully recommend it as occasion may require.

E. GRIFFITHS, *Teacher of Mathematics, Philadelphia.*

Philadelphia, Nov. 12, 1838.

The undersigned entirely concur in the opinions expressed by Messrs. Engles, Alexander, and Griffiths, respecting Mr. Greenleaf's Arithmetic.

JOHN W. FAIRES,
B. P. HUNT,   } *Teachers in Philadelphia*
JAMES P. ESPY,

I have examined Mr. Greenleaf's National Arithmetic with some care, and am much pleased with its arrangement ; his examples, under each rule, are numerous and appropriate : I am so well satisfied, that I intend to introduce it into my Seminary.   THOMAS McADAM.

Philadelphia, Nov. 14, 1838.

We fully concur with the gentlemen, who have already given recommendations of the National Arithmetic, considering the work well calculated to give youth a correct knowledge of the principles of Arithmetic.

WM. VOGDES,   } *Philadelphia Centre High School.*
E. O. KENDALL,

Copy of a letter from G. W. Harby, Esq., Principal of Harby's Academy, New Orleans, addressed to the Publishers.

Gentlemen : Viewing the publication of School Books of the first importance, it was with much pleasure that I received Greenleaf's National Arithmetic. For fifteen years, and upwards, I have devoted my life to the instruction of youth, during which time many Arithmetics have fallen under my inspection. I take a strong interest in every work that pertains to mathematical learning, and unhesitatingly pronounce Greenleaf's Arithmetic an important treasure to Academies ; it is wrought with a great deal of care, and in an easy, plain, and uniform style. His Geometrical, Mechanical, and Astronomical Problems are concise and clear : they lead the youthful mind to the exercise of a little patience,—not so arduous as to fatigue, but sufficiently laborious to call the mental faculties into exercise, and to create a taste for mathematical knowledge, and for scientific discovery and invention, — which has lately so conspicuously crowned some of our countrymen with brilliant success. I shall make it the standard book in my Institution, and recommend it to others of my profession.

I remain, gentlemen, your obedient servant,

New Orleans, August 22, 1839.   GEORGE W. HARBY.

3

GREENLEAF'S NATIONAL ARITHMETIC.

Portsmouth, Aug. 5, 1838.

Benjamin Greenleaf, Esq. Dear Sir : Having examined, and, to some extent, introduced into our Schools the National Arithmetic, of which you are the author, we deem it a duty we owe to the public, no less than to yourself, to express our decided approbation of its merits. The method, arrangement, and quantum of matter it contains, the clear and lucid manner in which its rules are demonstrated, together with its adaptation to the wants of the community, entitle it, in our humble belief, to the patronage of every lover of scientific investigation. Signed,

HAZEN PICKERING,   JOHN T. TASKER,
A. M. HOYT,   JOHN J. LANE,
JAMES HOYT,   EDWARD J. LAIGHTON.
C. E. POTTER,

School Teachers of Portsmouth, N. H.

From Rev. Dr. Hopkins, President of William's College.

My opinion of Greenleaf's Arithmetic is, that it is adapted to give a more thorough knowledge of that science, than any other that I have seen.
Respectfully, yours, **M. HOPKINS.**
Williamstown, Dec. 30, 1837.

Poughkeepsie Institute, Jan. 9, 1839.

We have carefully examined the National Arithmetic, and do not hesitate in pronouncing it the best work of the kind which has come under our notice. The deduction of the rule from the operations is, in our opinion, the proper method ; and the copious examples, under the various rules, are well selected and arranged. We hope it may meet with its merited success. We shall endeavour to extend and establish its use. Yours, respectfully,
J. L. DUSINBERY, } *Principals.*
A. H. TOBEY,

I have examined Greenleaf's Arithmetic, and consider it, in many respects, preferable to any work of the kind with which I am acquainted. I am particularly pleased with his illustration of the Square and Cube Roots, and the Rule of Proportion, and with the introduction of practical instruction on the subject of Banking, Custom-House Duties, Assessment of Taxes, &c. I think its introduction into Schools and Academies will prove of general interest to all who wish to acquire a knowledge of Arithmetic.

A. B. BULLOCK.

Hudson, Dec. 7, 1838.

I fully concur in the above, and shall use my influence to introduce it into my school. C. GREENE.

From the Principal of the Dutchess County Academy.

After a careful and comparative examination of Greenleaf's Arithmetic, I unhesitatingly say, I think it superior to any other Arithmetic within my knowledge. I shall with pleasure use my influence to give it a circulation in the Schools of this vicinity. WM. JENNOY
Poughkeepsie, Jan. 9, 1839.

I fully accord with Mr Jennoy in his opinion of Mr. Greenleaf's Arithmetic, and shall esteem it a privilege to recommend its use whenever an opportunity presents. O. M. SMITH,
Newburgh, January, 1839. *Principal of Newburgh High School.*

4

PROGRESSIVE EXERCISES IN ENGLISH COMPOSI-TION. By R. G. Parker, A. M., Principal of the Franklin Grammar School, Boston. Thirty-fifth Stereotype Edition.

☞ The reputation of this little Manual is now so well established as to render it unnecessary to present many of the numerous testimonials in its favor, received from teachers and others of the first respectability.

The School Committee of Boston authorized its introduction into the Public Schools of the city, soon after the first edition was issued, and it is now the only work on Composition used in them. It has also been adopted as a text-book in a large number of the best schools and higher seminaries in various sections of the United States, having been highly commended by all intelligent teachers, who have used it, and the demand is constantly increasing.

To show the high estimate of the work in England, the fact may be stated, that it has been republished and stereotyped in London, and *nine* large editions have been sold there; which, together with its favorable reception throughout the United States, furnishes sufficient evidence of its practical utility.

Among the public notices of the work in England, are the two following:

The design of this work is unexceptionably good. By a series of progressive exercises the scholar is conducted from the formation of easy sentences to the more difficult and complex arrangement of words and ideas. He is, step by step, initiated into the rhetorical propriety of the language, and furnished with directions and models for analyzing, classifying, and writing down his thoughts in a distinct and comprehensive manner. — *London Jour. of Education.*

Of the Exercises in Composition, by Parker, we can speak with unmingled praise. It is not enough to say, that they are the best that we have, for we have none worth mention. The book is fully effective both in suggesting ideas or pointing out the method of thinking, and also in teaching the mode of expressing ideas with propriety and elegance. — *English Monthly Magazine.*

From Mr. Walker, Principal of the Eliot School, Boston.

This work is evidently the production of a thorough and practical teacher, and in my opinion it does the author much credit. By such a work all the difficulties and discouragements which the pupil has to encounter, in his first attempts to write, are in a great measure removed, and he is led on, progressively, in a methodical and philosophical manner, till he can express his ideas on any subject which circumstances or occasion may require, not only with sufficient distinctness and accuracy, but even with elegance and propriety. An elementary treatise on composition, like the one before me, is certainly much wanted at the present day. I think this work will have an extensive circulation, and I hope the time is not distant, when this branch of education, hitherto much neglected, will receive that attention which in some degree its importance demands.

From J. W. Bulkley, Esq., Principal of an Academy, Albany.

I have examined "Parker's Exercises in Composition," and am delighted with the work; I have often felt the want of just that kind of aid, that is here afforded: the use of this book will diminish the labor of the teacher, and greatly facilitate the progress of the pupil in a study that has hitherto been attended with many trials to the teacher, and perplexities to the learner.

If Mr. Parker has not strewed the path of the student with flowers, he has "removed many stumbling-blocks out of the way, made crooked things straight, and rough places smooth." It is certainly one of the happiest efforts that I have ever seen in this department of letters, — affording to the student a beautiful introduction to the most important principles and rules of rhetoric; and I would add, that if carefully studied, it will afford a *"sure guide"* to written composition. I shall use my influence to secure its introduction to all our schools.

THE CLASS BOOK OF ANATOMY, explanatory of the first principles of Human Organization, as the basis of Physical Education; with numerous Illustrations, a full Glossary, or explanation of technical terms, and practical Questions at the bottom of the page. By J. V. C. Smith, M. D., formerly Professor of General Anatomy and Physiology in the Berkshire Medical Institution. Seventh, Improved Stereotype Edition.

☞ This work has received the highest testimonials of approbation from the most respectable sources, and has already been adopted as a text book in many schools and colleges in various sections of the United States.

The estimation in which it is held in other countries may be inferred from the fact, that a translation of it has recently been made into the Italian language, at Palermo, under the supervision of the celebrated Dr. Placido Portel. It is also in the progress of translation into the Hawaiian language, by the American missionaries at the Sandwich Islands, to be used in the higher schools, among the natives; and the plates are soon to be forwarded, with reference to that object, by the American Board of Commissioners for Foreign Missions; which furnishes conclusive evidence of its value and utility.

From Rev. Hubbard Winslow, Pastor of Bowdoin St. Church, Boston.

Boston, Nov. 7, 1836.

I have examined the Class Book of Anatomy, by Dr. Smith, with very great satisfaction. For comprehensiveness, precision, and philosophical arrangement, it is surpassed by no book of the kind which I have ever seen. The study of Anatomy and Physiology, to some extent, is exceedingly interesting and useful as a branch of common education; and it is to be desired that it should be more extensively adopted in all our higher schools. To secure this end, there is no other book before the public so well prepared as the one under remark. It is also a convenient compend to lie upon the table of the scientific anatomist and physician, and a very valuable family book for reference, and for explanation of terms which often occur in reading.

H. WINSLOW.

We are gratified to see the attempt to introduce a new subject to ordinary students. It is wonderful that civilized man has been so long willing to remain ignorant of the residence of his mind, and the instruments by which it operates. The book before us abounds in information in which every adult reader will feel a deep interest, and from which all may derive valuable lessons of a practical kind. We are gratified to see frequent references to the Great First Cause of life and motion. We cordially wish success to his enterprise in a path almost untrodden.—*American Annals of Education.*

Copy of a Communication from Mr. C. H. Allen, of the Franklin Academy, Andover, Mass.

North Andover, Dec. 10, 1836.

Mr. R. S. Davis. Dear Sir: During my vacation, I have had time to examine Smith's Class Book of Anatomy, the second edition of which you have recently published. I do not hesitate to speak of it as the very work which the public have long demanded. It contains knowledge which should be widely diffused. The author is remarkably clear in his explanations and descriptions, and very systematic in his arrangement. So that he has rendered this neglected branch of useful knowledge highly interesting to all classes.

Yours, respectfully,

CHAS. H. ALLEN.

14

ALGER'S MURRAY'S GRAMMAR; being an abridgment of Murray's English Grammar, with an Appendix, containing exercises in Orthography, in Parsing, in Syntax, and in Punctuation; designed for the younger classes of learners. By Lindley Murray. To which Questions are added, Punctuation, and the notes under Rules in Syntax copiously supplied from the author's large Grammar, being his own abridgment entire. Revised, prepared, and adapted to the use of the "English Exercises," by Israel Alger, Jr., A. M., formerly a teacher in Hawkins Street School, Boston. Improved stereotype edition.

As a cheap and compendious elementary work for general use, this is probably the best Grammar extant, which is indicated by its introduction into many Schools and Academies, in various sections of the United States. Though furnished at a moderate price, it is so copious, as, in most cases, to supersede the necessity of a larger work.

☞ By a vote of the School Committee, this work was introduced into all the Public Schools of the city of Boston.

ALGER'S MURRAY'S ENGLISH EXERCISES: consisting of Exercises in Parsing, instances of false Orthography, violations of the rules in Syntax, defects in Punctuation, and violation of the rules respecting perspicuous and accurate writing, with which the corresponding rules, notes, and observations, in Murray's Grammar are incorporated; also, References in Promiscuous Exercises to the Rules by which the errors are to be corrected. Revised, altered and particularly adapted to the use of Schools, by Israel Alger, Jr., A. M. Improved stereotype edition.

Extract from the Preface.

It is believed that both teachers and pupils have labored under numerous and serious inconveniences, in relation to certain parts of these Exercises, for the want of those facilities which this volume is designed to supply. Those rules in Mr. Murray's Grammar which relate to the correction of each part of the Exercises in Orthography, Syntax, Punctuation and Rhetorical construction, have been introduced into this manual immediately preceding the Exercises to which they relate. The pupil being thus furnished with the principles by which he is to be governed in his corrections, may pursue his task with profit and pleasure. In this edition, more than forty 18mo. pages of matter have been added from Mr. Murray's Grammar.

ALGER'S PRONOUNCING INTRODUCTION TO MURRAY'S ENGLISH READER, in which accents are placed on the principal words, to give Walker's pronunciation. Handsomely printed, from stereotype plates.

ALGER'S PRONOUNCING ENGLISH READER: being Murray's Reader, accented by Israel Alger, Jr. Printed from handsome stereotype plates, on good paper, and neatly bound.

☞ These editions of Murray's books are in the highest repute of any other published in the United States, and are sold at a cheap price.

BOSTON SCHOOL ATLAS. Embracing a Compendium of Geography. Containing seventeen Maps and Charts. Embellished with instructive Engravings. Twelfth edition, handsomely printed, from new plates. One volume, quarto.

The Maps are all beautifully engraved and painted; and that of Massachusetts, Connecticut, and Rhode Island, contains the boundaries of every town in those states.

☞ Although this book was designed for the younger classes in schools, for which it is admirably calculated, yet its maps are so complete, its questions so full, and its summary of the science so happily executed, that, in the opinion of many, it contains all that is necessary for the pupil in our common schools.

From the Preface to the Sixth Edition.

The universal approbation and extensive patronage bestowed upon the former editions of the Boston School Atlas, has induced the publishers to present this edition with numerous improvements. The maps of the World, North America, United States, Europe, England, and Asia, have been more perfectly drawn, and re-engraved on steel; and the maps of Maine, of New Hampshire and Vermont, and of the Western States, also, on steel, have been added; and some improvements have been made in the elemental part.

It has been an object, in the revision of this edition, to keep the work, as much as possible, free from subjects liable to changes, and to make it a *permanent Geography*, which may hereafter continue to be used in classes without the inconvenience of essential variations in different editions.

From R. G. Parker, author of " Progressive Exercises in English Composition," and other popular works.

I have examined a copy of the Boston School Atlas, and have no hesitation in recommending it as the best introduction to the study of Geography that I have seen. The compiler has displayed much judgment in what he has omitted, as well as what he has selected; and has thereby presented to the public a neat manual of the elements of the science, unencumbered with useless matter and uninteresting detail. The mechanical execution of the work is neat and creditable, and I doubt not that its merits will shortly introduce it to general use. Respectfully yours,

R. G. PARKER.

From E. Bailey, Principal of the Young Ladies' School, Boston.

I was so well pleased with the plan and execution of the Boston School Atlas, that I introduced it into my school, soon after the first edition was published. I regard it as the best work for beginners in the study of Geography which has yet fallen under my observation; as such I would recommend it to the notice of parents and teachers.

From the Principal of one of the High Schools in Portland.

I have examined the Boston School Atlas, Elements of Geography, &c., and think it admirably adapted to beginners in the study of the several subjects treated on. It is what is wanted in all books for learners—simple, philosophical, and practical. I hope it will be used extensively.

Yours, respectfully, JAS. FURBISH.

I have perused your Boston School Atlas with much satisfaction. It seems to me to be what has been needed as an introduction to the study of Geography, and admirably adapted to that purpose.

Very respectfully, yours, &c. B. D. EMERSON.

GREENLEAF'S NATIONAL ARITHMETIC.

I have examined, with considerable care and entire satisfaction, the System of Arithmetic by B. Greenleaf. I can say, without hesitation, I think it the most complete and well-arranged School System, in this branch of science, extant, and better calculated than any other to prepare our youth for active usefulness in all those pursuits where a knowledge of Arithmetic is requisite. I might speak of the happy combination of the Analytic and Synthetic methods of operation, and the still happier union of clearness with brevity in all the Rules and Definitions; but all this will be seen and pleasingly felt by those who peruse or study this truly valuable book. I shall do what I may, in my limited sphere of influence, to promote its introduction into the Schools of our State.

Albany, Dec. 1838. S. STEELE, *Teacher.*

I have examined Greenleaf's National Arithmetic, and am of opinion, from its practical character and the order of the arrangement, that it is well calculated to induct the inquiring pupil into the useful business operations of the community, for which the study of Arithmetic is designed. I shall not hesitate to recommend it to my own pupils and to the teachers of other Schools.

EDWARD SMALL,
Albany, Dec. 1, 1838. *Teacher of the Lancaster School, Albany.*

Mr. Greenleaf. Sir: I have examined your National Arithmetic and am glad to say, it meets my approbation; and I think I shall introduce it into my School, to the exclusion of all others. A. P. SMITH,
Albany, Nov. 28, 1838. *Teacher of the Second Public School, Albany.*

Mr. Greenleaf. Dear Sir: I have examined your System of Arithmetic, and am happy to state, that it meets with my unqualified approbation, and that I shall immediately introduce it into my School. Yours, respectfully,
Albany, Nov. 27, 1838. THOMAS McKEE.

We fully concur in the above.
NEWMAN & WALLACE, *Teachers, Mechanics Academy, Albany.*
D. E. BASSETT, *Principal of an Academy,* Do.
JOEL MARBLE, *Principal of District School, State Street,* Do.
J. W. BULKLEY, *Principal of an Academy,* Do.

From Dr. Fox, Principal of the Boylston School, Boston.

B. Greenleaf, Esq. Dear Sir: I have just been examining your new Arithmetic, and think it an excellent work. I like the plan of it much. Among its many excellences I perceive the following, viz. — The Tables of Money, Weights, and Measures carried out to the lowest denomination; the great variety of examples under each Rule, and likewise your method of treating several parts of the science, as Fractions, Proportion, Evolution, and Exchange, — every thing concerning them must appear clear, I think, to the student. The Geometry, Philosophical Problems, Mechanical Powers, and Book-keeping, seem also to be handled in a perspicuous manner. The Rules of Cross Multiplication and Position, I am happy to see have place in the work; for, after all, they are too useful, the latter especially, to be omitted in our arithmetical treatises. On the whole, the work appears to me well calculated to lead youth to a clear and thorough knowledge of the various branches of this Science, and I doubt not it will be sought after, as an improvement on former works of the kind, and obtain an extensive circulation. Yours, respectfully, CHARLES FOX.

A thorough examination of Mr. Greenleaf's Arithmetic has induced me to introduce it into the Academy with which I am connected. The arrangement is excellent, and much valuable matter is found in the National Arithmetic, not contained in others now in use. Very respectfully, yours,
Barnstable, Dec. 9, 1837. 7 F. A. CHOATE

WALKER'S BOSTON SCHOOL DICTIONARY. Walker's Critical Pronouncing Dictionary, and Expositor of the English Language. Abridged for the use of Schools throughout the United States. To which is annexed, an Abridgment of WALKER'S KEY to the pronunciation of Greek, Latin and Scripture Proper Names. Boston stereotype edition.

☞ This handsome and correct edition, prepared for the Boston schools, with great care, has so long been used, that it is only necessary for the publisher to keep it in a respectable dress, to ensure it a general circulation.

The price of the work, neatly bound in leather, is reduced to 50 cts. single, $5,00 a dozen.

THE CLASSICAL READER. A Selection of Lessons in Prose and Verse, from the most esteemed English and American Writers. Intended for the use of the higher classes in Public and Private Seminaries. By Rev. F. W. P. Greenwood and G. B. Emerson, of Boston. Tenth stereotype edition.

This work is highly approved, as a *First Class Reader*, and has received many commendable notices from Public Journals throughout the United States, from which the following are selected.

From the Visiter and Telegraph, Richmond, Va.

This work is a valuable acquisition to our schools. It is a work purely national and modern. It has many valuable historical facts and anecdotes in relation to the early history, the character, manners, geography and scenery of our country. In the matter it contains, it is well adapted to the taste, feelings, and habits of the present age. It embodies many of the brightest and most sparkling gems of Irving, Webster, Everett, Jefferson, Channing, Sparks, Bryant, Percival, &c.

From the American Journal of Education.

We are happy to see another valuable addition to the list of reading books, —one which has been compiled with a strict regard to the tendency of the pieces it contains, and which bears the stamp of so high a standard of literary taste. In these respects the Classical Reader is highly creditable to its editors.

Extract from the North American Review.

The Classical Reader is selected from the very best authors, and the quantity from each, or the number of pieces of a similar character, by different authors, affords all that can be required for classes, and in sufficient variety, too, of manner, to facilitate greatly the formation of correct habits of reading, and a good taste. From each of those considerations, we give it our cordial recommendation.

☞ *The Publisher respectfully solicits the attention of Teachers, School Committees, and all interested in the cause of Education, to the foregoing list of School Books,—feeling confident that an examination of the works will lead to a conviction of their merits,—copies of which will be furnished for this purpose, with a view to their adoption, without charge.*